K M Cannell
19·12·91

# The Ecology of
# Mixed-Species Stands of Trees

# The Ecology of Mixed-Species Stands of Trees

SPECIAL PUBLICATION NUMBER 11 OF THE
BRITISH ECOLOGICAL SOCIETY

EDITED BY

## M.G.R. CANNELL

Institute of Terrestrial Ecology (NERC)
Edinburgh Research Station
Penicuik

## D.C. MALCOLM

Institute of Ecology and
Resource Management
University of Edinburgh

## P.A. ROBERTSON

The Game Conservancy Trust
Fordingbridge

OXFORD

BLACKWELL SCIENTIFIC PUBLICATIONS

LONDON   EDINBURGH   BOSTON

MELBOURNE   PARIS   BERLIN   VIENNA

1992

© 1992 by the British Ecological Society
and published for them by
Blackwell Scientific Publications
Editorial offices:
Osney Mead, Oxford OX2 0EL
25 John Street, London WC1N 2BL
23 Ainslie Place, Edinburgh EH3 6AJ
3 Cambridge Center, Cambridge,
　　Massachusetts 02142, USA
54 University Street, Carlton
　　Victoria 3053, Australia

Other Editorial Offices:
Librairie Arnette SA
2, rue Casimir-Delavigne
75006 Paris
France

Blackwell Wissenschafts-Verlag
Meinekestrasse 4
D-1000 Berlin 15
Germany

Blackwell MZV
Feldgasse 13
A-1238 Wien
Austria

First published 1992

Set by Excel Typesetters Company, Hong Kong
Printed and bound in Great Britain by
The University Press, Cambridge

DISTRIBUTORS

Marston Book Services Ltd
PO Box 87
Oxford OX2 0DT
(*Orders*: Tel: 0865 791155
　　　　　Fax: 0865 791927
　　　　　Telex: 837515)

USA
　Blackwell Scientific Publications, Inc
　3 Cambridge Center
　Cambridge, MA 02142
　(*Orders*: Tel: (800) 759-6102)

Canada
　Oxford University Press
　70 Wynford Drive
　Don Mills
　Ontario M3C 1J9
　(*Orders*: Tel: (416) 441-2941)

Australia
　Blackwell Scientific Publications
　(Australia) Pty Ltd
　54 University Street
　Carlton, Victoria 3053
　(*Orders*: Tel: (03) 347-0300)

British Library
Cataloguing in Publication Data

The ecology of mixed-species stands of trees.
　1. Trees. Ecology
　I. Cannell, M.G.R.　II. Malcolm, D.C.
　(Douglas C.)
　III. Robertson, Peter *1961–*
　574.52642
　ISBN 0–632–03148–4

Library of Congress
Cataloguing-in-Publication Data

The ecology of mixed-species stands of trees / edited by
　M.G.R. Cannell, D.C. Malcolm and P.A. Robertson.
　　　　p.　　　　cm.—(Special publication number
　11 of the British Ecological Society)
　　　ISBN 0–632–03148–4
　　1. Forest ecology—Congresses.　2. Forest
　ecology—Great Britain—Congresses.　3. Forests and
　forestry—Congresses.　4. Forests and forestry—
　Great Britain—Congresses.　I. Cannell, M.G.R.
　(Melvin G.R.).　II. Malcolm, D.C. (Douglas C.).
　III. Robertson, P. (Peter).　IV. Series: Special
　publication . . . of the British Ecological Society:
　no. 11.
　QH541.5.F6E27　1992
　561.5'2642—dc20.

# Contents

# Contents

# Preface

This book is the proceedings of a symposium, held at the Heriot-Watt University, 2–6 September 1990, organized by the Forest Ecology Group of the British Ecological Society. The meeting was also held under the auspices of the International Union of Forest Research Organizations, Division S2.01.

During the 1960s and 1970s large areas of the British uplands were planted with conifer monocultures in order to establish a UK commercial forest resource. The objective was to produce timber and woodpulp as profitably as possible. During that time it became increasingly evident that these forests were having substantial impacts on UK soils, waters, flora and fauna, and many studies were done to quantify those impacts. In 1978, Edinburgh hosted a meeting entitled *The Ecology of Even-Aged Forest Plantations* (eds E.D. Ford, D.C. Malcolm & J. Atterson, published by the Institute of Terrestrial Ecology). During the 1970s and 1980s increasing attention was paid to the non-commercial benefits of forests, and to ways of ameliorating possible harmful effects of forests on water quality and quantity and on wildlife. Attention was focused on the role of broad-leaved woodlands in a meeting in 1982 (*Broadleaves in Britain*, eds D.C. Malcolm, J. Evans & P.N. Edwards, published by the Institute of Chartered Foresters) and during the 1980s encouragement has been given to planting and managing broad-leaved trees in both the British lowlands and uplands.

Foresters, ecologists and scientists from different disciplines have often claimed that tree species mixtures have advantages over monocultures. The purpose of this symposium was to bring those people together to discuss the various advantages, including aspects of timber production, synergism between trees with different life habits, the use of $N_2$-fixing tree species, the diversity of habitats offered to plants and animals, and the balance of biotic and physical factors that may diminish the risks of catastrophe from wind or pest outbreaks. Many of these advantages are accepted by practitioners, but some are controversial and the scientific basis for many of them is only now becoming clear. This symposium provides a review of the current state of knowledge and the current scientific issues.

Following an introductory review by Dr O. Rackham, this symposium first considers the growth, yield and silviculture of mixed-species stands from a forester's viewpoint. Secondly, the interactions among tree species in mixtures are considered, particularly the below-ground interactions and factors that affect tree nutrition. Thirdly, current information and some new studies are presented on the benefits, or otherwise, of tree species mixtures as habitats for herbaceous flora, birds, mammals and invertebrates. Finally, reports are presented of discussion groups which were formed at the end of the meeting to fill any gaps in the programme and to highlight the issues and priorities for further research. Throughout, the meeting was focused on the biophysical sciences; no attempt was made to cover the socio-economic or policy aspects. There was, understandably, a bias in this British Ecological Society meeting towards the UK scene, but we are grateful to the many overseas participants who helped greatly to raise the level of discussion, and who were able to bring information from countries with extensive and long-established mixed-species woodlands.

The meeting was attended by 105 participants from twelve countries and was planned and managed by an organizing committee. Special thanks are due to Mrs C. Morris, Mr J.D. Deans, Mr G. Patterson, Mr F.J. Harvey and Dr A.F. Harrison for helping to plan and run the meeting, and to Dr D.A. Burdekin, Dr J. Anderson, and Dr G. Peterken for chairing sessions.

M.G.R. CANNELL
D.C. MALCOLM
P.A. ROBERTSON
Edinburgh 1991

# Mixtures, mosaics and clones: the distribution of trees within European woods and forests

O. RACKHAM

*Botany School, University of Cambridge, Downing Street, Cambridge CB2 3EA, UK*

## SUMMARY

Most natural woods (or forests) in Europe consist of combinations of tree species, except in some relatively extreme environments. Sometimes the species form mixtures and sometimes mosaics. Natural mosaics may result from the patchy nature of the environment or from patchy regeneration; but much of this type of variation appears to be inherent in the behaviour of certain species of tree, especially (but not wholly) those that reproduce vegetatively to form clones.

In this paper the stability of mixtures and mosaics will be discussed as well as processes that add or subtract tree species. I will speculate on the mosaic nature of the original wildwood, dealing mainly with natural woodland, but also with plantations, hedges and wood-pasture.

## INTRODUCTION

Most natural woodland consists of mixtures of trees. This is commonly regarded as a contrast between natural woodland and plantation forestry. In Peterken's analysis (1981) of the tree communities of Great Britain, forty-five out of the fifty-nine 'stand types' are defined as mixtures of trees, and only six are typically pure stands. Other examples, taken at random (Table 1), show that this distribution is not atypical of west Europe. Ecologists and phytosociologists probably take less notice of simple stands, on the plea that they are either dull or artificial, but even so there can be little doubt that, except in some relatively extreme environments, most of the more nearly natural types of woodland in Europe are mixed.

I distinguish between plantations, natural woods and wildwood. Natural woods (or forests) are those in which the trees have not been planted. Wildwood (*Urwald*, 'primeval forest'), in Europe, is the past vegetation of prehistoric times, before the impact of civilized human activities. These activities began in the Neolithic period 6000–7000 years ago.

Many papers in this volume are concerned with plantations. This is partly because of the peculiarity of British (and Irish) forestry which, for over 100 years, was concerned almost exclusively with planted trees. But it is logical to investigate mixtures of species in the simplified ecosystem of a plantation, avoiding many of the irregularities and complexities of natural woods.

I

TABLE 1. Approximate numbers of single-species and mixed-species types recognized in a sample of regional classifications of European native woodland. Note that although each classification was meant to be comprehensive for its region, different authors had different standards of classification. It should not be inferred that Belgium and Catalonia have roughly the same number of types of woodland but that S. Germany is four times as diverse. (*Peterken (1981). †Lebrun *et al.* (1949). ‡Jovet (1949). §Oberdorfer (1957). ‖ Folch i Guillèn (1981).)

|                                                   | Britain* | Belgium† | N. France‡ | S. Germany§ | Catalonia ‖ |
|---------------------------------------------------|----------|----------|------------|-------------|-------------|
| Defined in terms of two or more distinct tree species | 45       | 15       | 32         | 172         | 17          |
| Defined in terms of two or more very closely related species | —        | 3        | 1          | 1           | —           |
| Two or more species not included in definition, but can occur | 8        | 13       | 1          | 15          | 21          |
| Single species                                    | 6        | 18       | 7          | 12          | 9           |
| Total number of types                             | 59       | 49       | 41         | 200         | 47          |

In Europe, most woods have been managed, for example by coppicing; indeed for millennia or centuries management has been almost as integral a feature of natural woodland as it has been of natural grassland. A question to be asked is which characteristics of mixed natural woods have been modified, or have resisted modification, by management. Nearly all natural woods that have not been managed are secondary woods that have arisen within the last 100 years or so on former farmland, heath, etc.—too recently for management to have been imposed.

## MOSAICS AND CLONES

The distinction between single- and multi-species stands is not always clear-cut. If a wood of 30 ha consists of two trees with *A* being twice as abundant as *B*, it may be that the whole wood is a mixed stand of *A* plus *B*; or that there are two single-species stands with 20 ha of *A* and 10 ha of *B*. What often happens is that the wood is an irregular mosaic of patches in which, let us say, 2 ha of pure *A* alternate with 1 ha of pure *B*.

The scale of the mosaic varies by more than three orders of magnitude. At one extreme, more than half of the Forêt de Chantilly, north-east of Paris, is a single patch of some 2200 ha of pure *Tilia cordata*. At the other extreme, in the woods of north Essex (Fig. 1) and south Suffolk (England), every hectare differs from every other hectare.

Especially at the lower end of the scale of variation, mosaic woods pose problems, both of concept and of practice, in classifying woodland and in defining plant communities. In a series of woods of diminishing mosaic size, at what point along the scale are we to stop treating the mosaic as an aggregate of different stand-types and to regard it, instead, as part of the structure of a single, complex, stand-type? In gathering data, what size of relevé do we use, and what do we do as the scale of the mosaic becomes smaller than the relevé? (In many mixed

(a)                          (b)

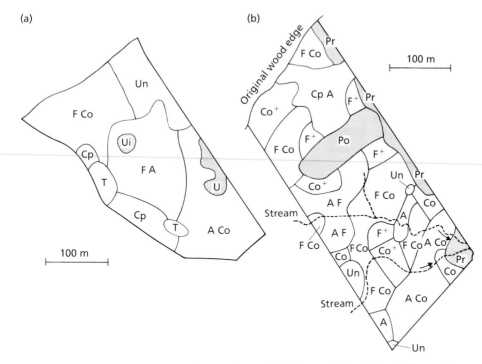

FIG. 1. Mosaic structure of two Essex (England) woods. (a) Tarecroft Wood, Rivenhall: a small ancient wood. (b) Hales Wood, Ashdon: a fragment of a large ancient wood. A, *Acer campestre*; Co, *Corylus avellana*; Cp, *Carpinus betulus*; F, *Fraxinus excelsior*; Po, *Populus tremula*; Pr, *Prunus spinosa*; Ui, *Ulmus minor* (invasive); Un, *Ulmus* 'Lineage' group (non-invasive); T, *Tilia cordata*. F$^+$ means a mixed stand with *Fraxinus* the most abundant tree. Trees which are obviously clonal are shaded.

stands, a relevé of at least 10 ha would be needed to catch significant but rare components of the mixture, for example *Sorbus torminalis*, *Malus sylvestris* or *Frangula alnus*. In small-scale mosaics, a relevé of less than 0.5 ha would be needed to avoid including more than one stand-type.) These questions are posed but answers cannot be suggested here. Nor can this paper try to deal with the question of whether the ground vegetation (undershrubs, herbs, bryophytes) should be integrated with the tree communities or should be treated as a separate set of plant associations (Rackham 1980; Ellenberg 1986).

## Causes of mosaics

### Planting

The first question to ask about a tree mosaic is how much of it is artificial. Areas of planted trees, within an apparently natural wood, are easy to detect if they are of non-native species. If of native species, they are usually detectable if the trees are in straight lines or if they stop suddenly at rides, straight edges or other management boundaries. Planted native trees often differ genetically from wild populations; for instance, *Tilia* is usually the hybrid *T.* × *vulgaris* rather than a native species.

Although some mosaic variation within native woodland passes these tests of artificiality, most of it does not. The patches, although they may have sharp edges, are irregular and unrelated to management boundaries. Planting a stand of trees within an existing wood is, even today, an operation which often fails; before the invention of chemicals for poisoning the existing trees, it would usually have failed.

The consequence of planting trees in an existing wood is usually more complex than the mere replacement of the existing trees by planted trees. Such an area is sometimes now marked by a patch of wild trees (especially birch) which came in after the planted trees failed or were felled. These patches, too, are excluded from the following discussion of mosaics as a natural phenomenon.

*Variation in environment*

In the east English Midlands, on chalky boulder-clay soils, there is a type of woodland of *Fraxinus excelsior, Acer campestre* and *Corylus avellana* lying on ridge-and-furrow — regular undulations, typically 11 m in wavelength and 0.3–0.8 m in height, which result from medieval strip-cultivation (Rackham 1986a). These woods probably arose in the late fourteenth century and are now very similar to the original woods of the area, which they often adjoin; they contain giant coppice stools (Rackham 1980) like any other ancient woodland. In detail, the ash stools lie mainly in the furrows and the maple and hazel stools on the ridges. The herbaceous vegetation also shows a zonation, the well-drained ridges being dominated by *Mercurialis perennis*, which is relatively sensitive to waterlogging, and the furrows by less sensitive species such as *Deschampsia cespitosa* and *Primula elatior* (Fig. 2).

All this can be explained in terms of the flooding of the furrows in spring (Rackham 1975). A probable mechanism is that poor drainage favours ash by repressing *Mercurialis*, which competes with its seedlings (Wardle 1959; Martin 1968).

This very small-scale mosaic is of two highly distinct plant communities, differing both in trees and in ground vegetation. They correspond roughly to Peterken's Stand Types 1Ba ('wet ash–wych-elm woods, heavy soil form') in the furrows and 2Ab ('wet maple woods') on the ridges. Because they occur in strips of only 5–6 m wide, no routine method of phytosociology could be expected to separate them; any relevé of reasonable size would contain a mixture of both. Had the determining feature of the environment been less conspicuous they would probably not have been detected at all.

Another example of a small-scale mosaic is the mosaic of *Pinus brutia* and *Cupressus sempervirens* in the White Mountains of Crete. Where these tree

FIG. 2. *Fraxinus–Acer* woodland on ridge-and-furrow. Buff Wood, Cambridgeshire (England), May 1987.

species occur together the cypresses tend to grow on bare rock and the pines on soil and fine scree.

### Gregarious trees

Much mosaic variation is connected with the gregarious nature of certain trees. If one finds a *Carpinus* tree in a wood, for example, the probability is that the next tree will also be *Carpinus*. This is not true of *Acer campestre*, which tends to be scattered apparently at random through those woodland types that contain it; it is still less true of *Malus sylvestris*, which though widespread is never abundant and seldom, if ever, occurs next to another *Malus* tree (Table 2).

Some trees are gregarious because they live in extreme habitats; for example, *Alnus glutinosa*, the chief tree of flushed sites. But many species, including *Carpinus* and *Tilia cordata*, have wide ecological ranges which defy such a simple explanation.

### Clones

Some European trees reproduce by suckering; these include *Populus alba*, *canescens* and *tremula*, *Prunus avium* and *spinosa*, *Sorbus torminalis*, *Ulmus* spp. (except *U. glabra* and a few others) and *Zelkova cretica*. This property, although

TABLE 2. Trees of English woods, arranged in descending order of gregariousness

Strongly gregarious    *Ulmus procera* group (clonal)
*U. minor* group (clonal)
*Carpinus betulus*
*Tilia cordata*
*Alnus glutinosa*
*Quercus petraea*
*Ulmus* 'Lineage' group (Rackham 1980)
*Castanea sativa* (ancient introduction)
*Prunus avium* (clonal)
*Populus tremula* (clonal)
*Corylus avellana*
*Acer campestre*
*Fraxinus excelsior*
*Quercus robur*
*Ulmus glabra*
*Crataegus monogyna*
*Sorbus torminalis* (weakly clonal)
*Malus sylvestris*
Anti-gregarious    *Pyrus pyraster*

doubtless favoured in managed woodland, is not the product of genetic selection through management, for it occurs also in American species of *Populus* and *Ulmus*, as well as in the American beech (*Fagus grandifolia*).

Clonal trees, which may have difficulty in reproducing by seed, are often gregarious. Those that have large clones and are strongly dominant, such as elms and *Prunus avium*, tend to form pure stands up to 5 ha in extent; the elms in particular invade existing woods and convert them into elm woods.

*Regeneration in gaps*

Another way in which woods might become patchy is by gaps arising from the death of one or a group of trees, and being colonized by a different species. A classic (though somewhat unusual) example of a regular gap-phase succession is provided by the beech woods of England, in which a gap is often colonized by *Fraxinus* (on calcareous soils) or *Betula* (on acid soils) and later replaced by new *Fagus* (Watt 1924). Such a predictable process is familiar in America (see bibliography in Peterken & Jones 1987). In Europe, succession is more chancy; it may depend on which of the surrounding trees happened to be fruiting well in the year in which a gap became available.

In the Forest of Białowieża (Poland), a mixed stand of *Carpinus, Tilia, Quercus* and *Ulmus glabra* has a very small-scale mosaic of patches of different species and ages of young to middle-aged trees. Some of these species, however, regenerate not only in gaps but also as saplings arising from seed beneath the canopy of trees of other species (Pigott 1975).

Different kinds and stages of gap cycles, perhaps 'frozen' at a particular moment by the introduction of coppicing, seem an attractive way of accounting for the apparent randomness of much mosaic variation. But there are arguments against it. Gaps left by the death of single trees are too small to account for any but the smallest-scale mosaics. In America, larger gaps are attributed to fires or hurricanes. Fires are unlikely in Europe except in pine woods or mediterranean vegetation. The English Channel storm of 16 October 1987 shows that hurricanes can occur in Europe. However, most of the trees uprooted or broken by that storm have not died and will not leave gaps. The damage occurred mostly in swathes which, if the trees were to die, would leave long narrow gaps; this is not a typical pattern in woodland mosaics.

*Conclusions*

Only some of the mosaic formation in woodland can be identified as artificial, clonal, environmental or developmental. A further possibility is that strongly gregarious trees which do not sucker, such as *Tilia cordata* and *Carpinus*, might be 'crypto-clonal', forming clones in some other way; for example, by taking root from fallen stems. The practice of propagating coppice stools by layering (bending shoots over and pegging them to the ground to take root), which is often mentioned in books on coppicing and occasionally found in practice, would have the same effect. This possibility can be checked with *Tilia*, in which individual clones can be recognized from the variable shape and pigmentation of the leaves of young coppice shoots. In English lime woods, for example Chalkney Wood (Essex) and Groton Wood (Suffolk), three or four stools at the most, but more often only a single stool, constitute a clone. There is no possibility of a clone extending to an entire element in the tree mosaic. The crypto-clonal theory thus fails for *Tilia*, although it could explain the gregarious nature of the 'lineage' group of non-*glabra* elms which appear not to sucker.

Much of the remaining mosaic variation may depend on features of the environment less easily observed than ridge-and-furrow or rock versus scree. Woodland soils, in particular, show much variation (especially fossil periglacial features) which can be detected only with difficulty if at all (Rackham 1980, 1986a). The very variable soils of the north Essex woods, full of ice-wedges, sand-lenses, etc., contrast with the remarkably uniform soils of the Forêt de Chantilly, which consist, over long stretches, of 70 cm of fine sand overlying soft limestone.

## MIXTURES: BORDERLINE CASES

*Mixtures of a common and a rare species*

Some mixtures are overwhelmingly of one species. Often, the rare component can be dismissed as a contaminant or recent invader (e.g. *Betula* in *Carpinus* woods), but sometimes it appears to be significant. For example, many otherwise pure oak woods, from Cornwall (England) to Macedonia (Greece), have *Sorbus torminalis* as a minor constituent—typically with well under one *Sorbus* to 1000 *Quercus*. Although rare, this is a highly characteristic tree of this group of woods, which it would be perverse to ignore.

*Mixtures of very closely related species*

Another borderline case is the oak woods which cover several thousands of square kilometres in the lower Pindos Mountains of N.W. Greece (Grevenà province). These woods have a long history of management: woodcutting, wood-pasture and cutting for leaf-fodder. They vary vastly in structure, often forming a mosaic with steppe, and displaying many different combinations of coppicing, browsing and pollarding. For the most part they are composed of oak with no other tree except sometimes an understorey of *Juniperus oxycedrus* and rare individuals of *Sorbus torminalis*.

These woods are an intimate mixture of at least seven oaks: *Quercus cerris, trojana, frainetto, brachyphylla, pubescens, pedunculiflora* and *virgiliana*. These are evidently highly interfertile and strongly introgressed; many intermediate oaks defy identification. There is no obvious ecological difference between the species except that, for example, *brachyphylla*-type oaks are commoner among the pollards and *frainetto*-type oaks are commoner among the coppice stools.

Whether or not these woods constitute a mixture depends on the weight one gives to the taxonomic differences between the oaks. It is conventional to treat them as separate species, but in the hands of a taxonomist of the 'lumping' school most of them would be demoted to subspecies. (This example should disquiet those ecologists who measure the diversity of vegetation by counting the number of species, regardless of how diverse those species are.)

## MIXTURES IN PLANTATIONS

Whether plantations are intended to be of single species or mixtures of two or more species is a matter of taste, in which fashions have changed. Many early plantations in England, from the seventeenth century onwards, were made by sowing a mixture of the seeds of various trees in imitation of natural woods. One species, usually oak or pine, was intended for timber and others, such as hazel or hornbeam, to be coppiced as underwood. Less often it was the practice to sow a single species, mainly oak or *Castanea*.

As time went on single-species plantations became more common, a development perhaps connected with the growing fashion for transplanting trees instead of sowing them on the spot. Records of the medals awarded by the Society of Arts (London) for tree-planting in the later eighteenth century show that by this time single-species plantations were becoming predominant (Rackham 1991). However, there are many surviving mixed plantations from the nineteenth century. There next appears the practice of planting two species, not for timber and underwood, but for two successive timber crops, or with the intention that one would act as a nurse to the other. Very often the intention was not realized and both crops were left standing so that they now form a mixture.

With the great expansion of plantation forestry in Britain and Ireland in the twentieth century, mixed plantations became unfashionable. Most plantations from the 1920s onwards are of single species, usually *Pinus nigra* subsp. *laricio, P. sylvestris, P. contorta, Picea sitchensis, Larix europæa, L. leptolepis* or *Populus* cultivars. The vogue for single-species plantations was one of the main points of dispute between foresters and conservationists. Chiefly for this reason, fashion has swung the other way since 1975 to such an extent that there is now a 'Standard Broad-leaf Mixture' used for amenity plantations which includes *Tilia cordata, Acer campestre, Prunus avium* and *Alnus glutinosa*. This is often thought of as a re-creation of natural woodland, although few, if any, natural woods actually have this whimsical composition.

Plantations can become mixed by default. The creator of a plantation may forget, or may lack the energy or the means, to prevent the natural processes of succession. The pine plantations of the Sandlings of east Suffolk (England) developed, after some 40 years, an understorey of oak; the poplar plantations of the Fens developed an understorey of alder. Both of these were on former farmland. In Britain in the 1950s and 1960s there was a fashion for destroying ancient woods and replacing them with plantations. This has had a poor record of success. The original trees (especially *Tilia*, which is tolerant of poison and shade) often turn out to be difficult to kill and have grown again from the stumps. What was intended for a simple plantation has become a complex mixture of surviving original trees, surviving planted trees and trees naturally colonizing (Fig. 3).

## SECONDARY WOODLAND

New woodland arising on former farmland, heath, fen, etc. is determined especially by the nearest seed source. Some recent woodland thus consists of one species, such as the *Crataegus* woods of the English Midlands and Essex (Ross 1936; Rackham 1986c), and some of two or more species, such as the *Quercus–Crataegus* woods of the same areas, or the majority of recent woods in the eastern United States.

In mainland Greece a common type of recent woodland is a mixture of *Quercus coccifera*, already present as a bitten-down bush and now growing up into a tree, and *Pinus halepensis* colonizing from outside (Rackham 1983). Gávdhos, the southernmost island in Europe, was almost treeless in the nineteenth century but now is more than half wooded; the trees are *Pinus brutia, Juniperus macrocarpa* and *J. phoenicea*, which tend to form separate stands according to the soils present but overlap to some extent.

FIG. 3. Mixture inadvertently created by attempted replanting of an ancient wood. Only a few planted trees (*Picea*) survive. *Populus tremula*, already present, has prospered. Some invasion by *Crataegus* and *Salix* species. Tiddesley Wood, Worcestershire (England), May 1985.

The original species in a recent mixed wood may remain in balance for many years, for example *Quercus robur* and *Crataegus monogyna*, or they may be invaded by a later successional species and converted to a pure stand; a classic English example is the succession from 'chalk scrub', a complex community of shrubs and small trees, to pure beech woods described by Watt (1934).

## HEDGES

Mixed and single-species stands occur among hedges as well as woods. In England, the number of species in a hedge tends to be proportional to its age (Pollard *et al.* 1974), but there are various exceptions to this rule (Rackham 1986a).

Hedges can be deliberately planted, can arise spontaneously at the edges of fields or (rarely) can represent the edge of a grubbed-out wood. Planted hedges (the commonest kind of modern hedge in England) are usually of a single species but slowly acquire further species with time.

Spontaneous hedges are usually mixed from the start. These are probably common among ancient hedges in England but rare among modern hedges. For example, a hedge near Hayley Wood (Cambridgeshire) which cannot be older than 1862 has seven or eight tree and shrub species. In America, spontaneous hedges are the norm; thus, in a hedge near Waco (Texas), which was not there in 1960 and is in a land where hedges until recently were unknown, I have found *Ulmus crassifolia*, *Celtis laevigata*, *Quercus velutina*, *Fraxinus texensis*, *Juniperus ashei*, *Rhus lanceolata* and *R. toxicodendron*.

## WILDWOOD: MIXTURES OR MOSAICS?

Early pollen analysts generally regarded the pre-neolithic wildwood of Europe as a more or less undifferentiated 'mixed oak forest', quercetum mixtum or *Eichenmischwald*; a mixture of *Quercus*, *Ulmus*, *Tilia*, *Fraxinus*, *Fagus*, etc. with *Corylus* as an understorey shrub. This was presumably influenced by the idea, then current, of a climax vegetation determined mainly by climate rather than by soil, clonal behaviour or the activities of animals or of mesolithic men.

It might be possible to work out the tree communities of certain kinds of wildwood by mapping the distribution of fossil trees in peat (bearing in mind that some trees, notably *Tilia* and *Ulmus*, seem not to be preserved). As far as I know this has not been done.

In one respect, the pollen record proves that wildwood cannot have been a uniform mixture. *Corylus avellana* can be a considerable tree as well as an understorey shrub, but it cannot compete in height with *Tilia*, *Ulmus* or probably *Quercus*. Although it can survive in shade, it produces pollen only where it forms part of the canopy. If this present behaviour of *Corylus* is any guide to the past, it follows that when a pollen sample contains large quantities of *Corylus* pollen as well as that of other taller trees this implies patches of *Corylus*-dominated woodland distinct from the rest of the wildwood.

The pre-neolithic pollen record of England can be divided into regions and subregions, but also shows local variation. Contemporary samples from similar sites a few kilometres apart, or even from different parts of the same site, for example the meres of south Norfolk (Rackham 1986b), show wide variation in the proportions of the 'mixed oak forest' trees.

Large quantities of wood, probably not from wildwood but from the earliest managed woodland, are preserved in the neolithic trackways laid across the peat of the Somerset Levels. The earliest, the Sweet Track of 3900 B.C., is a complex structure including poles of *Quercus*, *Fraxinus*, *Tilia*, *Corylus*, *Alnus* and *Ilex*. Some of the later tracks are composed of *Corylus*, sometimes with a little *Fraxinus* (Coles *et al.* 1975–85). These are sophisticated structures and we must not suppose that each is necessarily a random selection from a single stand of trees, but they suggest that there were both mixed and single-species woods, not very different from the ancient woods that still survive in the region.

This evidence, such as it is, works against the theory of a uniformly mixed wildwood; it indicates that the wildwood was a mosaic of different mixtures with some pure stands.

## NATURAL WOODLAND: STABILITY OR CHANGE IN MIXTURES AND MOSAICS

### *Gains and losses of species due to management*

In countries where modern forestry is a development from traditional woodland management, forestry methods are often intended to encourage a single species. Where the same species has remained desirable for a long period this can produce pure stands at the expense of mixtures and mosaics. There can be little doubt that the monotonous stands of beech, typical of many of the large forests of France, replace what had once been more varied woodland.

It has often been claimed that pure stands of one species, where not planted, result from the deliberate elimination of unwanted species in the course of management. For example, the oak coppice woods of Scotland were managed in the eighteenth century chiefly for bark for tanning, for which only oak will do, and there exist occasional prescriptions for extirpating other species, termed 'barren wood' (Monteath 1836). I am sceptical as to whether these attempts were widespread or persistent enough to have had much effect on the woods as they are today. To eliminate a species so completely that it has never returned would have called for sustained effort well beyond what normally went into the maintenance of woods. Except for a few specialized purposes, woodland records display remarkably little interest in the species comprising the underwood.

Management may add species that would not have been there otherwise. In west Europe, there can be little doubt that the ecological range of *Quercus* has been increased by the tradition, which in England goes back for well over 1000 years (Rackham 1990), of treating oak as a standard tree and allowing it to grow to timber size but coppicing other species as underwood. Oak already had a wide ecological range but has thus been encouraged, especially in types of woodland (e.g. those dominated by *Tilia*) in which it otherwise would not easily compete. (For the same process in Germany see Pott (1988).) More recently oak has been planted into existing woods, often successfully; such planted oaks can be recognized by their uniform genotype and age.

Composition can also alter as an incidental, often unwanted, consequence of management. Coppicing eliminates trees which do not respond, such as most conifers; it alters the competitive relations between other species. It may encourage trees such as *Corylus* and *Betula* which fruit within a coppice cycle. However, *Fraxinus*, *Acer campestre* and especially *Tilia*, though they may not fruit unless the coppice cycle is long, are not necessarily discouraged since a coppice stool lives indefinitely (Rackham 1990). A characteristic woodland type of north-east Suffolk is a mixture of *Carpinus* and *Corylus*. It is hard to suppose that hazel could withstand the competition of hornbeam unless both were coppiced. (For changes in non-mediterranean Italy, see Moreno (1971).)

In France and Germany there is a body of lore about which species are supposed to be encouraged by coppice cycles of different lengths (Ellenberg 1986). *Carpinus* is supposed to be favoured over *Fagus*, which coppices less vigorously;

this, however, is unlikely to account for the hornbeam woods of England in which *Tilia* is more likely as a predecessor of *Carpinus* (Rackham 1986c). Palatable trees are favoured if, as usual, browsing animals are kept out of a coppice wood lest they eat the young shoots. In general, therefore, coppicing might tend to increase the degree of mixture of a stand.

## Invading species

Introduced trees, originally planted, often invade existing woodland to form mixed stands with species already there; sometimes to such a degree (as with *Castanea* in England) that it is impossible to tell, except by pollen analysis, that they are not indigenous. Most species commonly naturalized in Europe, however, are highly competitive and tend to form pure stands—examples include *Acer pseudoplatanus* and *Rhododendron ponticum* outside their native ranges and *Robinia pseudacacia*, which is clonal in Europe.

  Indigenous species can also develop a capacity to invade. Two widespread changes in English woods over the last hundred years are the increases of *Fraxinus* and *Betula* (Rackham 1980, 1986a). *Betula* has become abundant even on soils (e.g. clays) which 70 years ago were thought unsuitable for it; this is part of a general increase beginning on disused heathland, derelict industrial land and felled plantations and which has greatly added to the seed available for colonizing less suitable habitats. *Fraxinus* has increased especially at the expense of *Corylus*; this increase, which has not been satisfactorily explained, is a warning against devising facile explanations for such changes in pre-history.

## Loss of species by disease

Chestnut-blight, caused by the fungus *Endothia parasitica* introduced in 1904, has very effectively subtracted *Castanea dentata* from mixed woods all over the eastern United States. The twentieth-century epidemics of Dutch elm disease in Europe have not usually had this effect. Woodland elms are rarely killed altogether but sucker from the roots or coppice from the base of the stem, unless the wood is browsed. However, an epidemic of *Ceratocystis ulmi* is the most likely cause for the elm decline which occurred suddenly throughout west Europe in the early neolithic period. It may not have killed the elms but would have reduced their pollen production; it is the only cause so far proposed that is sufficiently powerful and specific to elm to have had this effect (Rackham 1980, 1986a).

## Browsing and wood-pasture

A still more effective subtractor of species is browsing. For example, when coppicing was resumed in Hayley Wood (Cambridgeshire) in 1962, the regrowth of many species was not as expected because of browsing by fallow deer (*Dama dama*). These deer have a strong and definite preference for species: *Ulmus* >

*Fraxinus* > *Crataegus* > *Corylus* > *Acer campestre* > *Populus tremula*. Repeated browsing of their favourite trees, especially ash, kills the coppice stool. Where deer browsing is severe, one cycle of coppicing is enough to change a *Fraxinus*-dominated into a *Populus*-dominated area (Rackham 1975). The browsing order varies from one animal to another; thus muntjac deer (*Muntiacus reevesi*) prefer *Corylus* to *Fraxinus*.

The matter is not always simple. Deer activity in Hayley Wood varies from year to year; when browsing slackens, *Fraxinus* seedlings spring up in abundance in gaps created by previous browsing. Even if browsed later, saplings are not easily killed and in areas subsequently fenced may even create a pure stand of ash. Intermittent browsing may therefore introduce patches of pure ash into a mosaic. It may be significant that the woods of Hatfield Forest (Essex), which have a history of intermittent enclosure, contain much ash but few old stools of it (Rackham 1989).

In England there is a tradition, going back well over 1000 years, of separating wood-pasture browsed by domestic animals from ordinary woodland. Part of the wood-pasture tradition was concerned with deer (including the introduced *Dama*) which, for centuries, were treated as semi-domestic animals. For this reason Hayley Wood has not been exposed to deer until recently. Wood-pastures differ from ordinary woodland in their tree composition and often have few tree species or only one, usually oak or beech.

When wood-pastures cease to be browsed, they often revert to mixtures as palatable species (notably *Ilex aquifolium*) either colonize or grow up out of bitten-down bushes. Very similar changes to these in England are described in N.W. Germany by Pott and Burrichter (1983).

In part of Staverton Park (Suffolk) this process has generated a mixture with a peculiar structure. This medieval deer-park, last intensively browsed about 170 years ago (Peterken 1969), has close-set ancient pollard oaks and immense hollies which appear to be younger. The hollies grow in a ring around each oak; they have often overtopped and killed the oak (Fig. 4). When taking students there I invite them to theorize on how this could have happened; I now invite the reader of this paper to do the same. (The same happens with *Quercus robur* and *Ilex aquifolium* in the New Forest (England) and with *Q. laurifolia* and *I. opaca* on Cumberland Island, Georgia.)

In Wales, Scotland and Ireland the commonest type of ancient woodland is of oak (*Quercus petraea*), often with very few other species present. Most of these woods are accessible to sheep. Where other species occur, for example *Ulmus glabra* and *Populus tremula*, these are often on cliffs where the sheep cannot reach them. Although these woods were probably always strongly dominated by oak, owing to the infertile soils, it is likely that they would be mixtures had sheep not subtracted the more palatable species.

Throughout Europe, cliffs may answer the question of which trees have been subtracted from the local woodland by browsing. In Crete, for example, the many gorges contain *Quercus ilex, Pistacia terebinthus* and *Laurus nobilis*, which

FIG. 4. Ancient pollard oaks overtopped and partly killed by *Ilex aquifolium*. Staverton Park, Suffolk (England), July 1984.

are ill-adapted to browsing. *Cupressus sempervirens*, the most historically famous tree of Crete, is very resistant to browsing. Overgrazing in this island has probably not been limited to the 8000 years or so during which there have been sheep and goats. Like other mediterranean islands, Crete had a Pleistocene fauna of herbivores (elephants, hippopotami, deer) but no effective carnivore; vegetation 'impoverished' by browsing should therefore be regarded as the normal state.

## SOME HISTORICAL EXAMPLES

Historical studies of the tree composition of mixed natural woods are comparatively few. Peterken and Jones (1987) review what has been done on the repeated recording of permanent plots in Europe and America. Their longest example goes back 140 years. Historical records, made for other purposes, often mention tree species more or less casually (see, for instance, the medieval records of Gamlingay Wood (Cambridgeshire) in Rackham (1975) and of Hindolveston Wood (Norfolk) in Rackham (1980)); but they seldom give the total composition of a wood, for they are rarely interested in underwood species.

Fortunately, this meagre field includes three widely differing examples: a 'typical' coppice wood in eastern England; a mountain wood from the border of Wales; and a wood-pasture in southern England. (There are historical studies of woods from other parts of the British Isles, for instance Tittensor (1970) and Steven & Carlisle (1959), but the woods in these studies are not, or hardly, mixed.)

### Chalkney Wood (Essex, England)

This is one of the few woods which (i) are mixed, (ii) have historical evidence on the distribution of trees (including underwood) more than 200 years ago, and (iii) are still extant. For a full discussion see Rackham (1980).

The wood is on acid soils (loess and sand overlying London Clay). It consists (apart from a recent attempt at coniferization) of mosaics and mixtures principally of *Tilia cordata* and *Carpinus*, with some *Castanea*, *Fraxinus* and other trees.

The documentation consists of the wood-sale records of Roger Harlackenden, its owner from 1603 to 1610. He took some note of the underwood composition, especially of whether an area consisted wholly or partly of 'prye' (*Tilia*). The present mosaic of patches of *Tilia* and patches of other trees already existed, in general terms, by this date. The admixture of *Castanea* (which is sometimes an ancient introduction in England) appears to result from invasion since that date. There is some evidence for a decline in *Fraxinus*.

### Lady Park Wood (Herefordshire, England)

This wood in the Wye Gorge is an ancient coppice wood; its name indicates a wood-pasture episode in the distant past. It is an example of the less extreme, mixed woods of the Highland Zone. Part of it was last felled in *c*.1902 and is the subject of 40 years' observations on permanent plots by Peterken and Jones (1987).

The wood lies on very steep terrain. The soils are very varied, derived partly from the underlying limestone and partly from acidic loams and clays. The trees are a complex mosaic of three mixtures of species including *Fagus*, *Ulmus glabra*, *Tilia cordata*, *T. platyphyllos*, *Fraxinus*, *Corylus* and *Taxus*.

According to Peterken and Jones, beech gradually increased, for the first 70 years after the end of coppicing, at the expense of other species. Since 1970, there has been a series of sudden events: Dutch elm disease, then the great drought of 1976 which killed many beeches, and later storms which have produced gaps. These disturbances have suddenly halted the march towards a pure stand of beech, and have produced a patchy wood in which lime and elm are likely to fill the gaps.

This example gives colour to the theory that woodland mosaics are derived from random changes following patchy disturbance, in the American manner. However, Lady Park is an unusual wood. It is on very steep topography: crumbling cliffs are part of the disturbance, and it involves beech, a tree more sensitive to catastrophe than most.

## Avon Gorge (Bristol, England)

This amounts to a long-term experiment on browsing. The wood lies partly on a plateau and partly on steep slopes and cliffs. It is largely on hard limestone with mainly calcareous soils.

This wood was divided in the middle ages into a coppice part and a wood-pasture part. The coppice is a highly mixed wood including *Tilia*, *Fraxinus* and *Ulmus glabra*, of a kind frequent in and around the Mendip Hills and corresponding closely in composition to the local wildwood. The wood-pasture was grazed until the mid-nineteenth century. The present trees show that this part lost virtually all its ash and elm. During the wood-pasture period the principal trees were pollard oaks (a tree ill-adapted to these soils, but which resists browsing) with a few limes surviving on inaccessible cliffs. These oaks are now surrounded by a mixture of ash and elm, palatable species which returned after browsing ceased (Rackham 1982; Lovatt 1987).

Here browsing converted a mixture to a nearly pure stand; since browsing ceased, a mixture has arisen which is different from the original one.

## CONCLUSIONS

The division of woodland into mixtures, mosaics and clones has its parallel in other vegetation. For example, among grassland there are (besides artificial pure stands and artificial mixtures) natural mixtures, mosaics and pure stands. Mosaics may reflect variation in the environment brought about, for example, by anthills, periglacial features (stripes and polygons) or ridge-and-furrow. Or they may be the consequence of clonal behaviour by certain species, for example *Brachypodium pinnatum* invading chalk grassland. The invasion of salt-marshes by clones of the new species *Spartina anglica* is a close parallel to that of woodland by clones of species of *Ulmus* formed, at least in part, as a response to human intervention (Rackham 1980).

In a paper about mixtures of trees, rather than about the totality of species in a wood, I am reluctant to be drawn into questions of diversity theory. However, for what it is worth, I find that the total number of tree and shrub species in a wood is much less affected by the size of the wood than is the number of species of herbs and undershrubs. For separate woods it is commonly found that the number of species, $S$, is related to the area, $A$, by the regression

$$S = aA^b.$$

For five separate groups of woods in eastern England, I find that the value of the parameter $b$, for species of trees and shrubs, is relatively constant at 0.18–0.23. For the total of flowering plants and ferns, $b$ is higher and varies from 0.27 to 0.47 (Rackham 1980, 1986c). (A value of 0.50 means that a wood with four times the area has twice the number of species; a value of 0.20 means that thirty-two times the area is needed to double the number of species.) I leave others to explain this observation. (I have been unable to confirm the theory of Hooper (1971) that the value of $b$ depends on the woods' history of isolation.)

Among the natural woods of Europe, single-species stands are often in extreme environments: for example, the woods of *Pinus mugo*, *P. cembra* and *Alnus viridis* at high altitudes, of *Betula* in the far north, of *Cupressus sempervirens* on the south-facing limestone rock of Crete, of *Quercus petraea* in the Atlantic climate of west Britain, of *Alnus glutinosa* in peaty valleys, and of *Quercus* and *Fagus* in areas with a history of severe browsing.

In non-extreme environments, many single-species stands can be shown to be either artificial or reduced to the single-species state by management. Although natural single-species stands do occur, especially among gregarious trees, mixtures and mosaics are more usual. Some of this variation is environmental, some is related to changes with time, and some is due to intrinsic properties of the trees which are still poorly understood.

In eastern North America, where there are surviving wildwoods as well as managed woodland and vast areas of secondary woodland, these patterns are not exactly repeated. Single-species stands for the most part are either in the far north, or result from recent human intervention replacing more diverse tree communities (rather than modifying them as in Europe). For example, the vast and monotonous birchwoods (*Betula papyrifera*) in the Northwoods of Minnesota replace more varied pine-dominated woods; the monotonous *Pinus palustris* woods of Florida are artificially maintained by forestry.

In non-extreme environments the number of tree species is greater than in Europe, but not the number of tree communities. From what I have seen of them, eastern American woods typically consist of few communities, each of many species. Detailed analysis might reveal distinctions not apparent to the visitor, but American phytosociology seems not to have produced the many communities, each of few trees, that are regularly described by investigators in Europe.

American woods differ from European in many other ways. I have failed to find a well-marked permanent mosaic in American woods. There seem to be few or no strongly gregarious trees even among those which are clonal, such as *Fagus* and *Ulmus*. There are many more shade-bearing species than in most of Europe, with greater opportunities for cyclical succession. Shrub species are vastly more numerous and abundant than in European woods. American woods seem to change much more rapidly than is typical for Europe. We must not suppose that European woods were once less of a mosaic, like American woods today, nor that the differences are solely due to their different cultural histories.

As far as we can tell European woods have always consisted mainly of mixtures and mosaics, with some single-species stands. Any tree community which was confined to what (by prehistoric standards) was good agricultural land is unlikely to survive. Apart from this, those forms of human intervention and neglect which tend to increase the degree of mixture would seem to be roughly balanced by those which tend to reduce it.

## ACKNOWLEDGMENTS

I am indebted to many colleagues and friends who have introduced me to places mentioned in this paper. I would particularly mention Dr D.E. Coombe for the Cambridgeshire woods; Dr Susan Bratton for the S.E. United States; and Dr Jennifer Moody for America and Greece. The Greek work includes projects due to the University of Minnesota (Grevenà Archaeological Survey) and the European Community (Desertification in Southern Europe—Crete).

## REFERENCES

Coles, J.M. *et al.* (1975–85). *Somerset Levels Papers, passim.*

Ellenberg, P. (1986). *Vegetation Mitteleuropas mit den Alpen.* Ulmer, Stuttgart. (Translated by G.K. Strutt (1988) as *Vegetation Ecology of Central Europe.* Cambridge University Press, Cambridge.)

Folch i Guillèn, R. (1981). *La Vegetació dels Països Catalans.* Ketres, Barcelona.

Hooper, M.D. (1971). The size and surroundings of nature reserves. *The Scientific Management of Animal and Plant Communities for Conservation* (Ed. by E. Duffey & A.S. Watt), pp. 555–61. Blackwell Scientific Publications, Oxford.

Jovet, P. (1949). *Le Valois: Phytosociologie et Phytogéographie.* Société d'Edition d'Enseignement Supérieur, Paris.

Lebrun, J., Noirfalaise, A., Heinemann, P. & Vanden Berghen, C. (1949). Les associations végétales de Belgique. *Bulletin de la Société de Botanique Royale de Belgique,* **82,** 105–207.

Lovatt, C.M. (1987). The historical ecology of Leigh Woods, Bristol. *Proceedings of Bristol Naturalists' Society,* **47,** 3–19.

Martin, M.H. (1968). Conditions affecting the distribution of *Mercurialis perennis* L. in certain Cambridgeshire woodlands. *Journal of Ecology,* **56,** 777–93.

Monteath, R. (1836). *The Forester's Guide and Profitable Planter,* 3rd edn. Tegg, London.

Moreno, D. (1971). La Selva d'Orba (Appennino Ligure): note sulle variazioni antropiche della sua vegetazione. *Rivista Geografica Italiana,* **78,** 311–45.

Oberdorfer, E. (1957). *Süddeutsche Pflanzengesellschaften.* Fischer, Jena.

Peterken, G.F. (1969). Development of vegetation in Staverton Park, Suffolk. *Field Studies,* **3,** 1–39.

Peterken, G.F. (1981). *Woodland Conservation and Management.* Chapman & Hall, London.

Peterken, G.F. & Jones, E.W. (1987). Forty years of change in Lady Park Wood: the old-growth stands. *Journal of Ecology,* **75,** 477–512.

Pigott, C.D. (1975). Natural regeneration of *Tilia cordata* in relation to forest-structure in the forest of Bialowieza, Poland. *Philosophical Transactions of the Royal Society (London),* **B270,** 151–79.

Pollard, E., Hooper, M.D. & Moore, N.W. (1974). *Hedges.* Collins, London.

Pott, R. (1988). Entstehung von Vegetationstypen und Pflanzengesellschaften unter dem Einfluß des

Menschen. *Düsseldorfer Geobotanische Kolloquien*, **5**, 27–54.

Pott, R. & Burrichter, E. (**1983**). Der Bentheimer Wald: Geschichte, Physiognomie und Vegetation eines ehemaligen Hude- und Schneitelwaldes. *Forstwissenschaftliches Centralblatt*, **102**, 350–61.

Rackham, O. (**1975**). *Hayley Wood; its History and Ecology*. Cambridgeshire & Isle of Ely Naturalists' Trust, Cambridge.

Rackham, O. (**1980**). *Ancient Woodland: its History, Vegetation and Uses in England*. Edward Arnold, London.

Rackham, O. (**1982**). The Avon Gorge and Leigh Woods. *Archaeological Aspects of Woodland Ecology* (Ed. by M. Bell & S. Limbrey), British Archaeological Reports International Service, **146**, 171–6.

Rackham, O. (**1983**). Observations on the historical ecology of Boeotia. *Annual of the British School of Archaeology at Athens*, **78**, 291–351.

Rackham, O. (**1986a**). *The History of the* [British and Irish] *Countryside*. Dent, London.

Rackham, O. (**1986b**). The ancient woods of Norfolk. *Transactions of Norfolk & Norwich Naturalists' Society*, **27**, 161–77.

Rackham, O. (**1986c**). *Ancient Woodland of England: the Woods of South-East Essex*. Rochford District Council, Rochford.

Rackham, O. (**1989**). *The Last Forest: the Story of Hatfield Forest*. Dent, London.

Rackham, O. (**1990**). *Trees and Woodland in the British Landscape*. Dent, London.

Rackham, O. (**1991**). Landscape and the conservation of meaning. *Journal of the Royal Society of Arts*, **139**, 903–15.

Ross, R. (**1936**). The ecology of hawthorn scrub in south-west Cambridgeshire. In A.G. Tansley, *The British Islands and their Vegetation*, pp. 480–4. Cambridge 1939.

Steven, H.M. & Carlisle, A. (**1959**). *The Native Pinewoods of Scotland*. Edinburgh.

Tittensor, R.M. (**1970**). History of the Loch Lomond oakwoods. *Scottish Forestry*, **24**, 100–8.

Wardle, P. (**1959**). The regeneration of *Fraxinus excelsior* in woods with a field layer of *Mercurialis perennis*. *Journal of Ecology*, **47**, 483–97.

Watt, A.S. (**1924**). The development and structure of beech communities on the South Downs. *Journal of Ecology*, **12**, 145–204.

Watt, A.S. (**1934**). The vegetation of the Chiltern Hills, with special reference to the beechwoods and their seral relationships. *Journal of Ecology*, **22**, 230–70, 445–507.

# Predictions from growth and yield models of the performance of mixed-species stands

H.E. BURKHART* AND Å. THAM†

*Department of Forestry, Virginia Polytechnic Institute and State University, Blacksburg, VA 24061-0324, USA; and †Department of Forest Yield Research, Swedish University of Agricultural Sciences, S-770 73 Garpenberg, Sweden

## SUMMARY

Natural plant communities are usually mixtures of species, and forests are no exception. Even monocultures are often continuously invaded by other species. Despite the recognized advantages of mixtures of tree species (increased ecological diversity, diminished risks of catastrophe from pest outbreaks, wind and fire, etc.), there has been only limited effort devoted to quantifying growth and yield relationships for mixed-species stands as compared to the effort devoted to modelling pure stands.

Models of growth and yield for mixed-species stands are essential for making sound management decisions. The infinite variety of possible species mixtures, coupled with the range of environmental conditions under which mixtures might be grown, necessitates a modelling approach. It is not possible to observe all possible combinations of interest. Any modelling effort must be based on sound biological and quantitative principles in order to make the range of predictions and inferences required from a relatively small number of observed conditions.

Past efforts directed at modelling mixed-species stands have focused on stands of natural origin. Modelling approaches have included the use of differential equations, stand-level diameter distribution models, and individual-tree based increment equations. Future efforts should be aimed at developing data bases and modelling approaches that will enable interspecific and intraspecific competition effects to be quantified, so that the effects of a broad range of management alternatives can be predicted.

## INTRODUCTION

Natural plant communities, including forests, are usually mixtures of species. Even forest plantations generally experience continuous invasion by other species. Despite the recognized advantages of mixtures of tree species, most of the emphasis in commercial timber production has been on pure stands. There has been only limited effort devoted to quantifying growth and yield relationships for mixed-species stands.

Smith (1986) defined a *stand* as '. . . a contiguous group of trees sufficiently uniform in species composition, arrangement of age-classes, and condition to be a

21

distinguishable unit.' The internal structure of stands varies with regard to the degree that different species and age-classes are interspersed. The simplest structure consists of even-aged stands of a single species, but the range of complexity can extend to a variety of combinations of age-classes and species in various vertical and horizontal arrangements. As a result of differences in height-growth patterns, even-aged mixtures of tree species usually segregate into different canopy strata to form stratified mixtures. The focus of this paper is on even-aged stands (of varying composition) in temperate regions.

Although mixed-species stands are common, there is no universally accepted definition of 'mixed forest'. For the purposes of this paper, a mixed stand is defined as a stand of trees with two or more species comprising the usable volume. The question then arises: what proportion of the stand must be composed of the minor species to be regarded as a mixed-species stand? Partially as a result of the lack of commonly accepted definitions, few European countries provide statistics on the proportion of mixed stands in their forests; and the minimum composition of the minor species varies from country to country. For instance, in Sweden and Norway the composition limit for the minor species is 30% of the basal area. In central Europe, the limit is commonly taken to be 10% of either basal area or volume; however, in eastern Germany the limit is 20%. Thus, an exact quantification of the proportion of mixed stands in European forests is not possible. Similar circumstances exist elsewhere. In the south-eastern United States, where natural mixtures of pine and hardwood species are commonly found, pine stands are often defined as those with 75% or more of the basal area in pine, pine–hardwood stands consist of 25–75% pine by basal area, while hardwood stands contain less than 25% pine by basal area. It should be noted, however, that these definitions are not universal.

## Historical perspective

There has been more experience with creation and management of mixed-species stands in Europe than in other temperate regions. Hence, a brief review of historical trends in Europe will be given.

With the exception of the remote areas in the north and east, the species composition of European forests has been greatly affected by civilization. In nature, species richness decreases when the site is harsh. Therefore natural monocultures of pine or spruce occur close to the northern mountainous area. In central Europe there have been large areas of pure beech stands, some of which remain.

From the end of the eighteenth century, silviculturalists started to characterize the species composition of the forest stands (Thomasius 1973). At first, oak was a main species replanted but regeneration occurred also with other broad-leaved trees. Later, most of the harvested beech and oak forests were replaced with monocultures of spruce or pine as the reforestation techniques with conifers improved (Schwappach 1913). After the middle of the nineteenth century a reaction against man-made monocultures and a move back to natural mixed forests took place in central Europe (Heyer 1854). The monocultures had been exposed to heavy windthrow and insect damage and therefore multi-storied and mixed natural stands were regarded as safer. Today, the clear-felling system which aims at uniform stands is prevalent in the United Kingdom and the Nordic countries. The natural forest ideal with non-uniform stands and the selection system of silviculture is more prominent in central Europe.

## Planting mixed stands

Most mixed-species stands are the result of natural regeneration, but mixed stands can be created by planting. According to Smith (1986), the most successful mixed plantings are stratified mixtures composed of faster-growing intolerant species above slower-starting tolerants. If the trees in the upper canopy are not too dense and numerous, they grow more rapidly in diameter than they would if crowded into the single canopy of a pure plantation; the lower-stratum species can influence stem form and self pruning of the upper-stratum species similar to that of a pure stand.

If there are species that differ sufficiently in rate of growth and tolerance it is possible to create stands with several different strata. Properly selected species can be planted simultaneously and the stratified mixture will develop automatically. Sometimes the fastest growing species can be planted several years after those destined for the lower strata. Pioneer species are the most satisfactory for the overstorey because they grow rapidly in height and cast light shade. The best species for the lower strata are those that are shade tolerant (Smith 1986).

## Relative merits of pure and mixed stands

The choice of mixed versus pure stands is complex and depends on many considerations. Mixed stands generally require more skill to manage. In general, the yields and patterns of development of mixed stands are less predictable than those of pure stands. In some circumstances mixed stands may be more resistant to damage, more productive and more attractive than pure stands.

Mixtures of species are usually less susceptible to biotic pests than pure stands. The physical separation of susceptible plants inhibits the spread of many pests. Nevertheless, as Smith (1986) points out, there are important instances in which pure stands are more resistant to certain pests than stands of the same species mixed with highly susceptible ones.

The relative resistance of mixed or pure stands to damage from fire, wind, ice and snow depends on the physical characteristics of the individual trees or stands and the particular agent involved. Regardless of the agent of damage, however, a mixed stand of both resistant and non-resistant species should be more secure against injury than a pure stand of a non-resistant species, but less so than a pure stand of a resistant species (Smith 1986).

The question of whether mixed stands are more productive than pure depends on (i) the units of volume or value on which comparisons are based, (ii) the relative vulnerability to losses, and (iii) the relative degree of utilization of the growth factors of the site (Smith 1986).

In terms of total dry-matter production, single-canopied mixtures, with species mixed in the horizontal dimension, will generally yield less than pure stands of the most productive member. The productive capacity of the fastest growing species is generally diluted by its intermingling with the less productive species. Exceptions would be expected if the mixed stand suffered less loss or had some offsetting characteristic that improved growth, such as improved nitrogen fixation.

Stratified mixtures, however, are likely to exhibit greater total productivity than pure stands of the intolerant overstorey species. The tolerant species of the lower strata are, by nature, more efficient in photosynthesis than the overstorey species. It is, however, possible that a stratified mixture might be less productive than a pure stand of a tolerant species adapted to grow in the understorey. The situation may depend on whether soil factors or light are most limiting, or whether production is measured in terms of weight or volume of wood (Smith 1986).

Foresters are usually more concerned about merchantable yield than total production. In this regard, the observations of Smith (1986) are pertinent:

> One highly important consideration is the rate of diameter growth, which in turn affects rotation length. This often means that the typical overstory species may have some valuable attributes even though it is fundamentally less pro-ductive than some associate normally in the understory. The wood of one species is often inherently more valuable than that of another without respect to its ecological status. If the stratified mixture has virtue with respect to yield, it is often because of the opportunity to combine the quick growth of the individual trees of the overstory species with the slow-starting but high and long-sustained production that the understory species achieves after its release.

The choice between pure and mixed stands is often one between pure conifers and mixtures of conifers and hardwoods. Conifers generally have a higher propor-tion of their biomass in the usable bole than hardwoods. Furthermore, conifers are generally not as sensitive to unfavourable site conditions as hardwoods. Therefore, establishment of pure conifer stands on moderately productive sites is usually prudent. However, on some of the best sites, where hardwoods grow well and where the competing vegetation is hard to control, it may be desirable to grow mixtures.

One of the main disadvantages of growing mixed stands is the complexity of the harvesting operations. In stratified mixtures, the intolerant overstorey is typically removed well in advance of the rotation age of the more tolerant understorey. It is desirable, of course, to keep logging damage to the understorey stand to a minimum, but this can be difficult to achieve and expensive to execute in the field.

## GROWTH AND YIELD OF MIXED-SPECIES STANDS

It has been proposed that, for some circumstances, certain tree species will have higher yield if grown in mixed stands rather than pure stands. The search for this 'mixed species effect' has resulted in many investigations where the yields of two species in mixture are compared. The verification of the mixed species effect is difficult, however, because it is difficult to obtain comparative yields of pure stands of the species of interest, without confounding with other factors. Most studies have concentrated on comparing the mixed stand with a monoculture stand of the highest producing species. In spite of the difficulties with comparing mixed-stand to pure-stand yields, there have been a number of investigations which have produced interesting and useful results.

### Studies of mixed-species stands

Kennel (1965) in Bayern, Germany, found that spruce had a higher yield in mixture with beech than in monoculture. But beech, on the other hand, grew better alone. Jensen (1983) found that in Denmark spruce had a higher yield in mixture with silver fir than by itself.

The investigations made by Schilling (1925) and Busse (1931) in central Europe indicated that the volume-production of mixed spruce and pine stands exceeded their production in pure stands. Lappi-Seppälä (1930) found that a mixture of 40% birch and 60% Scots pine produced a higher total yield than an area divided in the same percentage ratio with the two species kept as monocultures. When Mielikäinen (1980) investigated different blends of pine and birch he found that birch grew better with pine than alone. However, when birch constituted a large proportion of the mixture, the yield of the pine decreased. Jonsson (1962) found that mixtures of Norway spruce and Scots pine, on intermediate sites, yielded more than the species grown separately.

Norway spruce and birch stands in the Nordic countries have been studied by Frivold (1982), Mielikäinen (1985), Agestam (1985) and Tham (1988). Mielikäinen, who separated silver birch and European birch found that a higher yield was obtained when Norway spruce and silver birch were mixed in the proportions 75:25%. A mixture of Norway spruce and European birch did not increase the total yield. Neither Agestam nor Frivold could indicate that the mixed stand had a higher yield than a monoculture of Norway spruce. Tham (1988) found that a mixture with between 600–800 birches and 2000–3000 Norway spruces per ha produced a higher total yield than a corresponding stand of spruces without birch-shelter, if the birches were removed at age 25–30 years. Frivold's, Mielikäinen's and Agestam's investigations were carried out as survey studies on temporary plots, while Tham analysed data from permanent plots.

In the south of eastern Germany Fiedler (1966) found that an admixture of birch in spruce stands did not reduce the volume production of spruce. The volume production of the stand as a whole was greater than the volume production of spruce alone. A growth and yield study from north-west Soviet Union (Cuprov, 1976) showed that the yield of a mixed stand of birch and spruce was between 135 and 160% of the yield of a monoculture birch stand and 120–150% of the yield of a monoculture spruce stand. The birch, however, needed to be cut before age 50–60 years.

Experience in central Europe has shown that a slight admixture of beech in spruce stands increases the yield on good sites compared with spruce alone (Kantor 1981). Yield tables for mixed stands of Norway spruce and beech (Wiedemann 1943) show that mixed stands have a higher yield than Norway spruce grown in monoculture on good sites. The opposite is true for poor sites where the yield of the mixed stands are less than for pure spruce stands. Several investigations have compared mixed and pure stands, but often it is not possible to judge whether the compared stands really are growing under the same conditions on equal sites (Auclair 1978).

In his seminal work *The Principles of Forest Yield Study*, Assmann (1970) devotes considerable attention to problems of studying the growth of mixed stands and to reviewing past studies of mixed-stand production. He presents results from studies with mixtures of (i) oak and beech, (ii) Scots pine and beech, (iii) Scots pine and Norway spruce, (iv) larch and beech, (v) beech and high-grade hardwoods, (vi) beech and spruce, and (vii) silver fir, Norway spruce and beech. Although adequate control plots for the yield of pure stands of the species involved in the mixtures were often not available, the general conclusion from most of the studies reviewed by Assmann is that mixtures result in increased total productivity.

Douglas fir, the premier timber species in north-west United States, often grows in association with red alder. There has been much interest in growth and yield of mixed Douglas fir/red alder stands because nitrogen availability often limits growth and red alder has the ability to fix nitrogen. Miller and Murray (1978) reported the long-term effects of red alder that were interplanted in 1933 within a 4-year-old Douglas fir plantation on a nitrogen-deficient site in the south-western part of the State of Washington. Red alder increased height and diameter of the associated dominant Douglas firs. By age 48 years, Douglas fir volume in the mixed stand averaged $217 \, m^3 \, ha^{-1}$ compared to $203 \, m^3 \, ha^{-1}$ in the pure stand. Red alder volume was $175 \, m^3 \, ha^{-1}$. Miller and Murray concluded that maintaining red alder in Douglas fir stands can increase merchantable yields on nitrogen-deficient sites.

## Modelling mixed-species stands

Models of growth and yield for mixed-species stands are essential for evaluating biological potential and making sound management decisions. The infinite variety of possible species mixtures, coupled with the range of environmental conditions under which mixtures might be grown, necessitates a modelling approach. Direct observation of all possible combinations of interest obviously is not possible. Any modelling effort must be based on sound biological and quantitative principles in order to make the range of predictions and inferences that are desired from a relatively small number of observed conditions. Interspecific competition effects are surely different than intraspecific effects; the quantification of these competition effects remains a major stumbling block to the development of reliable growth and yield models for mixed-species stands.

In spite of the recognized need for growth models for mixed-species stands, relatively few have been developed. Turnbull (1963) used a system of differential equations to study species competition in multiple-species forest stands, but he did not develop a yield simulator with the equations. Somewhat later, Leary (1979) indicated that differential equations can be used to model the dynamics of two or more forest tree species, but again an empirical growth and yield model based on the approach was not presented.

Lynch and Moser (1986) specified a system of differential equations to construct a mixed-species yield model that provides stand table predictions (i.e. information by diameter class). Their system of equations relates stand conditions at a particular time (initial conditions) to rates of change in these conditions. Ingrowth, mortality and survivor growth rates are used to determine net change rates. Because the equations in the system do not depend on age explicitly, the system can be used to model uneven-aged as well as even-aged stands. Lynch and Moser fitted their equation system to permanent plot data grouped into two species categories: (i) softwood (primarily white pine), and (ii) hardwood (primarily paper birch and quaking aspen).

Bowling et al. (1989) developed a stand-level growth and yield model for multi-species, natural stands of Appalachian hardwoods in the United States. Their model provides estimates by species groups and DBH (diameter at breast height) classes. Five species groups (red oak, white oak, intolerant, tolerant and miscellaneous), based on similarity in economic value, growth rate and shade tolerance, were defined. Equations were specified to predict stand attributes by species group and for the whole stand. Future DBH distributions were obtained by assuming that the diameters are Weibull distributed and employing the first two non-central moments of the DBH distribution to generate Weibull parameters. Future DBH distributions were generated for the whole stand and every species group except one; the diameter distribution of the remaining species group was obtained by subtraction.

Individual tree-based growth models for mixed-species stands have also been developed. Harrison et al. (1986) presented an individual tree, distance independent growth and yield model for Appalachian mixed-hardwood stands. Given a tree list (i.e. stand table), along with inputs of stand age and site index, the model applies species-specific individual tree equations to predict tree basal area increment and total tree height. Because it is an individual tree model, it provides both species and size-class resolution.

The models discussed, up to this point, are for naturally occurring mixtures of

*Growth and yield models*

species. While the modelling approaches and constructs should help to quantify the development of deliberately created (planted) mixtures, to date there has been only limited development of models for managed mixed stands. An exception is the model of Tham (1988) for yield prediction after thinning of birch mixed with Norway spruce. Tham developed a stand growth simulator to evaluate a range of alternatives, from eliminating self-propagated birches from Norway spruce stands to thinning the birches to different densities at various ages. The stand growth simulator consists of a method for estimating the spatial position of trees, a diameter distribution model for generating tree DBH values, a height–diameter function for estimating heights of trees from their diameters, and equations for predicting individual tree growth as a function of growth conditions including competition and thinning effects. The results of simulations show that a stand will produce a higher total yield if some birch cover is included. For

TABLE 1. Yield table for an initial stand of 1600 Norway spruces per ha without birch shelter, based on data from Sweden (from Tham 1988)

| | | Stand after thinning | | | Removed | | | Total yield | |
|---|---|---|---|---|---|---|---|---|---|
| | Age (years) | No. of stems (No. ha$^{-1}$) | Basal area (m$^2$ ha$^{-1}$) | Vol. (m$^3$ ha$^{-1}$) | No. of stems (No. ha$^{-1}$) | Basal area (m$^2$ ha$^{-1}$) | Vol. (m$^3$ ha$^{-1}$) | Basal area (m$^2$ ha$^{-1}$) | Vol. (m$^3$ ha$^{-1}$) |
| Spruce | 15 | 1600 | 0.1 | 0.7 | 0 | 0.0 | 0.0 | 0.1 | 0.7 |
| Birch | 15 | 0 | 0.0 | 0.0 | 0 | 0.0 | 0.0 | 0.0 | 0.0 |
| Total | | 1600 | 0.1 | 0.7 | 0 | 0.0 | 0.0 | 0.1 | 0.7 |
| Spruce | 20 | 1600 | 1.9 | 5.2 | 0 | 1.9 | 5.2 | 1.9 | 5.2 |
| Birch | 20 | 0 | 0.0 | 0.0 | 0 | 0.0 | 0.0 | 0.0 | 0.0 |
| Total | | 1600 | 1.9 | 5.2 | 0 | 1.9 | 5.2 | 1.9 | 5.2 |
| Spruce | 25 | 1600 | 6.8 | 22.1 | 0 | 0.0 | 0.0 | 6.8 | 22.1 |
| Birch | 25 | 0 | 0.0 | 0.0 | 0 | 0.0 | 0.0 | 0.0 | 0.0 |
| Total | | 1600 | 6.8 | 22.1 | 0 | 0.0 | 0.0 | 6.8 | 22.1 |
| Spruce | 30 | 1600 | 13.7 | 56.7 | 0 | 0.0 | 0.0 | 13.7 | 56.7 |
| Birch | 30 | 0 | 0.0 | 0.0 | 0 | 0.0 | 0.0 | 0.0 | 0.0 |
| Total | | 1600 | 13.7 | 56.7 | 0 | 0.0 | 0.0 | 13.7 | 56.7 |
| Spruce | 35 | 1600 | 22.1 | 118.7 | 0 | 0.0 | 0.0 | 22.1 | 118.7 |
| Birch | 35 | 0 | 0.0 | 0.0 | 0 | 0.0 | 0.0 | 0.0 | 0.0 |
| Total | | 1600 | 22.1 | 118.7 | 0 | 0.0 | 0.0 | 22.1 | 118.7 |
| Spruce | 40 | 1256 | 23.3 | 156.9 | 344 | 5.9 | 40.1 | 29.2 | 197.0 |
| Birch | 40 | 0 | 0.0 | 0.0 | 0 | 0.0 | 0.0 | 0.0 | 0.0 |
| Total | | 1256 | 23.3 | 156.9 | 344 | 5.9 | 40.1 | 29.2 | 197.0 |
| Spruce | 45 | 1256 | 27.9 | 217.7 | 0 | 0.0 | 0.0 | 33.8 | 257.8 |
| Birch | 45 | 0 | 0.0 | 0.0 | 0 | 0.0 | 0.0 | 0.0 | 0.0 |
| Total | | 1256 | 27.9 | 217.7 | 0 | 0.0 | 0.0 | 33.8 | 257.8 |
| Spruce | 50 | 1256 | 32.3 | 260.1 | 0 | 0.0 | 0.0 | 38.2 | 300.2 |
| Birch | 50 | 0 | 0.0 | 0.0 | 0 | 0.0 | 0.0 | 0.0 | 0.0 |
| Total | | 1256 | 32.3 | 260.1 | 0 | 0.0 | 0.0 | 38.2 | 300.2 |

TABLE 2. Yield table for an initial stand of 1600 Norway spruces and 600 silver birches per ha with the birch shelter removed at age 25 years, based on data from Sweden (from Tham 1988)

| | Age (years) | Stand after thinning | | | Removed | | | Total yield | |
|---|---|---|---|---|---|---|---|---|---|
| | | No. of stems (No. ha$^{-1}$) | Basal area (m$^2$ ha$^{-1}$) | Vol. (m$^3$ ha$^{-1}$) | No. of stems (No. ha$^{-1}$) | Basal area (m$^2$ ha$^{-1}$) | Vol. (m$^3$ ha$^{-1}$) | Basal area (m$^2$ ha$^{-1}$) | Vol. (m$^3$ ha$^{-1}$) |
| Spruce | 15 | 1600 | 0.1 | 0.6 | 0 | 0.0 | 0.0 | 0.1 | 0.6 |
| Birch | 15 | 600 | 4.0 | 21.3 | 0 | 0.0 | 0.0 | 4.0 | 21.2 |
| Total | | 2200 | 4.1 | 21.9 | 0 | 0.0 | 0.0 | 4.1 | 21.8 |
| Spruce | 20 | 1600 | 1.3 | 3.7 | 0 | 0.0 | 0.0 | 1.3 | 3.7 |
| Birch | 20 | 600 | 7.2 | 44.3 | 0 | 0.0 | 0.0 | 7.2 | 44.3 |
| Total | | 2200 | 8.5 | 48.0 | 0 | 0.0 | 0.0 | 8.5 | 48.0 |
| Spruce | 25 | 1600 | 4.6 | 14.6 | 0 | 0.0 | 0.0 | 4.6 | 14.6 |
| Birch | 25 | 0 | 0.0 | 0.0 | 600 | 10.2 | 72.1 | 10.2 | 72.1 |
| Total | | 1600 | 4.6 | 14.6 | 600 | 10.2 | 72.1 | 14.8 | 86.7 |
| Spruce | 30 | 1600 | 8.8 | 33.9 | 0 | 0.0 | 0.0 | 8.8 | 33.9 |
| Birch | 30 | 0 | 0.0 | 0.0 | 0 | 0.0 | 0.0 | 10.2 | 72.1 |
| Total | | 1600 | 8.8 | 33.9 | 0 | 0.0 | 0.0 | 19.0 | 106.0 |
| Spruce | 35 | 1600 | 18.1 | 91.4 | 0 | 0.0 | 0.0 | 18.1 | 91.4 |
| Birch | 35 | 0 | 0.0 | 0.0 | 0 | 0.0 | 0.0 | 10.2 | 72.1 |
| Total | | 1600 | 18.1 | 91.4 | 0 | 0.0 | 0.0 | 28.3 | 163.5 |
| Spruce | 40 | 1280 | 22.3 | 144.7 | 320 | 5.7 | 36.5 | 28.0 | 181.2 |
| Birch | 40 | 0 | 0.0 | 0.0 | 0 | 0.0 | 0.0 | 10.2 | 72.1 |
| Total | | 1280 | 22.3 | 144.7 | 320 | 5.7 | 36.5 | 38.2 | 253.3 |
| Spruce | 45 | 1280 | 28.4 | 203.9 | 0 | 0.0 | 0.0 | 34.1 | 240.4 |
| Birch | 45 | 0 | 0.0 | 0.0 | 0 | 0.0 | 0.0 | 10.2 | 72.1 |
| Total | | 1280 | 28.4 | 203.9 | 0 | 0.0 | 0.0 | 44.3 | 312.5 |
| Spruce | 50 | 1280 | 33.3 | 257.3 | 0 | 0.0 | 0.0 | 39.0 | 293.8 |
| Birch | 50 | 0 | 0.0 | 0.0 | 0 | 0.0 | 0.0 | 10.2 | 72.1 |
| Total | | 1280 | 33.3 | 257.3 | 0 | 0.0 | 0.0 | 49.2 | 365.9 |

example, simulation of a precommercially thinned mixed stand with 1600 Norway spruces and 600 birches per ha had a total yield of 365.9 m$^3$ ha$^{-1}$ at age 50 years. The birch shelter, in this case, was removed at age 25. A simulation of a corresponding stand without a birch shelter indicated a total yield of 300.2 m$^3$ ha$^{-1}$. Tables 1 and 2 contain detailed production data for these two scenarios.

Tham (1988) also found that Norway spruce grown under a birch shelter has a different growth pattern than spruce grown as a monoculture, especially with regard to growth in height. The dominant height of the Norway spruce grown under the birch shelter was retarded, compared to that of spruce in pure stands, until after the birches were cut. However, following removal of the birch shelter, the height growth of the released spruce exceeded that of spruces grown in pure stands. By age 50 years, the total dominant height of the released spruce exceeded that of the spruces grown without birch shelter (Fig. 1). The subtleties of

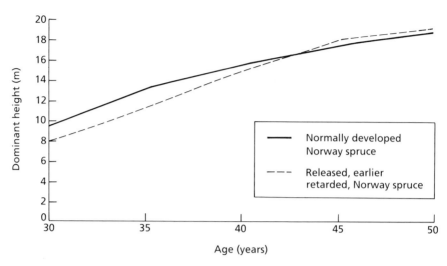

FIG. 1. Height development of Norway spruce grown in monoculture and after removal of birch shelter (from Tham 1988).

growth-pattern differences for various mixtures must be included to accurately predict total production.

Models for conifer plantations in which hardwood invaders are considered as 'weeds' have also been constructed. An example is the model of Burkhart and Sprinz (1984). Their model was developed to predict pine survival, growth and yield for unthinned loblolly pine plantations in the south-eastern United States with varying levels of hardwood competition in the main canopy. Inputs for the model are number of loblolly pine trees planted per unit area, site index for loblolly pine, per cent of hardwood basal area in the main canopy of the stand, and age(s) at which output is desired. From these inputs the model computes, by DBH class, the number of pine trees surviving, basal area, and volumes per unit area. No estimates of hardwood volume are included because the value of the hardwood component is typically very low in relation to the value of the pine 'crop' trees.

## NEEDS FOR FUTURE WORK

There are many areas in which lack of knowledge continues to act as an impediment to the establishment and management of mixed-species stands. Basic ecological relationships in mixed-species stands must be better understood—this is especially true for mixtures that might be created by planting. Quantitative models of mixed-species stand dynamics are essential for evaluating biological productivity and economic viability. Workable harvesting systems for various types of mixed-species stands, in particular where the components may have different rotation ages, must also be developed.

Because of the great variety of species and mixtures of species that might be grown, new and innovative approaches to data collection and modelling will be required in order to develop reliable growth and yield models for making informed silvicultural prescriptions and management decisions. The typical approach taken in past growth and yield studies was to define a population of interest, obtain a sample from the defined population (the sample could consist of temporary plots, permanent plots or both), and estimate coefficients (invariably with least squares) in specified equation forms. Obviously, it will not be possible to sample all species mixtures of interest. Thus, carefully designed data collection efforts will be required in order to model adequately the full range of conditions of interest. Plots in existing stands will be required, but such plots alone will not likely be fully adequate for modelling purposes. Designed experiments, with controlled mixtures (while holding other factors constant), will be needed to test hypotheses and to develop robust prediction equations.

Model populations—with varying spatial relationships and proportions of species—can be created experimentally. As noted by Harper (1977), the development of individuals in mixed or pure populations can be partly predicted from a knowledge of spatial relationships, provided that germination and establishment are synchronous. However, if the times of establishment of two neighbours differ, this can have a profound effect on their relative performance and overwhelm other influences such as differences in growth rate and spatial arrangement. Relationships for various mixtures and timing of establishment of mixed-species plantations of trees are not well understood or quantified. The growth of mixed-species stands can be studied by observations in naturally-occurring populations and by conducting experiments with the additive and substitutive designs described by Harper (1977).

Trees in stands share resources, such as light, water and nutrients, necessary for their survival and growth. Because resources are generally limiting, there is 'interference' between plants. The stress created by the proximity of neighbouring trees may result in mortality for whole plants or their parts, or reduced growth rates. A full understanding of neighbour tree effects—how these effects vary depending on whether the competition is inter- or intraspecific—is essential to evaluate mixed-species management options.

Although there have been a number of 'competition indices' developed for growth and yield modelling purposes, most of the work has been done for single-species stands and interspecific competition effects have not been studied thoroughly (see Daniels, Burkhart & Clason (1986) for a review and comparison of competition measures used to predict individual tree growth). Most of the competition indices, while logically constructed, have been highly empirical; their adequacy has been judged on the basis of correlation with observed tree growth. A more theoretical approach will likely be required to generalize competition relationships to multiple species and times of establishment. In this context, Tilman's (1982, 1986) model of differential resource utilization may prove useful. Tilman has attempted to explain the coexistence by competing plants on the basis of differential resource utilization. A key feature of Tilman's approach is that he explicitly considers the dynamics of the resources as well as the dynamics of the competing species. Ecological field theory (EFT) may provide further useful insights into this complex competition quantification problem. In EFT plant spatial influences are quantified as pulsating geometric zones about individual plants (Walker *et al.* 1989). It provides a methodology to include spatial interactions between plants of different size, function and growth form in models of plant community dynamics. Because the methodology includes variable amounts of spatial interaction as a function of plant size and prevailing climatic and environmental conditions, EFT should be helpful to generalize competitive relationships in mixed-species forests.

Levins (1966) argued that modellers of population biology strive to maximize simultaneously three desirable properties of a given model: generality, reality and precision. In any one model, he asserted, one may sacrifice one of these desired properties to achieve a higher level of the other two. In traditional growth and yield models, generality is sacrificed for increased reality and precision (with the primary emphasis being on precision). Given the usual objectives of growth and yield modellers, this is a reasonable strategy. However, for models of mixed-species stands, it may be desirable, at least initially, to emphasize generality and reality at the expense of precision. Models that provide realistic estimates for a wide range of options seem a reasonable starting place, as opposed to models of high precision for a narrower range of conditions.

## CONCLUSION

In conclusion, mixed-species stands, under some conditions, are attractive alternatives to monocultures. Evaluation of silvicultural and management options for mixed-species stands requires reliable growth and yield models. The development of models of mixed-stand dynamics will require innovative approaches to data collection and modelling. Designed experiments, in addition to data collection in field plots, will be required to develop the data bases needed. These data must be interpreted in light of relevant ecological theory, and models fitted to the data should strive for increased generality, even if some precision is sacrificed.

## REFERENCES

Agestam, E. (1985). *En produktions modell för blandbestånd av tall, gran och björk i Sverige.* (Summary: A growth simulator for mixed stands of pine, spruce and birch in Sweden.) Swedish University of Agricultural Sciences Department of Forestry Yield Research Report No. 15, 150 pp.

Assmann, E. (1970). *The Principles of Forest Yield Study.* Pergamon Press, New York.

Auclair, D. (1978). *La sylviculture des forêts mélangées.* Etude bibliographique. Document, Centre de recherches forestières d'Orléans 78/30. Olivet, 51 pp.

Bowling, E.H., Burkhart, H.E., Burk, T.E. & Beck, D.E. (1989). A stand-level multispecies growth model for Appalachian hardwoods. *Canadian Journal of Forest Research*, 19, 405–12.

Burkhart, H.E. & Sprinz, P.T. (1984). *A model for assessing hardwood competition effects on yields of loblolly pine plantations*. Virginia Polytechnic Institute & State University, School of Forestry and Wildlife Resources, Publ. No. FWS-3-84.

**Busse, J. (1931).** Ein Kiefern-Fichten-Mischbestand in Sachsen. *Tharandter forstliches Jahrbuch*, **82**, 595–601.

**Cuprov, N.P. (1976).** Vozrasty spelosti i rubok berenznjakov. *Lesnoe hozjajstvo 1976*, **6**, 49–54.

**Daniels, R.F., Burkhart, H.E. & Clason, T.R. (1986).** A comparison of competition measures for predicting growth of loblolly pine trees. *Canadian Journal of Forest Research*, **16**, 1230–7.

**Fiedler, F. (1966).** Zuwachs und Ertrag im Fichten-Birken-Mischbestand. *Archiv für Forstwesen*, **15**, 283–91.

**Frivold, L.H. (1982).** *Bestandsstruktur og produksjon i blandingsskog av bjork* (Betula verrucosa *Ehrh.*, B. pubescens *Ehrh.) og gran (Picea abies (L) karst.) i Sydost Norge.* (Summary: Stand structure and yield of mixed stands of birch (*Betula verrucosa* Ehrh., *B. pubescens* Ehrh.) and spruce (*Picea abies* (L) Karst.) in South East Norway.) Agricultural University of Norway, Department of Silviculture Scientific Report No. 18, 108 pp.

**Harper, J.L. (1977).** *Population Biology of Plants*. Academic Press, London.

**Harrison, W.C., Burkhart, H.E., Burk, T.E. & Beck, D.E. (1986).** Growth and yield of Appalachian mixed hardwoods after thinning. Virginia Polytechnic Institute & State University, School of Forestry and Wildlife Resources, Publ. No. FWS-1-86.

**Heyer, C. (1854).** *Der Waldbau oder die Forstproduktenzucht*. B.G. Teubner, Leipzig. 390 pp.

**Jensen, A.M. (1983).** *Aedelgranens* (Abies alba *Mill.*) *vekst sammenlignet med rodgranens* (Picea abies (L) *Karst.) i henholdsvis rene og blandede bevoksninger på sandede jorder i Midt- og Vestjylland.* (Summary: Growth of Silver fir (*Abies alba* Mill.) compared with the growth of Norway spruce (*Picea abies* (L) Karst.) in pure and mixed stands on sandy soils in the western parts of Denmark.) Meddelolser fra skovbruksinstitutett Series 2, No. 14, 498 pp.

**Jonsson, B. (1962).** Om barrblandskogens volymproduktion. (Summary: Yield of mixed coniferous forest.) *Meddelanden Fran Statens Skogsforskningsinstitut*, **50**, 1–143. Stockholm.

**Kantor, P. (1981).** Vliv buku na produkci smrkových porosto v horskych polochach. *Zpravy lesnickeho vyzkumu*, **26**, 7–11.

**Kennel, R. (1965).** Untersuchungen über die Leistung von Fichte und Buche im Rein- und Mischbestand. *Allgemeine Forst- und Jagd-Zeitung*, **136**, 149–61, 173–89.

**Lappi-Seppälä, M. (1930).** Untersuchungen über die Entwicklung gleichaltriger Mischbestände aus Kiefer und Birke. *Communicationes Instituti Forestalis Fenniae*, **15**, 1–243.

**Leary, R.A. (1979).** Design. *A generalized forest growth projection system applied to the lake states region*, pp. 5–15. USDA Forest Service General Technical Report NC-49.

**Levins, R. (1966).** The strategy of model building in population biology. *American Scientist*, **54**, 421–31.

**Lynch, T.B. & Moser, J.W. (1986).** A growth model for mixed species stands. *Forest Science*, **32**, 697–706.

**Mielikäinen, K. (1980).** Mänty-koivusekametsikböiden rakenne ja kehitys. (Structure and development of mixed pine and birch stand.) Communicationes Instituti Forestalis Fenniae, **99**, 1–82.

**Mielikäinen, K. (1985).** Koivousekoituksen vaikutus kuusikon rakenteeseen ja kehitykseen. (Summary: Effect of an admixture of birch on the structure and development of Norway spruce stands.) Communicationes Instituti Forestalis Fenniae, **133**, 79 pp.

**Miller, R.E. & Murray, M.D. (1978).** The effects of red alder on growth of Douglas-fir. *Utilization and management of red alder*, pp. 283–306. USDA Forest Service General Technical Report PNW-70.

**Schilling, L. (1925).** Ostpreussische Kiefern-Fichtenmischbestände. *Zeitschr. Forst unt Jagdwesch*, **57**, 257–96.

**Schwappach, A. (1913).** Forstgeschichte. In *Handbuch der Forstwissenschaft* (Ed. by C. Wagner), Vierter Band, Ausgabe A. pp. 1–190. H. Laupp, Tubingen.

**Smith, D.M. (1986).** *The Practice of Silviculture*. John Wiley & Sons, New York.

**Tham, Å. (1988).** *Yield prediction after heavy thinning of birch in mixed stands of Norway spruce* (Picea abies (L) *Karst.) and birch* (Betula pendula *Roth and* Betula pubescens *Ehrh.).* (Sammanfattning: Produktionsförutsägelser vid kraftiga gallringar av björk i blandbestånd av gran (*Picea abies* (L.) Karst.) och björk (*Betula pendula* Roth and *Betula pubescens* Ehrh.). Swedish

University of Agricultural Sciences, Department of Forestry Yield Research Report No. 23, Garpenberg. 36 pp.

**Thomasius, H. (1973).** *Wald, Landeskultur und Gesellschaft*. Steinkopff, Dresden. 439 pp.

**Tilman, D. (1982).** *Resource Competition and Community Structure*. Princeton University Press, Princeton.

**Tilman, D. (1986).** Resources, competition and the dynamics of plant communities. *Plant Ecology* (Ed. by M.J. Crawley), pp. 51–75. Blackwell Scientific Publications, Oxford.

**Turnbull, K.J. (1963).** *Population dynamics in mixed forest stands. A system of mathematical models of mixed stand growth and structure*. Ph.D. thesis, University of Washington.

**Walker, J., Sharpe, P.J.H., Penridge, L.K. & Wu, H. (1989).** Ecological field theory: the concept and field tests. *Vegetatio*, **83**, 81–95.

**Wiedemann, E. (1943).** Der Vergleich der Massenleistung des Mischbestandes mit dem Reinbestand. *Allgemeine Forst- und Jagd-Zeitung*, **119**, 123–32.

# Silviculture and yield of mixed-species stands: the UK experience

G. KERR,* C.J. NIXON† AND R.W. MATTHEWS*

*Forestry Commission, Forest Research Station, Alice Holt Lodge, Wrecclesham, Farnham, Surrey GU10 4LH, UK; and †Forestry Commission, Northern Research Station, Roslin, Midlothian EH25 9SY, UK

## SUMMARY

The term 'mixture' is used to mean a formal planting pattern to establish a plantation. 'Mixed-species stand' implies that more than one species is present and includes a wide range of situations from ancient semi-natural woodland to a plantation of Corsican pine with an intrusion of birch. Mixed species stands account for 26% of the Forestry Commission's estate.

Conifer/broad-leaved mixtures are planted in the lowlands, usually with the objective of securing a final crop of quality broad-leaves. Three benefits of such mixtures are often cited: the nursing effect, increased profitability and flexibility. The nursing effect has been much discussed by foresters; however, in experiments this effect has often been difficult to prove. Other benefits and problems and the importance of effective management are discussed.

Mixtures have been widely used in the uplands to alleviate difficulties associated with poor soils and testing climatic conditions. In contrast to the lowlands, the benefits associated with nursing mixtures have been clearly demonstrated. The relative vulnerability of mixed stands to wind damage is less well understood, although empirical work suggests aspects of stand structure which may be important. Mixed-species crops, particularly spruce with pine or larch, have also been of major importance in the afforestation of upland heaths and peatlands with nutritionally deficient soils.

The use of yield models for predicting the growth of mixed stands is discussed.

## EXTENT OF MIXTURES IN BRITISH FORESTS

### Forestry Commission

The Forestry Commission maintains a comprehensive database of the species composition of all of its woodlands (Table 1). However, the primary objective of

The phrase 'mixed-species stand' includes a range of situations from ancient semi-natural woodland to a plantation of Corsican pine with an intrusion of birch. The term 'mixture' is used here to mean a formal planting of species together, either in lines or groups, to achieve silvicultural or economic objectives.

35

TABLE 1. Areas of UK Forestry Commission high forest with pure stands and species mixtures (1988) (thousands of hectares)

| Species | Pure stands | Mixed and two-storied |
| --- | --- | --- |
| Scots pine | 57.7 | 26.0 |
| Corsican pine | 25.2 | 9.2 |
| Lodgepole pine | 62.3 | 38.8 |
| Sitka spruce | 324.4 | 75.8 |
| Norway spruce | 39.2 | 12.4 |
| European larch | 3.3 | 3.6 |
| Japanese larch | 37.2 | 9.7 |
| Hybrid larch | 8.8 | 2.5 |
| Douglas fir | 17.0 | 6.4 |
| Western hemlock | 3.6 | 2.9 |
| Western red cedar | 0.8 | 1.3 |
| Grand fir | 1.5 | 0.9 |
| Noble fir | 1.1 | 0.6 |
| Others | 3.0 | 1.8 |
| All conifers | 585.1 | 191.9 |
| Oak | 6.5 | 6.3 |
| Beech | 6.9 | 6.8 |
| Sycamore | 0.5 | 1.2 |
| Ash | 0.3 | 1.1 |
| Birch | 0.3 | 0.7 |
| Poplar | 0.8 | 0.1 |
| Sweet chestnut | 0.2 | 0.2 |
| Others | 1.6 | 1.3 |
| All broad-leaves | 17.1 | 17.7 |
| Total | 602.2 | 209.6 |

the database is to produce production forecasts for valuation and timber supply estimates, so the data on mixtures has limitations.

1   The smallest territorial unit of the database is the subcompartment; if two or more species exist within this boundary it is classified as 'mixed'. This gives no indication of how the species are distributed; a subcompartment which has two blocks of pure species is classified as mixed.

2   The figures in Table 1 refer to productive high forest. Other areas of high ecological diversity, e.g. low-grade broad-leaves, coppice and scrub, are not included.

3   Figures cannot be separated for mixed and two-storied forests. The latter is usually of mixed-species composition, but is only a small proportion of the total area ($<1\%$).

However, the figures in Table 1 show that: (i) approximately 26% of the Forestry Commission's estate consists of mixed-species stands; (ii) broad-leaves are twice as likely to be found in mixture as conifers; and (iii) seven conifer and six broad-leaved species have significant (defined as $>30\%$ of area) areas of mixed-species stands. Thus, mixed-species stands constitute a substantial area of the Forestry Commission's estate.

## Private sector

The private sector now manages 58% of Britain's productive woodland (Forestry Commission 1989a). The species composition of these woodlands is given in the 1979–82 Census of Woodlands (Locke 1987). The results of this census are presented as 'mainly coniferous' or 'mainly broad-leaved' and therefore of limited value to establish the extent of mixed-species stands.

# THE SILVICULTURE OF MIXTURES IN THE LOWLANDS

Conifer/broad-leaved mixtures represent the most widespread and interesting silvicultural system in the lowlands of Britain. Mixtures of broad-leaves are increasingly being planted (Forestry Commission 1989b) and pose few problems: one species does not so readily dominate as occurs in conifer/broad-leaved mixtures, and even if dominance does occur there is little loss of amenity and wildlife interest. Mixtures of conifers are relatively unimportant in the lowlands. Uneven-aged species mixtures, such as two-storied forests are not common in Britain, although a few notable examples do exist (e.g. the Bradford–Hutt 'continuous cover forestry system').

## Benefits of mixtures

### Nursing

Brown (1953), Evans (1984), Wood and Nimmo (1962), Joslin (1982) and Wright (1982) refer to the nursing role of a conifer component in a conifer/broad-leaved mixture; in this context the word 'nurse' means 'help to grow'. There are at least two ways in which this can be achieved: (i) by protecting the broad-leaved species from unseasonal frosts, particularly in late spring; and (ii) by providing side shelter, essential to the growing of quality broad-leaves, by aiding upward growth and better tree form (Evans 1984).

The Forestry Commission has established a number of experiments to test the hypothesis that, in the lowlands, the incorporation of a conifer component into a broad-leaved crop will have a nursing effect upon it and a corresponding increase in growth will occur.

At Crumblands, in south-east Wales, an experiment was planted in 1931 to compare the growth of pure oak with that in line and group mixtures with European larch (Wood, Miller & Nimmo 1967). The site is the leeward side of a moderately exposed slope and there is considerable site variation. At age 30 years, the oak in the mixtures were 1.8–3.6 m taller and up to 5 cm greater in girth than the oak in pure stands. However, the experimental design did not permit statistically sound conclusions to be drawn.

At Friston, in south-east England, an experiment was planted on chalk downland in 1927 to compare pure plots of beech, ash and sycamore with plots of these species in mixture with birch, grey alder, European and Japanese larch. Each of the latter nurse species appears in separate plots as a 25% and as a 50% component of the stand. The experiment was established in two sections on different sites, the first being unploughed and the second ploughed. At age 25 years, there was little evidence for any nursing effect, partly because of poor establishment (Wood & Nimmo 1962). However, at age 30 years, at the ploughed site, ash and sycamore had made good growth in some of the mixtures with alder. Where alder had spread by root suckering into pure crops of ash, there was a marked response by the latter species, probably because of improved nitrogen supply. By contrast, the ash and sycamore in pure stands had not closed canopy.

However, statistically sound evidence for a nursing effect in lowland conifer/ broad-leaved mixtures is lacking. Nevertheless, there has been a growing recognition of the value on former woodland sites of natural regeneration as side shelter for planted trees and as a supplementary source of potential crop trees (Evans 1988). The significant silvicultural benefits from mixing broad-leaves with conifers on former woodland (restocking) sites may be equally effectively provided by other woody growth, where tree numbers reach about 2500 stems $ha^{-1}$.

*Financial criteria.* In lowland Britain, it is usually financially more rewarding to grow conifers than broad-leaved trees. The maximum mean annual increment of oak and beech is commonly in the range $6-8\,m^3\,ha^{-1}\,year^{-1}$, whereas Norway spruce, Douglas fir or western red cedar growing the same sites yield $14-24\,m^3\,ha^{-1}\,year^{-1}$. A conifer/broad-leaved mixture produces returns earlier than a pure broad leaved stand. This is a powerful argument to growers, who usually expect the first thinning return from conventionally managed pure oak stands between 30 and 40 years. The Forestry Commission (1985) suggests that up to 50% of a lowland mixture can consist of conifers without detriment to the quality of the final broad-leaved crop.

*Flexibility.* In Victorian times, it was common to plant a mixture of many species in the hope that at least one or two species would survive. We now have sufficient silvicultural and site-related knowledge to disregard this 'scatter and hope' approach. However, a mixture can be regarded as an insurance policy (Hiley 1967).

If one component of a mixture is attacked lethally or sublethally, the second component can be favoured for the final crop. For example, beech and sycamore can be bark-stripped by grey squirrels, and Norway spruce may be attacked by *Dendroctonus micans*.

The presence of two or more species on a site gives the forester greater marketing flexibility. However, the volumes of each product must be above a critical size to ensure that competitive prices are paid, and any delay in removing conifer thinning may be deleterious to the broad-leaved component.

*Wind stability.* Mixtures may confer greater wind stability, though sometimes this is only due to one species in a mixture failing to grow well. On clay soils in central and eastern England, Norway spruce within oak/spruce mixtures has blown over at the pole stage, whereas the oak, with its deeper rooting habit, remained firm. In the Forest of Dean, on heavy clay, pure Douglas fir is prone to blow over at 25–30 years, but when mixed with oak it is able to grow to maturity and achieve 40 m height.

*Disease resistance.* The effects of silvicultural practices on the rate of disease development are poorly understood (Watt, this volume). However, there is some observational evidence that a conifer nurse (excepting larch) can reduce the incidence of beech bark disease during the establishment period (Lonsdale & Wainhouse 1987).

## Problems of mixtures

*Compatibility.* Species grown in mixture must have similar rates of growth in height to avoid one component swamping the other(s), or alternatively, the most vigorous component must be removed at an early stage, which may not be economic. The forester must be able to anticipate the performance of the species (and provenances) on the sites to be planted. Evans (1984) stated that: 'the anticipated conifer yield-class should never be more than double that for the broad-leaved component, except for larch which should not be more than 50% better. For example, oak Yield-Class 6 ($m^3 ha^{-1} year^{-1}$) with European larch Yield-Class 8 and beech Yield-Class 8 with Western red cedar Yield-Class 16 are probably compatible, whereas oak Yield-Class 6 with Norway spruce Yield-Class 16 probably is not.'

The reasoning behind larch being treated differently is its very fast early height growth rate compared with other conifers (commonly about 8 m at age 10 years, Yield-Class 12, compared with 3.0–4.5 m for other conifers (Edwards & Christie 1981)).

*Planting patterns.* Planting patterns can be chosen to increase the compatibility of two or more species. Theoretically, if the two species have equal early growth rates they may be planted in an intimate mixture; however, this may bring operational problems later. Differential growth rates are more likely, in which case group or line mixtures are more likely to enable each component to survive until the time of first thinning. The minimum number of rows considered to be robust enough to survive until first thinning is three, and the recommended group size is 12–25 broad-leaved trees at 10–12 m spacing, planted in a matrix of conifers.

*Other problems.* Differing susceptibilities of the components of a mixture to herbicides may cause problems, and it has been reported that broad-leaves in mixtures with conifers suffer greater squirrel damage than pure broad-leaved crops (Aldhous 1981). However, published quantitative support for this is lacking; Pepper (pers. comm.) reports that a survey of pure beech, and beech in mixture with spruce and pine, showed no significant differences in damage.

## THE SILVICULTURE OF MIXTURES IN THE UPLANDS

### *History*

Mixtures of Scots pine, Norway spruce and European larch were widely planted in Scotland on privately owned estates during the nineteenth and early twentieth centuries. The Forestry Commission continued to use mainly these species well into the 1930s, until the more productive Sitka spruce became the most commonly planted tree in the southern and western uplands.

Historically, the planting of broad-leaves has been mainly confined to large private estates. Many mixture combinations have been established, and, although relatively few experimental examples exist in comparison with the lowlands, they provide invaluable information on appropriate silvicultural techniques for upland conditions.

### *Benefits of mixtures*

*Nursing*

*Conifer/conifer nursing mixtures.* As afforestation expanded on to the heather (*Calluna vulgaris*)-dominated heathlands of eastern Scotland and northern England in the 1920s, it was observed that Sitka spruce went into 'check' despite good early growth. It was hypothesized that this was as a result of competition with heather. This was proved correct when effective heather control improved tree growth for a number of years. However, the spruce still suffered from nutrient deficiency, particularly nitrogen. Later observations (Weatherell 1953, 1957; Zehetmayr 1960) showed that the spruce soon recovered when planted near pine or larch. The fast early growth of the pine and larch helped to suppress the heather, allowing the spruce to become established. The discovery of this 'nursing effect' led to the instigation of a major research programme by the Forestry Commission and, from 1945 to 1965, Sitka spruce was planted in mixture with Scots pine or Japanese larch on a large scale on eastern heathland sites.

For similar reasons, of heather suppression, mixtures of Sitka spruce and lodgepole pine were frequently established on peats, peaty gley and peaty ironpan

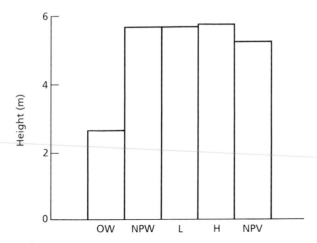

FIG. 1. Mean height of Sitka spruce at 20 years in a mixtures experiment at Strath Forest, Inverness-shire in north Scotland. OW = Sitka spruce + heather control; NPW = Sitka spruce + periodic inputs of nitrogen + heather control; L = Sitka spruce (25%) mixed with lodgepole pine (75%); H = Sitka spruce (25%) mixed with hybrid larch (75%); NPV = Sitka spruce + periodic inputs of nitrogen.

soils during the late 1950s and 1960s. Unfortunately, many of these mixtures failed, owing to the choice of unsuitable seed origins of lodgepole pine or unsuitable ratios or patterns of mixture. Southern lodgepole pine from the coasts of Oregon and Washington often grew too vigorously and suppressed the spruce, resulting in an open stand of pine. Other stands contained pine origins which had poor exposure resistance and low canopy density; these had little or no suppressive effect on the heather and were therefore equally unsuccessful as nurses.

The value of both pine and larch as nurses to Sitka spruce on peat has now been clearly demonstrated by experiment. One example is illustrated in Fig. 1. Because the long-term aim of these mixtures was to produce a final crop of Sitka spruce it was evident that the choice of nurse species and provenance should be made with some care. Problems of identifying suitable nurse species, combined with the benefits of improved cultivation, fertilizer application and weed control, led to a marked decline in the use of these nursing mixtures during the 1960s and 1970s. However, they are now widely used once again following the discovery of nutritional benefits to the spruce.

*Broad-leaved/conifer nursing mixtures.* Interest in nursing mixtures of broad-leaves and conifers has centred upon the potential of such crops to ameliorate the harsher climate of the uplands, thereby promoting faster growth and improved quality in the broad-leaved component.

To investigate the benefits to broad-leaves of nursing mixtures with pine and larch, an experiment was planted in 1955 on a typical heathland site in the North Yorkshire Moors. This experiment compared the growth of five main and four subsidiary broad-leaved species in both pure plots and plots containing a row

TABLE 2. Heights of trees in pure and mixed-species stands in an experiment situated on the North Yorkshire Moors in northern England. JL, Japanese larch; SP, Scots pine; GYC, General Yield-Class

| Main species | | Mean height (m) | | | | Dominant height (m) | | Top height (m) | Provisional GYC |
|---|---|---|---|---|---|---|---|---|---|
| | | 3 years | 6 years | 10 years | 15 years | 20 years | 25 years | 25 years | (m³ ha⁻¹ year⁻¹) |
| Sessile | pure | 0.3 | 1.3 | 1.6 | 2.1 | 4.6 | 5.3 | 6.9 | 4 |
| oak | with JL | 0.4 | 1.3 | 2.3 | 3.9 | 6.1 | 7.1 | 8.0 | 4 |
| | with JL and SP | 0.4 | 1.3 | 2.0 | 3.2 | 5.8 | 6.8 | 7.9 | 4 |
| Beech | pure | 0.3 | 1.0 | 1.6 | 1.7 | 3.5 | 4.6 | 5.3 | 4 |
| | with JL | 0.4 | 1.0 | 2.2 | 4.0 | 6.3 | 7.2 | 8.2 | 6 |
| | with JL and SP | 0.3 | 1.0 | 1.7 | 3.0 | 5.6 | 6.8 | 8.0 | 4 |
| Silver | pure | 1.0 | 2.2 | 3.1 | 4.1 | 7.2 | 8.2 | 9.6 | 4 |
| birch | with JL | 1.0 | 2.3 | 3.3 | 4.4 | 7.3 | 8.6 | 9.8 | 4 |
| | with JL and SP | 1.0 | 2.3 | 3.4 | 4.5 | 7.0 | 7.6 | 8.6 | 4 |
| Sweet | pure | 0.2 | 0.8 | 1.3 | 1.7 | 4.3 | 5.5 | 6.6 | 4 |
| chestnut | with JL | 0.3 | 1.2 | 2.3 | 4.4 | 7.3 | 9.4 | 10.3 | 6 |
| | with JL and SP | 0.3 | 0.9 | 1.5 | 2.7 | 7.0 | 8.5 | 9.4 | 6 |

Subsidiary species

| | | | | | | | | | |
|---|---|---|---|---|---|---|---|---|---|
| Red oak | with JL and SP | 0.2 | 0.9 | 1.5 | 2.4 | 5.8 | 6.6 | 7.6 | 4 |
| *N. procera* | with JL and SP | 0.6 | 2.1 | 2.8 | 4.3 | 7.1 | 7.9 | 9.7 | c.6 |
| *N. obliqua* | with JL and SP | 0.7 | 1.8 | 2.3 | 3.4 | 5.9 | 6.6 | 7.9 | c.4 |
| Downy birch | with SP only | 0.7 | 1.0 | 1.6 | 2.0 | 4.3 | NM | — | — |

mixture with Japanese larch and Scots pine as nurse species. The row mixtures consisted of three conifer trees followed by four broad-leaved trees planted alternately along each row (in the direction of ploughing). The five main species were sessile oak (*Quercus petraea*), beech (*Fagus sylvatica*), silver birch (*Betula pendula*), grey alder (*Alnus incana*) and sweet chestnut (*Castanea sativa*). Subsidiary species were: red oak (*Quercus borealis*), downy birch (*Betula pubescens*), raoul (*Nothofagus procera*) and roble (*Nothofagus obliqua*). The early survival rates of all the species were very good, with the silver birch lowest at 85%. The initial growth of all the species was also good and a suggestion of a nursing effect from the larch and pine was evident at age 3 years (Table 2). The subsequent growth of all the species, except grey alder and downy birch, has been maintained and the nursing effect was still evident at the time of the last assessment at age 25 years (Gabriel 1986).

This experiment demonstrated the significant improvements that can be obtained in the rate of growth and quality of a range of broad-leaved tree species by using nursing mixtures on a poor heathland site. Another notable benefit has been protection from frost damage, particularly on *Nothofagus procera*.

## Nutrition

Afforestation of unflushed peats in north Scotland began in the early 1970s. The increasing cost of fertilizer meant that it was no longer economic to establish spruce crops, which required very high inputs of fertilizer; so large areas were established with pure lodgepole pine, which is less site-demanding. The resultant loss in productivity, and also the generally poor form of the pine, led to a re-evaluation of 'nursing mixtures' with spruce. A series of experiments established in the mid 1960s, to investigate the benefits of *Calluna* suppression by a nurse species, were used to assess the value of these mixtures in terms of the nutrition of the spruce later in the rotation.

A representative experiment from this series, at Inchnacardoch Forest in Inverness-shire, contains Sitka spruce in intimate mixture with Scots pine and Japanese larch. In this, and in several other experiments on deep peat, the initial benefits of the nursing mixture were continued, at least to the pole stage (Fig. 2). In particular, the concentration of nitrogen in the spruce foliage has been greatly increased in the mixture plots (Table 3), whilst the pure plots required repeated nitrogen applications to maintain satisfactory growth (Taylor 1985). The effect is seemingly related to the root activity of the nurse species and substantial differ-

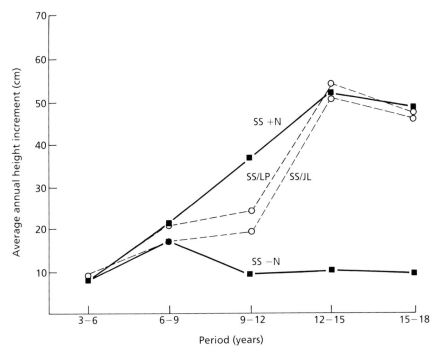

FIG. 2. Average annual height increment of Sitka spruce, pure and in mixture with lodgepole pine and Japanese larch, growing at Inchnacardoch Forest in Inverness-shire. SS = Sitka spruce; LP = lodgepole pine; JL = Japanese larch; +N, −N = with or without the addition of nitrogen fertilizer.

TABLE 3. Per cent of nitrogen in the foliage of Sitka spruce in a species mixture experiment at Inchnacardoch Forest (north Scotland). <1.2 = deficient nitrogen levels; >1.5 = optimum nitrogen levels; SS = Sitka spruce; LP = lodgepole pine; JL = Japanese larch; +N = additional nitrogen fertilizer

| Age (years) | SS | SS +N | JL/SS | JL/SS +N | LP/SS | LP/SS +N |
|---|---|---|---|---|---|---|
| 8 | 1.01 | 1.70 | 1.11 | 1.93 | 1.02 | 1.86 |
| 22 | 0.81 | 1.39 | 1.32 | 1.57 | 1.51 | 1.69 |

ences have been found in the mycorrhizal fungal flora on pure spruce compared with spruce in mixture with larch or pine (Carey, McCarthy & Miller 1988).

There is little evidence that the use of a broad-leaved component will improve the nutrition of a conifer in the British uplands. Nitrogen-fixing broad-leaved tree species such as alder show the most promise, but the fast early growth of red alder could suppress a conifer in an intimate mixture. There is evidence from North America of favourable effects of red alder on the nitrogen status of Douglas fir (Miller & Murray 1979; Binkley, this volume), but there has been limited experimentation on nutritional mixtures of this kind in upland Britain.

*Wind-throw*

Stand vulnerability to wind damage increases with age (height) regardless of the structure of the woodland. The onset of vulnerability is earlier in conifers and increases more rapidly with age than in broad-leaves (Quine & Miller 1990). It has also proven very difficult to isolate differences between species owing to the many combinations of site type, cultivation, stand management and structure.

Canopy roughness clearly influences the pattern of air flow over and through a stand, with a greater transfer of momentum to the trees occurring in stands with a more broken structure. Therefore, it is probable that a mixed-species stand in which the growth rates of component species are markedly different, and where the components are arranged in relatively large blocks or groups, will be more vulnerable to wind-throw.

There is also evidence to show that stem taper influences susceptibility to wind-throw and that trees with a low height:diameter ratio may be more stable (Neckelmann 1981). Stem taper can be increased by using a wide initial spacing, early respacing or by the use of a 'self-thinning' mixture. Patterns of self-thinning Sitka spruce/Scots pine mixtures have been observed in an experiment at Kielder Forest. The mixture plots remain stable at 50 years despite the occurrence of wind-throw in the surrounding pure Sitka spruce stand.

Although wide initial spacing will produce trees with greater stem taper, the increased light available to each tree will also allow greater crown development and the extra wind resistance of the trees may counteract the benefits of the stem form. The ideal mixture, therefore, for exposed sites would seem to be one composed of an intimate mixture (to avoid increased canopy roughness) of a matrix species which is more light-demanding and yet slower growing than the main final crop species. This mixture would help to ensure that the matrix (or nurse) trees were suppressed relatively quickly after the establishment phase and that the transition to a pure crop was more gradual than that which would result from respacing.

Little is known about the wind-firmness of self-thinning mixtures compared with pure stands although, at worst, they are likely to be comparable to a thinned stand. It may be that the gradual elimination of the matrix species might increase the canopy roughness and thus enhance the risk of wind-throw. Alternatively, the different rooting patterns of the component species, particularly in broad-leaved/conifer mixtures, might result in deeper rooting and hence a more stable stand structure.

There are few stands of self-thinning mixtures on high 'wind-throw hazard class' sites adjacent to comparable pure crops, making it difficult to draw objective conclusions. One example of a Sitka spruce/Scots pine mixture at Birkley Wood in Falstone forest remains relatively wind-firm after 70 years and appears to have resulted in a more stable stand than would be expected from a pure Sitka spruce crop.

## Long-term mixtures

In the early 1950s there was much debate about the value of mixed-species stands which were grown with the intention of retaining a mixture for the entire rotation. Apart from the possible benefits in terms of stability, the main advantages were thought to be the potential for increasing the total timber production from a site by using species which utilized different canopy and/or soil layers and the potential to improve soil fertility, in particular by the inclusion of a broad-leaved component.

One result of the debate was the establishment of an experiment at Gisburn (Bowland) Forest in 1955 (see Brown, this volume). The experiment was designed to investigate the long-term characteristics of several species mixtures and pure crops in terms of tree growth and their effects on the soil and ground vegetation. The component plots include Scots pine, Norway spruce, alder and oak planted pure, and in mixture, with the other species. There has been no apparent benefit to the growth in height of either of the conifers by being in mixture with the broad-leaved species (Table 4). This situation may not, however, continue to be the case in subsequent rotations. Scots pine has been shown to be the most effective nurse, particularly for Norway spruce, although the site is not one on which spruce would be expected to suffer seriously from nitrogen deficiency. However, although there may still have been a growth-enhancing effect from increased nitrogen availability in the pine/spruce mixture, competition for light in the presence of an overtopping nurse could also have contributed to the greater height growth of the spruce in mixture compared with the pure plots.

TABLE 4. The dominant height (m) of trees in mixtures at Gisburn Forest, northern England (see also Brown this volume). Figures in brackets refer to pure plots; * $P < 0.05$; ** $P < 0.01$. [†] Mean of mixed stands only

| Age (years) | Species | Pure | In mixture with | | | | |
| --- | --- | --- | --- | --- | --- | --- | --- |
| | | | NS | SP | Oak | Alder | Mean[†] |
| 7 | Norway spruce | 1.20 | (1.20) | 1.17 | 1.04 | 1.12 | 1.13 |
| | Scots pine | 1.96 | 2.15 | (1.96) | 2.20 | 2.03 | 2.09 |
| | Oak | 1.49 | 1.16 | 1.52 | (1.49) | 1.18 | 1.34 |
| | Alder | 1.67 | 1.53 | 1.68 | 1.55 | (1.67) | 1.61 |
| | Mean | | 1.51 | 1.58 | 1.57 | 1.50 | |
| | 5% L.S.D. | 0.27 | | | | | |
| 10 | Norway spruce | 1.85 | (1.85) | 2.18 | 1.51 | 1.73 | 1.82 |
| | Scots pine | 3.25 | 3.49 | (3.25) | 3.58 | 3.36 | 3.42 |
| | Oak | 2.03 | 1.71 | 2.52* | (2.03) | 1.87 | 2.03 |
| | Alder | 2.45 | 2.28 | 2.57 | 2.34 | (2.45) | 2.41 |
| | Mean | | 2.33 | 2.63 | 2.36 | 2.35 | |
| | 5% L.S.D. | 0.34 | | | | | |
| 20 | Norway spruce | 5.87 | (5.87) | 7.11* | 5.29 | 6.22 | 6.12 |
| | Scots pine | 8.04 | 8.30 | (8.04) | 8.05 | 7.94 | 8.08 |
| | Oak | 4.21 | 3.62 | 6.11** | (4.21) | 4.47 | 4.60 |
| | Alder | 5.77 | 4.88 | 6.54 | 4.99 | (5.77) | 5.55 |
| | Mean | | 5.67 | 6.95 | 5.64 | 6.10 | |
| | 5% L.S.D. | 0.90 | | | | | |
| 26 | Norway spruce | 8.80 | (8.80) | 10.62** | 8.76 | 9.84 | 9.51 |
| | Scots pine | 11.12 | 11.56 | (11.12) | 11.34 | 11.12 | 11.28 |
| | Oak | 6.58 | 5.67 | 8.82** | (6.58) | 7.29 | 7.09 |
| | Alder | 8.24 | 7.57 | 9.31* | 7.18 | (8.24) | 8.07 |
| | Mean | | 8.40 | 9.97 | 8.46 | 9.12 | |
| | 5% L.S.D. | 1.14 | | | | | |

## PREDICTING THE YIELD OF MIXED-SPECIES STANDS

To predict yield in pure stands, reference is made to the yield models published by the Forestry Commission (Edwards & Christie 1981). Each model describes expected growth for a range of species, Yield-Classes*, planting spacings and thinning treatments. These models have been constructed using periodic measurements from a national network of permanent sample plots and as such are primarily intended for forest management and planning at the national level.

The simplest way of assessing Yield-Class is to measure the average height of the dominant trees in a stand (the top height), as it has long been established that volume production is well correlated with top height (Hummel & Christie 1957). An estimate derived in this way is known as the General Yield-Class (GYC) of a stand. A better estimate of Yield-Class, known as Production-Class, can be obtained if the basal area or volume growth of the crop is also measured (Rollinson

* Yield-Class is the maximum average rate of timber volume production attained by a crop, and has units of $m^3 ha^{-1} year^{-1}$.

1986). Use of Production Class is thus recommended when making forecasts for forest management and planning at the local level.

To predict the yield of a mixture, adjustments are made to the appropriate pure species models for each component. The expected yield of each component is first worked out as if each formed a separate pure stand. The percentage of the canopy occupied by the crowns of each component is then estimated and the yield of the mixture computed by taking a weighted average, as illustrated in the following example. Suppose that the mixture consists of oak standards (aged 58, Yield-Class 8) underplanted with western hemlock (aged 25, Yield-Class 18), in which the oak occupies 0.1 of the crown space, with a predicted standing volume of 186 m$^3$ for a pure stand, and the western hemlock occupies 0.9 of the crown space, with a predicted standing volume of 161 m$^3$ for a pure stand. The standing volume of mixture $= (0.1 \times 186) + (0.9 + 161) = 163\,m^3\,ha^{-1}$.

This simple method, employed by the Forestry Commission to forecast production in mixtures, has proved adequate. However, it is unlikely that the main assumption—that each mixture component behaves like a pure stand—will hold in all cases. Evidence for change in the patterns of growth of trees in mixture can be sought using data from 103 Forestry Commission sample plots in mixtures. Compared with the 1444 sample plots in pure stands, this is a modest outlay, reflecting the dominance of pure species plantations in commercial forestry in

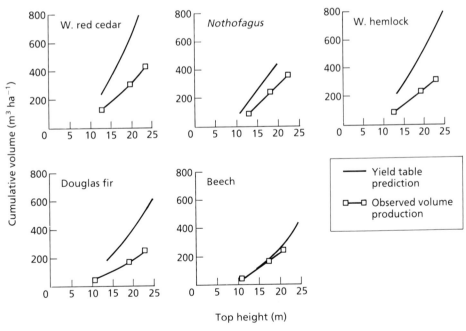

FIG. 3. Development of stand cumulative volume production with top height by five tree species planted under European larch standards. Comparison with volume production for pure stands as predicted by Forestry Commission yield tables.

Britain. While some plots occur as isolated yield plots, others form small mixture and underplanting experiments. One such experiment can be found at Haldon, near Exeter in Devon. A large plot of European larch was underplanted in 1937 with five different species, and sample plots were established in each.

Figure 3 shows cumulative volume production plotted against top height for each of the underplanted plots, adjusted to allow for the space occupied by the larch standards, and compared with the yield table function for a pure stand. All five species achieve rates of production similar to that predicted by the yield table, but it is interesting to note that volume production falls slightly lower than the yield table line in all three conifers. This observation conforms to silvicultural doctrine, suggesting that the underplants are being 'drawn up' by the overstorey, making them tall and thin. Unfortunately, it is impossible to tell whether this is a genuine effect as the experiment is unreplicated and has no control. Whether typical of underplantings or not, such deviations from normal growth are easily taken into account by estimating Production-Class for the stand.

However, there are other aspects of the behaviour of trees in mixture that are harder to predict, as illustrated by a series of plots at Beddgelert, in Gwynedd, Wales. The series consists of three pairs of plots at different altitudes on a hillside. Each pair is formed of one plot each of Japanese larch and lodgepole pine in mixture with Sitka spruce. The plots were thinned to maintain a 50/50 mixture, and Fig. 4 shows the proportion of each species remaining in each plot after 46 years. The results clearly show how attempts to maintain a 50/50 mixture failed. The Sitka spruce has largely suppressed the lodgepole pine and to a lesser extent the larch. The dominance of the Sitka spruce also becomes more pronounced at higher altitudes, such that the lodgepole pine has been completely excluded in the highest plot. Although this experiment has yielded a very clear result, it is unlikely that the same pattern would be observed generally. It is quite conceivable that the reverse result, with the larch or pine becoming dominant, might be observed at other sites. Without considerable supporting fieldwork it would be very difficult to predict this sort of behaviour with a yield model.

Future progress on the modelling of mixtures may be made using two approaches. First, the methods employed in constructing the existing empirical yield models could be extended to cover mixtures. Although data from permanent sample plots would be invaluable for such work, it is unlikely that all possible mixture combinations will ever be represented in sample plots. However, by combining data from pure-species plots with additional information from temporary sample plots in mixtures, it would be possible to produce reliable mixture models. For example, it is generally accepted that any mixture model must take into account the interaction between the crowns of the species involved. Temporary sample plots would be ideal for quickly collecting the necessary data.

The second approach is known as process modelling. While an empirical model makes direct assumptions about the patterns of growth of trees and stands, a process model seeks to predict those patterns by making more fundamen-

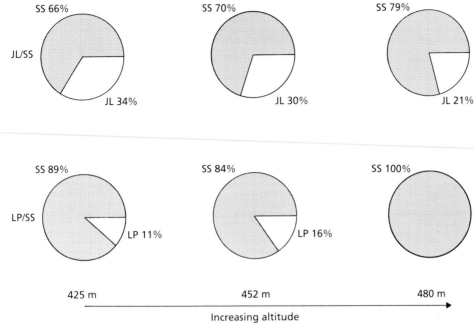

FIG. 4. Japanese larch and lodgepole pine planted in equal mixture with Sitka spruce. Relative survival of each species after 46 years. Top row: Japanese larch/Sitka spruce; bottom row: lodgepole pine/Sitka spruce.

tal assumptions about tree physiology. From the forester's point of view, it is irrelevant at what level the assumptions are made, provided sufficiently accurate and useful models can be produced. It will be some time before a process model which can predict the growth of mixtures is available but, once working, it should be capable of quickly producing results equivalent to many empirical models. Whichever method is used, sample plot data will always be needed to test the predictions made by the models.

While new models for mixtures may improve the accuracy of production forecasts at the national level, their value for local forecasts is less clear. As was the case at Haldon and Beddgelert, any change in growth rates due to interactions between species is likely to be masked by many other factors. Dewar (1988) observed that the prominant causes of forecasting errors are inaccurate descriptions of growing stock and assumptions about future treatment. Both empirical and process models require accurate field measurements in order to produce meaningful forecasts. It is arguable that, compared to the task of improving the models themselves, there is considerably more work to be done to improve the way in which the existing yield models are used. Production forecasts are as good as a forest manager needs them to be, provided the expenditure required to collect sufficiently intensive and accurate measurements can be justified.

## ACKNOWLEDGMENTS

Our thanks go to J. Evans, D. Paterson, J. Methley, T. Randle and S. Abbott for their useful comments and assistance when drafting this paper. We also thank P. Wright, H. Payne, B. Dickinson and J. Bell for conscientiously typing and amending the text.

## REFERENCES

**Aldhous, J.R. (1981).** Beech in Wessex—a perspective on present health and silviculture. *Forestry*, **54**, 197–210.

**Brown, J.M.B. (1953).** Studies on British beechwoods. *Forestry Commission Bulletin*, No. 20. HMSO, London.

**Carey, M.L., McCarthy, R.G. & Miller, H.G. (1988).** More on nursing mixtures. *Irish Forestry*, **45**, 7–20.

**Dewar, J. (1988).** Private sector production forecasting in Scotland. *Timber Grower*, **109**, 31–32.

**Edwards, P.N. & Christie, J.M. (1981).** Yield models for forest management. *Forestry Commission Booklet*, No. 48. HMSO, London.

**Evans, J. (1984).** Silviculture of broadleaved woodland. *Forestry Commission Bulletin*, No. 62. HMSO, London.

**Evans, J. (1988).** Natural regeneration of broadleaves. *Forestry Commission Bulletin*, No. 78. HMSO, London.

**Forestry Commission (1985).** *Guidelines for the Management of Broadleaved Woodland.* HMSO, London.

**Forestry Commission (1989a).** *Forestry Facts and Figures 1988–1989.* Polton House Press Ltd, Midlothian.

**Forestry Commission (1989b).** *Broadleaves Policy Progress 1985–1988.* Forestry Commission, Edinburgh.

**Gabriel, K.A.S. (1986).** Growing broadleaved trees on the North York Moors. *Quarterly Journal of Forestry*, **80**, 27–32.

**Hiley, W.E. (1967).** *Woodland Management.* Faber & Faber, London.

**Hummel, F.C. & Christie, J.M. (1957).** Methods used to construct the revised yield tables for conifers in Great Britain. *Forestry Commission Report on Forest Research 1957*, pp. 137–40. HMSO, London.

**Joslin, A. (1982).** Management of broadleaves in the Forest of Dean with special reference to regeneration. In *Broadleaves in Britain: Future Management and Research* (Ed. by D.C. Malcolm, J. Evans & P. Edwards), pp. 53–60. Institute of Chartered Foresters, Edinburgh.

**Locke, G.M.L. (1987).** Census of woodlands and trees 1979–82. *Forestry Commission Bulletin*, No. 63. HMSO, London.

**Lonsdale, D. & Wainhouse, D. (1987).** Beech bark disease. *Forestry Commission Bulletin*, No. 69. HMSO, London.

**Miller, K.F. & Murray, M.D. (1979).** The effects of red alder on growth of Douglas fir. Utilisation and management of alder. *USDA Forestry Services General Technical Report PNW-70*, 283–306.

**Neckelmann, J. (1981).** Stabilisation of edges and internal shelter-zones in stands of Norway spruce on sandy soil. *Dansk Skouforenings Tiddeskrift*, **56**, 296–314.

**Quine, C.P. & Miller, K.F. (1990).** Windthrow—A factor influencing the choice of silvicultural systems. In *Silvicultural Systems* (Ed. by P. Gordon). Institute of Chartered Foresters, Edinburgh.

**Rollinson, T.J.D. (1986).** Don't forget Production Class. *Scottish Forestry*, **40**, 250–8.

**Taylor, C.M.A. (1985).** The return of nursing mixtures. *Forestry and British Timber*, **14**, 18–19.

**Weatherell, J. (1953).** The 'checking' of forest trees by heather. *Forestry*, **26**, 37–40.

**Weatherell, J. (1957).** The use of nurse species in afforestation of upland heaths. *Quarterly Journal of Forestry*, **51**, 298–304.

Wood, R.F. & Nimmo, M. (1962). Chalk downland afforestation. *Forestry Commission Bulletin*, No. 34. HMSO, London.

Wood, R.F., Miller, A.D.S. & Nimmo, M. (1967). Experiments on the rehabilitation of uneconomic broadleaved woodland. *Forestry Commission Research and Development Paper 51*. Forestry Commission, Edinburgh.

Wright, T.W. (1982). The National Trust's approach to the regeneration of broadleaved woodland. In *Broadleaves in Britain: Future Management and Research* (Ed. by D.C. Malcolm, J. Evans & P. Edwards), pp. 77–83. Institute of Chartered Foresters, Edinburgh.

Zehetmayr, J.W.L. (1960). Afforestation of upland heaths. *Forestry Commission Bulletin*, No. 32. HMSO, London.

# Silviculture of mixed-species stands in Germany

G.K. KENK

*Forest Research Institute Baden-Württemberg, Wonnhaldestrasse 4, D-7800 Freiburg, Germany*

## SUMMARY

In the former FRG more than 50% of the forests consist of pure stands and the 1:3 ratio of broad-leaved to coniferous trees in the original forest has been turned the other way round. Economic arguments have been decisive in the move towards conifers and pure stands. However, the importance of mixtures was recognized more than 100 years ago, and the use of site-suited mixed forests has been in the foreground of silvicultural programmes of the last decade.

The extraordinary diversity of mixed stands (729 combinations) is rather confusing.

Clarity about the type of mixture required, together with a knowledge of site and of the growing rates of the species being grown are indispensable for silvicultural treatment and for the success of mixed stands. The selection forest is often seen as an ideal and an example of 'natural' forest management. But its possibilities tend to be overestimated.

The main subject of this paper is the silviculture of even-aged, mixed stands. Treatment of these stands is aimed at controlling the growing space of crop trees, and this is illustrated by the example of a spruce/beech mixture. The transformation of pure coniferous stands into mixed stands is realized by preliminary cultivation under the canopy of the mature stand or by open-field planting.

The importance of stable and diverse forest ecosystems and the steps needed to achieve a 'natural' silviculture are discussed.

## INTRODUCTION

The recently finished 1987 forest inventory in the FRG lists 729 combinations of mixed-species stands (Smaltschinski 1990). The diversity of mixed stands in south-western Germany (land Baden-Württemberg) is summarized in thirty-five types.

The manager of these mixed-species stands* is faced with a confusing multitude of mixtures whose composition results from historical considerations about profitability as well as from site conditions, growth dynamics and past silviculture. Attitudes to some of these factors have changed in the last few decades.

---

* A stand composed of two or more tree species, which significantly influence the ecology of the stand both by themselves and especially in combination. The tree species can be mixed in small clusters, in groves, or in groups; rows and lines are possible, too, as well as woodlots. A mixture is called temporary if a tree species is used before the end of the stand's cutting cycle (Brünig & Mayer 1980).

This contribution deals with three aspects of mixed-species stands, namely:

1  their historic development and present state;
2  the silvicultural treatment of stands, with an example (spruce/beech); and
3  the transformation of pure coniferous into mixed-species stands.


## HISTORICAL DEVELOPMENT AND PRESENT STATE

The biogeographical history of central Europe has lead to a dendroflora comparatively poor in species. In the dawn of history most of the ecosystems were
dominated by the very competitive beech (*Fagus sylvatica* L.) mixed with maple
(*Acer* spp.), ash (*Fraxinus excelsior* L.) and elm (*Ulmus* spp.) and in mountainous
regions with fir (*Abies alba* Mill.) and spruce (*Picea abies* (L.) Karst.). The oaks
(*Quercus* spp.) occupied the drier and warmer lowlands, where they were mixed
with hornbeam (*Carpinus betulus* L.), lime (*Tilia* spp.), cherry (*Prunus* spp.),
beam-tree berry (*Sorbus* spp.) etc. A few almost pure stands of spruce were
located on the peripheries of moorlands and on higher montane sites. Pine (*Pinus
sylvestris* L.) occurred on sand dunes and gravel ridges, as well as in the highlands
of the Black Forest (Firbas 1949, 1952; Rubner & Reinhold 1953; Ellenberg
1978).

'After the end of the Middle Ages an intentional or unintentional promotion
of coniferous trees took place' (Hausrath 1911). There were many reasons for
this: the great clearances from A.D. 700 to 1300 affected primarily the better sites
stocked with broad-leaved species, while natural restocking (for example after
the Thirty Years' War, 1618–1648) favoured the light seeded conifers. Further
advance of coniferous trees was favoured by successful artificial establishment,
higher biological and financial yields, and by site deterioration due to litter
removal, grazing in the forest by domestic animals, and mismanagement (Hausrath
1911). The wider timber uses, higher productivity and better profitability of
spruce were beyond doubt until the recent past: at the beginning of the twentieth
century, beech stands realized only 4–5% of the ground rent and 30% of the
wood rent achieved by spruce crops. Even in the 1970s, beech and some other
hardwoods were regarded as 'lost timber species' and 'forest weeds' (Brandl
1989). In consequence, they were continually decimated when they occurred in
mixtures.

Man's manipulation of German forests has reversed the balance between
broad-leaved and coniferous trees found in original woodland communities
(Hesmer 1938): from two-thirds of deciduous trees the proportion has decreased
to scarcely one-third.

More than half of the forests in western Germany are made up of 'pure stands',
i.e. stands in which one species occupies 90% or more of the area (Table 1).

In the original forest vegetation types, pure stands were rather uncommon; the
reasons lie in the dynamics of woodland succession and in the remarkably small-
scale diversity of site conditions.

TABLE 1. Tree species (%) in Germany (formerly FRG) by stand-type. Sources: 1-4 = Federal Forest Inventory 1987 (Smaltschinski 1990); 5 = estimated according to Hesmer (1938)

| Tree species group | Total | Share in area (%) Pure stands (admixture < 10 %) | Mixed stands (admixture ≥ 10 %) | Natural stands |
|---|---|---|---|---|
| 1 | 2 | 3 | 4 | 5 |
| Spruce | 42 | 66 | 34 | 20 |
| Fir | 2 | 18 | 82 | 8 |
| Pine | 20 | 63 | 37 | 5 |
| Douglas fir | 2 | 46 | 54 | — |
| Larch | 2 | 39 | 61 | — |
| Beech | 17 | 49 | 51 | 40 |
| Oak | 9 | 41 | 59 | 20 |
| Others | 6 | 37 | 63 | 7 |

Today the forests are dominated by spruce and pine. About two-thirds of the stands of these species contain no or very little admixture. Stands of larch (2%), Douglas fir (2%) and silver fir (3%) play a minor role in comparison.

The deciduous forests consist of beech, oaks and other broad-leaved trees (*Carpinus*, *Acer* spp., *Fraxinus*, *Betula* spp., *Prunus* spp., *Tilia* spp., *Salix* spp., *Sorbus* spp., etc.). More than half of these stands are mixed.

The advantages and disadvantages of pure and mixed forests have been the subject of discussion since forestry was first practised (Burger 1941). In Germany, pure stands were promoted in the second half of the eighteenth century, but subsequently mixed stands were preferred for some decades.

In the second half of the nineteenth and after the turn of the twentieth century the arguments against mixed forest were predominant. As a result, the loss of deciduous woodlands between 1883 and 1927 amounted to 1.2 million ha (Hesmer 1938).

However, as early as 1886, Gayer started the move back towards mixed forests with the publication of his work *The Mixed Forest, its Establishment and Care, in particular by Management with Groves and Groups*. He wished beech to be grown in spruce stands and to cover about 20% of the stand area (= 5–10% of volume) in order to ensure future natural regeneration. Demands that this target be met have been made again and again up to the present—above all because of the widespread impression that pure spruce stands suffer catastrophic damage by storm, snow and insects more often than mixed stands (Burschel 1987a).

Those managing mixed stands experienced failures and setbacks caused by unsuitable combinations of species, inadequate consideration of stand dynamics and growth behaviour, ignorance of site conditions and poor silvicultural treatment, as well as by excessive stocking of game animals (Burschel & Huss 1987). These led to further decreases of hardwoods and mixed stands.

Over the last two decades a growing movement to favour tree species of the

TABLE 2. Woodland communities of outstanding value for the protection of fauna (according to Arbeitskreis Forstliche Landespflege 1984)

| Woodland communities | Examples |
| --- | --- |
| Forests on wet and humid soils | Riparian forest, fen forest, swamp forest |
| Thermophilous deciduous forests | Stands with *Quercus pubescens*, dryness-tolerant oak-dominated mixed woodlands, orchid/beech stands |
| Natural pine forests | Pine stands with *Molinia* |
| In maple and ash dominated mixed stands | Stands of maple and ash, brook forests with ash, ravine forests with maple and phyllitis |
| Communities with original participation of fir | Fir/beech forests, spruce/fir forests |
| Historical forms of woodland utilization | Coppice with and without standards, oak forest in the zone of beech forests |
| Mature forest ecosystems | Mature stands with higher percentages of dead trees |

natural plant associations can be seen, especially in public forests. This has been made easier by a remarkable improvement in the profitability of hardwoods and by the increased importance attached to the stability and variety of forests. After all, at least a quarter and often more of the annual cut in former West Germany is of trees damaged by storm, snow or insects. For coniferous trees only, this proportion is significantly higher (Seitschek 1989).

Stability of forests depends to a large extent on the species composition of stands and their silvicultural treatment. Mixed-species stands, adapted to site conditions, are widely believed to be more stable than pure stands, and with good reason they are today in the foreground of silvicultural programmes (Burschel 1987a,b, 1990; Weidenbach 1988; Seitschek 1989). They are also of great importance as biotopes (Table 2).

Mixtures are not an end in themselves; they serve economical, ecological, and also visual objectives, the last in the context of the character and design of landscapes.

In Germany, beech is the most important species in mixed stands: it helps to stabilize spruce and pine stands, as well as to keep the stems of valuable oaks and larches free of epicormic branches. Furthermore it prevents the spread of ground vegetation which often forms an important obstacle to natural regeneration.

In beech stands the presence of ash, maple and cherry, etc., leads to a significantly-improved financial return. They attract much higher prices per unit volume.

Deciduous trees and shrubs arising from natural regeneration in plantations of spruce and oak may allow substantial reductions in the number of planted trees and hence of establishment costs; total production costs are also reduced and at the first thinning bigger dimensions of timber can be expected (Kenk 1988a;

TABLE 3. Examples of tree species mixtures in Germany (following Burschel & Huss 1987; Schütz 1989)

| Mixture | Examples |
| --- | --- |
| At least two dominant tree species grown on the same rotation ('mixture' in groups or single trees) | Spruce/beech<br>Spruce/fir/beech<br>Larch/beech<br>Spruce/pine<br>Pine/spruce<br>(/beech) |
| Tree species in secondary function (stem protection, site tending, diversity) ('mixed stand with secondary species') ('subordinate mixtures') | Oak with hornbeam<br>Oak with beech<br>Larch with beech<br>Pine with beech<br>Douglas fir with beech |
| At least two dominant tree species with different cutting cycles ('temporary mixture') ('subordinate mixtures') | Spruce in beech stands; maple, ash in beech stands<br>Cherry in ash stands |

Kenk, Schlör & Weise, 1989; Kenk 1990). Some examples of common mixtures are shown in Table 3.

The dominant tree species of the main 'mixture' (Schütz 1989) should have similar growth rates and rotation lengths. This ideal case occurs rarely; the necessary balance of growing conditions has to be achieved partly by silvicultural management (e.g. taking advantage of differences in age and growth rate, thinning to favour tree species with inferior growth).

## SILVICULTURAL TREATMENT OF MIXED STANDS

The growth of mixed stands has been the object of forest research since the turn of the century. The first 'Instruction for managing yield experiments in mixed stands' was published by the forest research institutes in the 1920s (Deutsche Forstliche Versuchsanstalten 1927). A substantial body of valuable results has been built up and can be found in the textbooks of Wiedemann (1951), Assmann (1961), Mitscherlich (1978) and Kramer (1988).

The first, and so far only, yield tables for mixed stands of spruce/beech and oak/beech were set up by Wiedemann (1942, 1949). They show the higher (+18%) and lower (−17%) volume increments of spruce/beech mixtures compared with pure spruce stands on good and bad sites, respectively. With a better water supply, lower alkali reserves in the soil and lower temperatures, the growth rate of spruce becomes superior to that of beech. However, differences in the detailed composition of mixtures, site conditions and silvicultural treatments leave open many questions about their productivity.

Those responsible for the silviculture of mixed stands should not be seduced by unrealistic proposals for promoting uneven-aged or selection forests with an abundance of species as an ideal and as an example of a type of management which is 'true to nature'. Such forests are often very different from the natural vegetation in our biogeographic conditions. The transformation of pure to mixed stands and their introduction into our silvicultural practice may be possible if there are clear, long-term management objectives appropriate for the site conditions and stand structures. There are some very convincing examples of the conversion of conventional forests (Hiller von Gäertringen 1989; Rotenhan 1988).

Selection forests with single tree or small group mixtures of silver fir, spruce and beech are often taken as the model for a more 'natural' silviculture which could be used with other species. But this interpretation is based on misunderstanding: selection forests are artefacts and very sensitive to silvicultural treatment. They depend on the maintenance of a finely-balanced mixture of shade and semi-shade trees, and today they play a very limited role in some natural silver fir forests. To perpetuate the structure of selection forests without planting and with comparatively few tending operations is possible if special conditions are met (regular harvesting, silvicultural knowledge, accessibility and a rather low game population). Their attraction to foresters is their potential as a biological and technical production forest with higher volumes per piece and therefore higher profits (Mitscherlich 1961; Siegmund 1974). Their aesthetic value is also higher than that of uniform high forests.

The conversion of even-aged into selection forests is very difficult and runs the risk of failure for decades after the process has started.

Today, some of the advantages of uneven-aged forests can also be achieved by silvicultural systems with regeneration phases up to 60 years, as well as by artificial establishment of stands with a low tree number (Kenk 1988b, 1990).

In the foreground of our silvicultural practice in western Germany are even-aged, more or less intensively-mixed stands which can be established by planting and/or natural regeneration. Their silvicultural management is intended to control the growing space of the final crop trees.

Of greatest importance for stand development are the first five decades of the stand's life. This period covers the phases of tending and early thinning. Early, heavy thinning to favour the crop trees is necessary to guarantee fully developed crowns (40–50% of total tree height) and the associated increases in individual tree value and stability.

Simple and effective means of controlling the species ratio in a mixed stand are based on the relationship between diameter at breast height and area of crown projection. Depending on production targets in terms of breast heights-diameters, Abetz (1974) deduced crop tree numbers and the distances between them (Table 4).

An example of the treatment of spruce/beech stands in Baden-Württemberg (MLR Baden-Württemberg 1988) is shown in Table 5. In their fundamental character, these recommendations correspond to the management instructions of various other regional forestry administrations. Modifications concern mostly the number of final crop trees.

TABLE 4. Numbers and spacing of crop trees* (Abetz 1974) for seven tree species and suitable mixtures in Germany. *Not identical with the total number of trees in the final stand

| Tree species | Spruce | Silver fir | Douglas fir | Pine | Larch | Beech | Oak |
|---|---|---|---|---|---|---|---|
| Crop trees ha$^{-1}$ | 400 | 300 | 100 | 200 | 100 | 110 | 60 |
| | | | Mean (minimum) distance between crop trees (m) | | | | |
| Spruce | 5.7 (4.0) | | | | | | |
| Silver fir | 6.2 (5.0) | 6.6 (4.0) | | | | | |
| Douglas fir | | | 11.5 (8.0) | | | | |
| Pine | 7.0 (6.0) | 7.3 (6.0) | | 8.1 (5.0) | | | |
| Larch | | 9.0 (7.0) | | 9.8 | 11.5 (8.0) | (5.0) | |
| Beech | 8.5 (7.0) | 8.7 (7.0) | | 9.5 (7.0) | 11.2 (8.0) | 11.0 (5.0) | |
| Oak | | | | 11.4 (8.0) | 13.1 (9.0) | 12.9 (9.0) | 14.8 (8.0) |

TABLE 5. Treatment of mixed spruce/beech stands

| | Spruce (mostly from plantation) | Beech (natural regeneration or precultivation) |
|---|---|---|
| If sheltered | In case of natural regeneration: releasing of vital trees only in a 3–5 m distance | Aim: small cluster or groups |
| ≤ 5 m height | Reduction to about 1500 (resp. 800–1000 on steep slopes) dominant trees ha$^{-1}$ Pruning of 200–400 suitable trees to 2.5 m | Regulation of mixture corresponding to defined targets; If risk of snow break: reduction of trees to a mean spacing of 1.5 m |
| 10 m height | Reduction to about 1000–1500 ha$^{-1}$ Pruning to 5 m | Reduction to a mean spacing of 2.5 m (maximum: 2000 dominants ha$^{-1}$) |
| ≥ 15 m height | Selection of 200–300 crop trees in distances of 5–9 m Heavy thinning to favour crop trees | Reduction to a mean spacing of 3 m (maximum: 1000 dominants ha$^{-1}$) Selection of up to 30 crop trees in spacings of 6–10 m |
| ≤ 25–30 m height | Thinnings in intervals of 3–4 m height increase, or all 5–10 years | |
| ≥ 35 m height | Beginning of a 20–50 years lasting regeneration period. First step: preregeneration or precultivation of beech in gaps; they should be ahead with about 20–30 years in age or 8–10 m in height. Second step: natural or artificial regeneration of spruce | |

# TRANSFORMATION OF PURE CONIFEROUS INTO MIXED-SPECIES STANDS

The first steps in the transformation of pure spruce stands into mixtures with beech were taken in the 1920s, and in public forests the process has been a focal point in silviculture over the last decade (Weidenbach 1988; Burschel 1987b, 1990; Seitschek 1989).

Transformation can be achieved in two ways.

1 Partial substitution of spruce (or pine) by beech (or on suitable sites by fir) by
   (a) planting in small groups in the mature crop, or
   (b) restocking of clear-cut areas.

In the first case (a) natural or artificial gaps in the mature crop are planted and then gradually enlarged according to the light requirements of the planted species. The duration of this phase depends on the rate of height growth of the planted trees; it is often 20–30 years, but can take up to 50 years if fir is involved. After the final extraction of the mature timber, spruce or other softwoods are planted on the remaining area, at appropriate distances from the established groups of hardwood or fir. Palmer (1985) gives a detailed description.

The second case (b) is one of expedience and is done when there is no time to establish groups of hardwood/fir before final felling. Such situations arise after hurricane damage (Zimmermann 1985), or if there is a danger of rapid deterioration of the mature timber by heart rot, or if there is a high risk of wind-throw.

In general, spruce shows better growth than beech and fir on restock sites, and mixtures will flourish only if the latter are planted in groups and/or as larger plants.

In some cases, beech can be replaced by the faster growing sycamore or lime. Where fir is required on difficult areas within its natural habitat (for example on wet sites), it may be protected from dangerous late frosts by a nurse crop of black alder.

Experiences with both practices (a and b) are often bad, mostly due to excessive stocking of game animals or poor silvicultural treatment. Establishing stands with lower plant numbers and/or more intensive early tending could help to improve the stability of the mixture (Kenk 1990).

2 Complete replacement of unsuitable spruce (and pine) after clear-cutting.

In this case hardwood crops are established on the restock area, especially oaks (with hornbeam or lime); on suitable sites mixtures of maple, ash, elm, etc., can also be planted.

Both practices—partial substitution and complete replacement of conifers—are state-subsidized for many forest-owners in former West Germany.

As well as these deliberate transformations of mature stands, use of natural regeneration of desirable mixture tree species can play an important role in stand treatment. In younger conifer stands in particular there are often hardwoods or other tree species compatible with the main species, and by judicious thinning they can be used to form a mixture. They may then be capable of producing seed for the next rotation.

## CONCLUSIONS

The stability of forests primarily depends on their species composition, but also on their silvicultural treatment. Mixed-species stands are widely believed to be more stable than pure stands, especially if they consist of tree species that naturally occur in the region.

Ignorance of the growth behaviour of trees in mixtures, particularly in the early stages of stand development, and mistakes in establishing and tending, have led to many failures in the management of mixed species stands.

Even if their present composition is often far from that of natural woodland, forests represent living spaces which are relatively 'close to nature'. There are those who consider that foresters have an ethical obligation to ensure that the characteristics and variety of forests are not only maintained, but are improved where possible (Weidenbach 1988; Ott 1989).

In many cases—in particular in public forests—this objective can be achieved without conflicts and in parallel with economic aims. Germany has high labour costs and a free market economy in which conventional forestry on many poor sites produces deficits and will do so for a long time to come. In such circumstances, the transition to extensively or unmanaged woodlands—and perhaps also the conversion towards stable tree communities closely matched to site conditions—might be easier. Some of these sites could stock a mosaic of climax forests. Woodlands of this type could meet some of the demands for biodiversity made by different sectors of society including the nature conservation lobby, today supported by all political groups (Pöppinghaus 1990).

These diverse objectives might best be achieved with a silviculture which takes maximum advantage of natural cycles. Essential features of such silviculture and management are the stability of single trees and stands, the adequate representation of native tree species, the production of large-sized, valuable timber, and the use of natural regeneration wherever it is possible. The last is possible only if a limited stock of game animals is present.

Today, our knowledge of the properties and dynamics of soils, natural floristic communities and of the growth of tree species, permit us to consider the problems of managing mixed-species stands in a purposeful way. This should enable us to contribute to the realization of the multiple functions of our forests. Thus, the prospects for the 'third period of mixed-species stands' (Huss 1987) are better than ever before.

## REFERENCES

Abetz, P. (1974). Zur Standraumregulierung in Mischbeständen und Auswahl von Zukunfts bäumen. *Allgemeine Forstzeitschrift*, **29** (41), 871–73.
Arbeitskreis Forstliche Landespflege (1984). *Biotoppflege im Wald. Ein Leitfaden für die forstliche Praxis.* KILDA-Verlag, Greven.
Assmann, E. (1961). *Waldertragskunde.* BLV Verlagsgesellschaft, München.
Brandl, H. (1989). Die Auseinandersetzungen um die Anteile der Buche am Bestockungsaufbau auf den Forstvereinstagungen in Baden-Württemberg seit 1839. *Forst und Holz*, **45** (1), 3–6.

**Brünig, E. & Mayer, H. (1980).** *Waldbauliche Terminologie.* IUFRO-Gruppe Ökosysteme Wien.

**Burger, H. (1941).** Beitrag zur Frage der reinen oder gemischten Bestände. *Mitteilungen der Schweizerischen Anstalt für forstliches Versuchswesen,* **22** (1), 164–203.

**Burschel, P. (1987a).** Karl Gayer und der Mischwald. *Allgemeine Forstzeitschrift* **42** (23), 587–603.

**Burschel, P. (1987b).** Der Wald von morgen. *Allgemeine Forstzeitschrift* **42** (45), 1162–5.

**Burschel, P. (1990).** Waldumbau. *Allgemeine Forstzeitschrift* **45** (3), 57–9.

**Burschel, P. & Huss, J. (1987).** *Waldbau.* Verlag Paul Parey, Hamburg.

**Deutsche Forstliche Versuchsanstalten (1927).** Anleitung zur Ausführung von Untersuchungen in gemischten Beständen (beschlossen auf der Versammlung des Vereins der Deutschen Forstlichen Versuchsanstalten 1926 zu Rostock). *Allgemeine Forst- und Jagd-Zeitung,* 76–7.

**Ellenberg, H. (1978).** *Vegetation Mitteleuropas mit den Alpen in ökologischer Sicht.* Ulmer-Verlag.

**Firbas, F. (1949).** *Spät- und nacheiszeitliche Waldgeschichte von Mitteleuropa nördlich der Alpen.* 1. Bd.: Allgemeine Waldgeschichte. Verlag Gustav Fischer, Jena.

**Firbas, F. (1952).** *Spät- und nacheiszeitliche Waldgeschichte von Mitteleuropa nördlich der Alpen.* 2. Bd.: Waldgeschichte der einzelnen Landschaften. Verlag Gustav Fischer, Jena.

**Gayer, K. (1886).** *Der gemischte Wald, seine Begründung und Pflege insbesondere durch Horst- und Gruppenwirtschaft.* Verlag Paul Parey, Berlin.

**Hausrath, H. (1911).** *Pflanzengeographische Wandlungen der deutschen Landschaft.* Verlag Teubner., Leipzig.

**Hesmer, H. (1938).** *Die heutige Bewaldung Deutschlands.* Verlag Paul Parey, Berlin.

**Hiller von Gäertringen, H. (1989).** Rationalisierung durch konsequente Vorratspflege. *Allgemeine Forstzeitschrift,* **9/10** (44), 232–5.

**Huss, J. (1987).** Mischwald zwischen Wunsch und Wirklichkeit. *Forstwissenschaftliches Centralblatt,* **106**, 114–32.

**Kenk, G. (1988a).** Fichtenwirtschaft ohne Vornutzung? Das Beispiel eines 115- jährigen Bestandes im Wuchsgebiet Schwäbische Alb. *Allgemeine Forstzeitschrift,* **43** (30), 837–9.

**Kenk, G. (1988b).** Der Volumen-und Wertzuwachs im Stadium der natürlichen Verjüngung eines Kiefern-Tannen-Bestandes durch den Schirmkeilschlag in Langenbrand, Nordschwarzwald. *Allgemeine Forst- und Jagd-Zeitung,* **159** (8), 154–64.

**Kenk, G. (1990).** Fichtenbestände aus Weitverbänden. Entwicklungen und Folgerungen. *Forstwissenschaftliches Centralblatt,* **109**, 86–100.

**Kenk, G., Schlör, P. & Weise, U. (1989).** Erntekosten und Nettoerlöse in Baden-Württemberg nach Brusthöhendurchmesser und Konsequenzen für die Bestandesbehandlung. *Allgemeine Forstzeitschrift,* **44** (27), 710–12.

**Kramer, W. (1988).** *Waldwachstumslehre.* Verlag Paul Parey, Hamburg.

**Mitscherlich, G. (1961).** Untersuchungen in Plenterwäldern des Schwarzwaldes. *Allgemeine Forst-und Jagd-Zeitung,* **132**, 65–73, 85–95.

**Mitscherlich, G. (1978).** Wald, Wachstum und Umwelt. In *Form und Wachstum von Baum und Bestand.* 1. Band: 144 p. 2. J.D. Sauerländer's Verlag. Aufl. Frankfurt.

**MLR (Ministerium für Ländlichen Raum, Ernährung, Landwirtschaft und Forsten) Baden-Württemberg (1988).** Jungbestandspflege in den wichtigsten Betriebszieltypen.

**Ott, W. (1989).** Überlegungen zur Ertragsentwicklung in der Forstwirtschaft. *Allgemeine Forstzeitschrift,* **44** (28), 735–42.

**Palmer, S. (1985).** Der Buchen-Vorbau. *Allgemeine Forstzeitschrift* **40** (45), 1217–20.

**Pöppinghaus, G. (1990).** Naturschutz im Walde. *Forst und Holz,* **46** (10), 261–6.

**Rotenhan, S. Frhr. v. (1988).** Stabiler Wald—stabile Forstwirtschaft. *Allgemeine Forstzeitschrift* **43** (27–28), 748–51.

**Rubner, K. & Reinhold, F. (1953).** *Das natürliche Waldbild Europas.* Verlag Paul Parey, Hamburg.

**Schütz, J.-P. (1989).** Zum Problem der Konkurrenz in Mischbeständen. *Schweizerische Zeitschrift für Forstwesen,* **140** (12), 1069–83.

**Seitschek, O. (1989).** Aufbau stabiler Wälder—eine zentrale Aufgabe des Waldbaus. *Forst und Holz* **44** (7), 163–9.

**Siegmund, E. (1974).** Aufwand und Ertrag bei waldbaulichen Betriebsformen (untersucht an Modellen von Tannen-Fichten-Buchen-Mischbeständen). *Dissertation München.*

**Smaltschinski, Th. (1990).** Mischbestände in der Bundesrepublik Deutschland. *Forstarchiv* **61** (4), 137–40.

**Weidenbach, P. (1988).** Grundsätze künftigen Waldbaus. *Allgemeine Forstzeitschrift*, **43** (51–53), 1405–8.

**Wiedemann, E. (1942).** Der gleichaltrige Fichten-Buchen-Bestand. *Mitteilungen der Preußischen Versuchsanstalt*. Verlag Schaper, Hannover.

**Wiedemann, E. (1949).** *Ertragstafeln der wichtigsten Holzarten*. Verlag Schaper, Hannover.

**Wiedemann, E. (1951).** *Ertragskundliche und waldbauliche Grundlagen der Forstwirtschaft*. J.D. Sauerländer's Verlag, Frankfurt.

**Zimmermann, H. (1985).** Zur Begründung von Mischbeständen mit Fichte und Buche auf Sturmwurfflächen im öffentlichen Wald Hessens. *Allgemeine Forstzeitschrift*, **40** (49), 1326–30.

# Nitrogen relations of mixed-species stands on oligotrophic soils

J.L. MORGAN,* J.M. CAMPBELL AND D.C. MALCOLM
*School of Forestry, Institute of Ecology and Resource Management, University of Edinburgh, Mayfield Road, Edinburgh EH9 3JU, UK*

## SUMMARY

1   The development of afforestation practice to establish stands of Sitka spruce on oligotrophic heath and peatland sites had to overcome problems caused by competition from *Calluna vulgaris*, inadequate phosphorus availability and poorly-aerated soils.

2   Despite site amelioration and control of *Calluna* pure spruce stands develop nitrogen-deficiency symptoms within 8 years of planting. Spruce planted in mixture with pine or larch species can maintain growth rates comparable to pure stands fertilized with nitrogen (N).

3   Field studies in a number of 15-year-old experiments established the extent of improvement of mixed species stands and eliminated atmospheric inputs, litterfall and dinitrogen fixation as sources for additional N in mixtures. Rates of N mineralization in soil organic layers of mixed stands could be double those in pure stands of spruce.

4   Measurements of microbial activity in mixture material and the effect of some tree species on this substrate accord well with mineralization rates measured in the field.

5   It is concluded that although tree species probably do not differ in ability to access soil organic-N, their differential rooting patterns and different mycorrhizal associations create conditions for enhanced availability of mineral N to spruce in mixture.

## INTRODUCTION

Since the end of the First World War the extent of forest in Britain has increased from 3 to 11% of the total land area. The majority of new plantations have been restricted to upland sites with low agricultural value which have presented a silvicultural challenge to the successful establishment of plantations on sites which were exposed and had infertile soils. Much empirical research has been undertaken to determine effective silvicultural techniques and the most suitable choice

---

* Present address: Forestry Commission, Northern Research Station, Roslin, Midlothian, EH25 9SY, UK.

of tree for timber production. Early trials devoted much attention to the choice of species; now, however, only a small number of the most productive coniferous tree species are selected for upland afforestation (Busby 1974). Sitka spruce (*Picea sitchensis* (Bong.) Carr.) is the most important coniferous tree grown in Scotland and the North of England because of its wide range of site tolerance, high yield potential and desirable properties for the wood processing industries (Henderson & Faulkner 1987). Sitka spruce comprises approximately 40% of coniferous high forest in Britain, a proportion which will increase, since about 65% of current planting or regeneration is with this species (Locke 1987).

Peaty soils, the main land resource for afforestation in upland Britain, with their physical and chemical characteristics present specific problems for establishment. Very early attempts at planting used notch planting, with no ground preparation or fertilization, but it became apparent that survival and growth rates of plants were very poor on peaty soils. Stirling-Maxwell (1907) reported the success of Belgian techniques for planting on peaty soils and recommended their use in Scotland. Usually, a system of drainage was created and trees were planted on turves with a top dressing of phosphorus as basic slag. Although widespread adoption of planting on Belgian turves progressed slowly, the method gave significant benefits to the establishment of all tree species. Ploughs were later developed to combine the benefits of drainage and a raised planting position and these techniques have been reviewed by Zehetmayer (1954, 1960).

### Afforestation of oligotrophic soils with spruce

In the 1920s, trials in Scotland showed that Sitka spruce did not grow as well as pines or larches on peats and heathland soils. Spruce growth rates deteriorated soon after planting and eventually height growth stagnated or 'checked'. Spruce often did not close canopy under these conditions whereas pines and larches continued to grow relatively well. Where spruce had been established in a mixed plantation with pine or larch the same growth check was usually not apparent. Macdonald (1936) noted that when pine was introduced into checked spruce plantations the growth rates of spruce gradually improved. Confirmation that pine could 'nurse' spruce through a period of check was obtained when trees had recovered and achieved canopy closure within the pine matrix (Macdonald & Macdonald 1952). Mixtures of spruce with a 'nurse' species became a common prescription for planting peaty soils during the 1930s. Mixed stands were intended to provide a 'fail safe' combination with spruce as the preferred final crop and the nurse as insurance against failure of the spruce.

The potential for remedial dressings of basic slag to release spruce from check was explored during the 1930s (Zehetmayer 1954, 1960). Improvement in spruce growth rates were apparent on certain soils where phosphorus was deficient, but spruce still remained in check where heather (*Calluna vulgaris* (L.) Hull) was a dominant component of ground vegetation. Mixed stands on heathland sites were more effective at nursing a final crop of spruce if a top dressing of slag had been applied. Where this treatment was omitted, spruce was often outgrown by the nurse species and eventually suppressed by shading. Suppression by the nurse component became more likely as lodgepole pine (*Pinus contorta* Dougl.) and Japanese larch (*Larix leptolepis* (Sieb & Zucc.) Gord.) were planted in preference to the slower growing species of pine.

Despite trials with intensive ploughing and top dressings of basic slag, Sitka spruce still experienced growth check on some heathland soils and peats. Check became apparent as *Calluna* recolonized sites after ploughing, and it was observed that the condition of spruce was worse where *Calluna* was more vigorous. Research was focussed on potential mechanisms which could explain the checking of spruce by *Calluna*, and Weatherell (1953) established that N deficiency was associated with the yellow–green needles of checked trees. On a podzolized soil he showed that check could be alleviated by killing *Calluna* surrounding the trees or by application of a nitrogenous fertilizer. Treatments, as simple as rotovating between rows of trees or applying a heather mulch, were sufficient to release spruce from check (Weatherell 1953). Leyton (1955) confirmed that control of *Calluna* by artificial shading increased the concentrations of N contained in spruce foliage. Leyton proposed that the benefits to spruce grown in nursing mixtures were largely a result of shading of *Calluna* by those trees which were tolerant of competition from *Calluna*. Later work by Handley (1963) and Robinson (1972) was to establish that extracts from live *Calluna* roots inhibited the development of symbiotic relationships between spruce roots and mycorrhizal fungi. It was also shown that mycorrhizal fungi, specific to infections with pines and larches, were unaffected by the same *Calluna* root extracts. It was concluded that this antagonistic mechanism restricted the uptake of nutrients by spruce, but did not affect the nutrition of trees used as nurse species.

### Early trials of mixed species stands

Weatherell (1957), reviewing the trials of spruce planted in mixed-species stands on heathland soils in the North Yorkshire Moors, concluded that Japanese larch was a more effective nurse species due to its rapid suppression of *Calluna*, compared with the pines which had slower crown development. Observations from these trials suggested that spruce roots travelled considerable distances to proliferate beneath larch litter layers. In contrast, roots growing beneath spruce had to compete with *Calluna* and were thinner, poorly branched and had smaller root tips. Zehetmayer (1960) observed that spruce did not show any response to adjacent stands of nurse species on deep peats when plots were separated by deep cross drains. Spruce planted next to nurse stands with no root barriers developed a canopy profile which indicated a marked response of edge trees to the nurse. Weatherell (1957) noted that only the roots of spruce at the edge of plots had proliferated beneath adjacent larch. Roots of trees from plot centres did not reach the larch, and these trees remained in check.

Weatherell (1957) and Zehetmayer (1960) both reported trials of growing spruce with leguminous plants as a nurse. Broom, gorse, tree lupin and laburnum were all tested for their potential to increase N availability for spruce by dinitrogen fixation. Both lupin and broom increased early growth rates of spruce but were short-lived and required intensive site preparation for establishment. The high costs of such treatments prevented their introduction as regular practice, although gorse and broom were accepted when they occurred as components of heathland vegetation. Trials with different varieties of *Alnus* had also proved unsuccessful (Zehetmayer 1954, 1960). The main problem associated with alder was frost damage and this genus was generally short-lived in acid soil conditions.

Better understanding of the nature of spruce check by *Calluna* led to modified prescriptions for the establishment of pure spruce plantations on peaty soils. Cultivation and drainage were routinely undertaken to provide well-aerated planting positions which increased the period when spruce was free from competition with *Calluna* (Taylor 1970; Neustein 1976). However, *Calluna* began to grow back onto plough ridges between 3 and 4 years after planting and herbicides often had to be applied to avoid checked plantations (Mackenzie 1974). Widespread afforestation with pure spruce created large areas requiring such treatment. Prescriptions for weed control considered both the potential for spruce to be released from check by nursing mixtures and the contribution of N-fixing species where they occurred (Mackenzie, Thompson & Wallis 1976).

Mackenzie (1974) reported that on peaty ironpan soils N and P deficiency were the two main factors limiting the growth rates of spruce. On deep peats, K deficiency was recognized as an additional factor to N and P deficiency. Experimental trials had helped refine fertilization techniques using rock phosphate and superphosphate (Dickson 1971; Mackenzie 1972; Everard 1974) and P fertilizer was considered essential for any planting on peaty soils.

Mackenzie (1974) also recommended *Calluna* control as a successful alternative to N fertilization of checked spruce on peaty podzols. Malcolm (1975) supported these views with results from experimental work on a deep peat. However, Dickson and Savill (1974) sounded a warning note from Northern Ireland by suggesting that pure spruce stands were likely to require repeated nitrogen applications up to canopy closure. Their observations from deep oligotrophic peats suggested that the availability of inorganic-N in peat only sustained satisfactory growth of spruce for 6–8 years after planting. Consequently, *Calluna* control and added P(K) was not sufficient to establish pure stands of spruce on oligotrophic peats. Even the response of spruce to N fertilization was short-lived and it was anticipated that N applications would need to be repeated at intervals of 3–4 years. Malcolm (1975) offered a choice of silvicultural systems to the forest manager for the establishment of trees on peaty soils dominated by *Calluna*:

1   pure spruce with N, P, K and no *Calluna* control on oligotrophic peats;
2   pure spruce with P(K) and *Calluna* control, where N was not limiting;
3   *Calluna*-tolerant species with P and no *Calluna* control; and
4   nursing mixtures with P(K) and no *Calluna* control.

McIntosh (1981) developed these principles with detailed fertilizer prescriptions for Sitka spruce in upland Britain. These recommended treatments had realized the earlier predictions of Dickson and Savill (1974) with the requirement for N applications to be repeated on heathland podzols and deep peats. In a later publication, McIntosh (1983) was to describe the characteristic N deficiency of spruce and the sites with which it was associated. The rate of N application recommended to correct this deficiency was $150\,kg\,N\,ha^{-1}$ usually applied as urea.

Garforth (1979) has described the experience of the Forestry Commission with Sitka spruce/lodgepole pine mixtures in South Scotland. Large areas of mixed stands had been planted in the 1950s and 1960s but had proved difficult to manage due to poor choice of pine provenance and inadequate provision of phosphate supply. The difficulty of implementing management operations on mixtures was considerably greater than for stands of one species and may explain why many foresters adopted the pure spruce option in the 1970s and early 1980s. The most common forestry practice on peaty soils was either to adopt high fertilizer-input regimes with Sitka spruce or low-input alternatives with lodgepole pine and the larches.

## Current developments in practice

In recent years there has been a renewal of interest in the use of nursing mixtures as a silvicultural option on oligotrophic peats (O'Carroll 1978; McIntosh 1983; Taylor 1985; Carey, McCarthy & Miller 1988). Observations from experiments in Scotland and the Irish Republic have confirmed that the production of Sitka spruce can be improved by its planting in admixture with pines and larches. Results for one such experiment at Mabie Forest are given in Table 1. Foliar analysis has clearly demonstrated that improvements in spruce growth are associated with better N status in mixed stands. The checking of spruce on similar peats described by Dickson and Savill (1974) and McIntosh (1983) was caused by a lack of available N and not solely by allelopathic effects of *Calluna*. Pure spruce developed N-deficiency symptoms despite *Calluna* control, whereas spruce in mixed stands continued to grow satisfactorily (O'Carroll 1978; Carlyle 1984; Taylor 1985). The foliar N concentrations at Mabie are shown in Table 2. It has been concluded that the benefits to spruce grown in mixtures were due to some mechanism which had increased the availability of N above levels occurring in pure spruce stands.

The most recent work to be published on N deficiency in Sitka spruce plantations (Taylor & Tabbush 1990) describes four site categories which are associated with different degrees of N deficiency. These categories have been defined from the response of pure spruce stands to combinations of N and *Calluna* control over a range of field experiments. A method for predicting the site category from soil type, lithology and *Calluna* cover has been described. The poorest site categories are oligotrophic peats which overlie Old Red Sandstone and quartzitic lithologies. Three to four N applications are required to establish pure stands of Sitka spruce

TABLE 1. Top height (m) and basal areas (m$^2$ ha$^{-1}$) of spruce and mixed species at Mabie Forest, Expt 7, aged 19 years. On an equal density basis pure SS−N basal area would be approximately 5 m$^2$. SS = Sitka spruce; HL = hybrid larch; LP = lodgepole pine

| Treatment | Height | Basal area | |
|-----------|--------|-----------|---|
| SS−N | 8.60 | 20.08 | |
| SS 25% | 9.60 | 12.70 | 20.97 |
| HL | 7.07 | 8.27 | |
| SS 25% | 10.31 | 5.36 | 36.13 |
| LP | 9.71 | 30.77 | |
| SS+N (620 kg) | 11.04 | 47.28 | |

TABLE 2. Spruce foliar N concentrations in pure and mixed species stands at Mabie Forest, Expt 7, P67 (% dry wt). SS = Sitka spruce; HL = hybrid larch; LP = lodgepole pine. Data kindly provided by Forestry Commission Research Division

| Age | SS−N | SS+HL | SS+LP |
|-----|------|-------|-------|
| 3 | 1.90 | 0.96 | 0.93 |
| 5 | 1.09 | 0.86 | 0.75 |
| 7 | 0.96 | 1.06 | 1.05 |
| 9 | 1.27 | 1.26 | 1.51 |
| 12 | 1.05 | 1.30 | 1.43 |
| 14 | 0.91 | 1.25 | 1.46 |
| 16 | 1.16 | 1.47 | 1.83 |
| 18 | 1.06 | 1.18 | 1.67 |
| 22 | 1.02 | 1.19 | 1.46 |

on these soils and nursing mixtures are advocated in preference to pure spruce in these circumstances. Spruce may be planted with Japanese larch or Scots pine on heathland soils whilst Alaskan provenances of lodgepole pine should be chosen as a nurse on deep peats. Recommendations are that spruce and nurse should be intimately planted as alternate pairs or triplets in each row. Stands on peaty soils with high wind-throw hazard rating are unlikely to be thinned and these mixed stands are intended to be 'self-thinning' with spruce shading out the nurse species early in the rotation.

## SITKA SPRUCE IN PURE AND MIXED-SPECIES STANDS

### *Field studies: above-ground biomass*

The observations of improved growth rates and N status of Sitka spruce growing in mixed plantations were followed by more intensive studies in field experiments which demonstrated the 'mixture effect'. The results of three research projects will be reviewed to present the most recent information on the N cycles of pure spruce and nursing mixtures planted on oligotrophic peats. A summary of the treatment types and site descriptions relating to the three investigations at Mabie Forest (Carlyle 1984), Avondhu in Ireland (Carey *et al.* 1986) and Inchnacardoch and Culloden (Miller *et al.* 1986) is given in Table 3. Each of these experiments was over 15 years old when investigated.

Mixed stands were planted with Sitka spruce and either lodgepole pine, Scots pine, Japanese larch or hybrid larch. The common similarity across all locations was for mixed stands to contain larger above-ground biomass than pure spruce with no N fertilization (Table 4). When the biomass of spruce was expressed as a function of stand area occupied, this was also larger than the equivalent measurement for pure spruce stands. On the deep peat at Inchnacardoch, spruce planted with larch represented an equivalent above ground biomass of $105\,t\,ha^{-1}$ compared with $23\,t\,ha^{-1}$ in pure spruce plots. The spruce component of the same spruce/larch stands contained a similar above-ground biomass to pure spruce which had been fertilized with N ($102\,t\,ha^{-1}$). However, comparisons between the above-ground biomass of each *treatment* (spruce + nurse) revealed that the weight of pure spruce with added N was always larger than the nursing mixtures. The comparison drawn from Inchnacardoch showed that *treatment totals* for spruce/larch mixtures represented approximately half the above-ground biomass of pure spruce with added N (Miller & Miller 1987). The magnitude of biomass differences between pure spruce ($-N$) and the treatment totals for mixed stands varied between locations (Table 4). Within the same experimental site, comparisons between pure spruce and mixtures were also influenced by the form of

TABLE 3. Experimental sites for mixed-species stands. All sites received P and/or K as required

| Location, planting year (elevation, rainfall) | Soil type | Peat depth (cm) | Stands/treatments studied |
|---|---|---|---|
| Avondhu, 1960 (225 m, 1250 mm) | Podzolized gley | 2–4 | Pure spruce ($-N$) <br> Spruce/J larch [1:1 alternate double rows] ($-N$) <br> Spruce/L pine [1:1 alternate double rows] ($-N$) |
| Culloden, 1969 (200 m, 750 mm) | Indurated peaty gley | 10–15 | Pure spruce ($-N$) <br> Spruce/S pine [1:1 alternate triplets] ($-N$) |
| Inchnacardoch, 1965 (300 m, 1250 mm) | Deep peat over gley | 30–200 | Pure spruce ($-N$) <br> Pure spruce ($+N$) <br> Spruce/L pine [1:1 alternate triplets] ($-N$) <br> Spruce/J larch [1:1 alternate triplets] ($-N$) |
| Mabie, 1967 (14 m, 1000 mm) | Deep peat (type 10a) | 300–500 | Pure spruce ($-N$) <br> Pure spruce ($+N$) <br> Spruce/L pine [1:3 alternate triplets in alt. rows] ($-N$) <br> Spruce/H larch [1:3 alternate triplets in alt. rows] ($-N$) |

TABLE 4. Comparative production of pure and mixed-species stands (t ha$^{-1}$). DT = deep spaced furrow ploughing; CDT = deep complete ploughing

| Site | Treatment | Pure spurce | Total Mixture | Spruce | Nurse |
|---|---|---|---|---|---|
| (a) *Above-ground biomass* | | | | | |
| Culloden | SS/SP(DT) | 9 | 49 | 19 | 30 |
| | SS/SP(CDT) | 40 | 68 | 32 | 36 |
| Inchnacardoch | SS/JL | 22 | 50 | 26 | 24 |
| | SS/LP | 22 | 77 | 20 | 57 |
| Avondhu | SS/JL | 15 | 53 | 35 | 18 |
| (b) *Below-ground biomass* | | | | | |
| Culloden | SS/SP(DT) | — | 20 | 8 | 12 |
| | SS/SP(CDT) | 18 | 21 | 10 | 11 |
| Avondhu | SS/JL | 6 | 20 | 12 | 8 |

initial cultivation and the choice of nurse species. Pure stands of spruce (−N) performed better on completely ploughed soils at Culloden (CDT—40 t ha$^{-1}$) than where single-furrow ploughing had been used (DT—9 t ha$^{-1}$). The margin for increases of biomass in mixed spruce/pine stands over pure spruce were lower in completely ploughed treatments (28 t ha$^{-1}$) than for single-furrow ploughing (40 t ha$^{-1}$). More effective control of *Calluna* (Thompson & Neustein 1973) and increased mobilization of soil organic-N (Ross & Malcolm 1988) could account for the benefits of complete cultivation to pure spruce. The total above-ground biomass for spruce/larch mixtures at Inchnacardoch was lower than for mixtures with lodgepole pine. Japanese larch did not perform as well as lodgepole pine on the deep peat site and was subject to deer damage (Miller *et al.* 1986). Similar observations had been made at Mabie where the rapid growth rates of lodgepole pine threatened to suppress the spruce it was nursing in contrast to hybrid larch, which was a much less competitive nurse (Table 1).

The above-ground biomass of all treatments was reflected by the total N capital contained in these stands (Table 5). Nitrogen capitals were larger in mixtures than in pure spruce with no added N at all locations investigated.

TABLE 5. Above-ground N capital in pure and mixed-species stands (kg ha$^{-1}$). DT = deep spaced furrow ploughing; CDT = deep complete ploughing

| Site | Treatment | Pure spruce | Total mixture | Spruce | Nurse |
|---|---|---|---|---|---|
| Culloden | SS/SP(DT) | 25 | 169 | 69 | 100 |
| | SS/SP(CDT) | 110 | 241 | 120 | 122 |
| Inchnacardoch | SS/JL | 68 | 147 | 80 | 67 |
| | SS/LP | 68 | 224 | 64 | 160 |
| Avondhu | SS/JL | 94 | 246 | 177 | 69 |
| Mabie | SS/HL | 54 | 87 | 31 | 56 |
| (foliage only estimate) | | | | | |

Similarly, spruce as a component of mixed stands contained more $N\,ha^{-1}$ than when grown as a monoculture. The differences between N capitals of pure and mixed stands were large. Total N in mixtures ranged between two and six times the N capital of pure spruce stands.

*Field studies: below-ground biomass and rooting intensity*

The below-ground contribution to total stand biomass was estimated by excavation of stumps and large lateral roots at Avondhu and Culloden (Table 4b). At both sites there was a greater proportion of spruce biomass below ground in pure stands than in mixtures. Within spruce/pine or spruce/larch mixtures, spruce and nurse both allocated a similar proportion of total biomass to large roots. McKay (1986) investigated the fine-root dynamics under pure spruce and spruce/pine treatments with single-furrow ploughing at Culloden. The mean standing crop of fine roots under pure spruce ($112\,g\,m^{-2}$) was approximately twice as large as those measured under mixed stands ($57\,g\,m^{-2}$). Spruce root contributed about two-thirds of the standing crop in mixed stands and this equated to about two-thirds of the rooting *intensity* of spruce in pure plots (standing crop/stocking density). Annual production rates of fine roots beneath pure spruce were also double those measured under mixed stands. Turnover rates of fine roots were faster under pure spruce, and the amount of dead root recovered was larger than under mixtures. The fine-root study supported the findings from excavations regarding the allocation of resources to roots. McKay and Malcolm (1988) calculated that only 10% of net primary production was required for growth and respiration of fine roots under mixed stands compared with 60% in pure spruce stands. McKay and Malcolm suggested that the differences in *spruce* carbon allocation to fine roots were a consequence of low productivity and nitrogen stress in pure spruce stands.

The vertical distribution of fine roots was different under pure spruce and spruce/pine stands (McKay 1986; McKay & Malcolm 1988). The majority of fine roots beneath pure spruce were found less than 6 cm below ground, whereas pine and spruce roots in mixtures were more evenly distributed to a depth of 9 cm. The relative frequency of pine roots was greater than spruce below a depth of 6 cm in mixed stands. High winter water table levels were suggested as the main factor controlling root distribution and it was considered that better aeration under mixed stands encouraged a greater depth of rooting (McKay & Malcolm 1988). Scots pine roots are not noted for high tolerance of soil anaerobism, although their distribution at Culloden may suggest greater tolerance than Sitka spruce roots. Spruce roots are known to be very sensitive to soil anaerobism and high root mortality is caused by rising water tables in peaty soils (Coutts & Phillipson 1987). Lodgepole pine and hybrid larch are two nurse species which are considered to be tolerant of anaerobic soil conditions and demonstrate the ability to root deeply in poorly-drained soils (Boggie 1972; Morgan 1984). The capacity for mixed stands to root deeper than pure spruce may be an important advantage to

TABLE 6. Annual litterfall and N content in pure and mixed-species stands (kg ha$^{-1}$ year$^{-1}$). DT = deep spaced furrow ploughing; CDT = deep complete ploughing

| Site | Treatment | Dry wt | N content |
|---|---|---|---|
| Culloden | SS−N(DT) | 896 | 3.3 |
| | SS/SP(DT) | 3497 | 12.8 |
| | SS−N(CDT) | 1061 | 3.9 |
| | SS/SP(CDT) | 4097 | 16.1 |
| Inchnacardoch | SS−N | 425 | 1.8 |
| | SS/JL | 1336 | 6.5 |
| | SS/LP | 2847 | 11.0 |
| Avondhu | SS/JL | 1384 | 11.1 |
| Mabie (only HL collected) | SS/HL | 987 | 7.0 |

nutrient uptake. At Culloden the rooted volume under mixed stands was effectively 30% larger than under pure stands.

## Field studies: litter accumulation

The accumulation of litter occurred at greater rates under mixtures than under stands of pure spruce (Table 6). Larch is deciduous and pine has a shorter period of needle retention than spruce (3 and 5–7 years, respectively). Larger crowns had developed in mixed stands and consequently litter was returned earlier and in greater quantities than by stands of pure spruce. The total N capital of annual litterfall from mixed stands was more than three times as large as litterfall-N under pure spruce. Litterfall-N from mixed stands containing larch was no greater than under mixtures with pine, despite larch being deciduous. Larch had poor crown development at Mabie and Inchnacardoch, whereas pine mixtures had reached canopy closure with associated increases in litterfall. The contribution of litterfall-N to soil could have been expected to occur earlier and in larger quantities under larch compared with pines in mixed stands. Weatherell (1957) demonstrated that larch litter-N did make a significant contribution to the availability of N for spruce uptake in mixed stands. In a subsequent investigation Leyton and Weatherell (1959) demonstrated that litter amendments from a range of tree species could improve the growth rates and N status of checked spruce on a heathland soil. They concluded that growth responses were proportional to the total nitrogen added in litter. Larch litter had been the most effective source of N due to larger litter-N concentrations than other species. From the results of this investigation Leyton and Weatherell recommended that larch should be used as a nurse in preference to pines.

Litterfall inputs of N would seem to be a plausible source from which spruce could gain access to N provided by the nurse species in mixed stands. However, the maximum N capital of annual litterfall was recorded as only 16 kg N ha$^{-1}$ for spruce/pine mixtures at Culloden (Table 6). This level of N input could not have accounted for the differences between spruce-N capital in pure and mixed stands ($130\,kg\,N\,ha^{-1}$), even if all N had been released from litter. Actual mineral-N release rates from litter are slow and usually follow a period of immobilization lasting for up to 1 year. The total quantity of nitrogen added as litter by Leyton and Weatherell (1959) actually ranged from 156 to $835\,kg\,N\,ha^{-1}$ over a 3-year period. This level of input would significantly increase the availability of inorganic-N for small trees. Even if nurse litter-N inputs were able to account for a proportion of the increased *spruce-N* uptake at the experimental sites the total *stand-N* uptake would still require explanation. Alternative processes which could have provided an input into the N cycles of mixed stands were therefore sought.

### Atmospheric inputs and nitrogen fixation

The larger leaf area of mixed-species stands may have been responsible for greater interception of precipitation and aerosols than by pure spruce. However, an increased flux of N was only observed in the throughfall and stemflow at Avondhu. At all other locations it appeared that canopies of mixed stands actually absorbed N, and if more N had been intercepted, then this was not measured in throughfall. At Avondhu the maximum gain to mixed stands from precipitation was $10\,kg\,N\,ha^{-1}\,year^{-1}$, a relatively small input in relation to stand-N capitals. Enriched throughfall was not considered to be responsible for the more effective accumulation of N in mixed stands (Miller *et al.* 1986).

Carlyle (1984) investigated the potential for dinitrogen fixation in the top 3 cm of soil beneath stands at Mabie. Acetylene reduction techniques failed to indicate the presence of any non-symbiotic nitrogen fixation in any of the stands tested. Rates of nitrogen fixation by free living heterotrophs are generally low in temperate forest ecosystems. Norhstedt (1988) reported that N fixation measured in older litter layers under birch was approximately $1.4\,kg\,N\,ha^{-1}\,year^{-1}$. Rates of N fixation in pine and spruce litter layers were much lower than those reported for birch. Nitrogen fixation was not considered to be an important input of N for trees growing on oligotrophic peats.

### Soil nitrogen and mineralization

Total nitrogen capitals contained in litter and native peat layers were measured for the soils of three experimental sites (Table 7). At Culloden there was more N in the litter layers under mixed stands than under pure spruce. Litterfall-N inputs could not account for the total amounts of N measured in these layers, and it is assumed that they are mainly derived from the original vegetation. *Calluna* was completely shaded beneath mixed stands but was still actively growing amongst spruce which had a more open canopy. At Mabie the litter layer-N capitals were much the same for both pure spruce and spruce/larch treatments. Soil organic layers contained reserves of N which reflected the depth of peat at the site

TABLE 7. Forest floor N capitals in mixture trials (kg ha$^{-1}$). LFH = organic matter derived from tree stands; DT = deep spaced furrow ploughing; CDT = deep complete ploughing. *Ridge and flat only

| Site | Treatment | | Total to 10 cm | LFH | Native organic matter |
|------|-----|-----|------|------|------|
| Culloden | DT | SS–N | 848 | 291 | 557 |
| | DT | SS/SP | 808 | 367 | 441 |
| | CDT | SS–N | 343 | 168 | 175 |
| | CDT | SS/SP | 660 | 447 | 214 |
| Avondhu | | SS−N | 1091 | 238 | 853 |
| | | SS/JL | 1143 | 417 | 726 |
| Mabie* | | SS−N | 1160 | 130 | 1030 |
| | | SS/HL | 1160 | 130 | 1030 |

TABLE 8. Net mineralization rates from soils of mixture trials (kg N ha$^{-1}$ year$^{-1}$). LFH = organic matter derived from tree stands; DT = deep spaced furrow ploughing; CDT = deep complete ploughing

| Site | Treatment | | Total to 10 cm | LFH | Native organic matter |
|------|-----|-----|------|------|------|
| Culloden | DT | SS−N | 51 | 6 | 45 |
| | DT | SS/SP | 66 | 25 | 41 |
| | CDT | SS−N | 13 | 8 | 5 |
| | CDT | SS/SP | 23 | 6 | 17 |
| Mabie | | SS−N | 22 | 9 | 13 |
| | | SS/HL | 32 | 19 | 13 |

investigated. There were no collective trends to suggest accumulation or depletion of peat organic-N by any treatment in the experimental trials.

The availability of mineral N for uptake was examined in detail for soils at Mabie and Culloden (Table 8). Net N mineralization rates were measured from isolated cores of soil incubated *in situ* and in the laboratory. Both studies revealed that the release of inorganic-N was occurring at a faster rate beneath mixed stands compared with pure spruce. However, the quantity and distribution of net N mineralization differed considerably between sites and within cultivation treatments at Culloden. At Mabie most mineralization occurred in the top 3 cm of soil beneath pure spruce and spruce/larch treatments. Carlyle (1984) reported annual net mineralization rates of 28 and 60 kg N ha$^{-1}$ for pure spruce and spruce/larch, respectively. The majority of additional N released beneath mixed stands was derived from the flat portion of the planting position. Carlyle found no evidence for any influence of microclimate or soil physical conditions on the mineralization rates in each treatment. He concluded that greater N release rates under mixtures were either due to changes in substrate quality or superior microbial decomposition in the rhizosphere of mixed stands.

At Culloden, Williams (1986) reported that N mineralization rates were greatest in the *organic layers* beneath spruce and spruce/pine stands. Differences in the distribution of nitrogen mineralization were most apparent in treatments with less intensive cultivation (DT). The total amounts of N released during incubations were also *larger* in less intensively cultivated treatments. Soils beneath spruce/pine stands released more N than under pure spruce and the differences between treatments originated from the LFH layers of DT plots and from organic layers of CDT treatments. Williams concluded that the differences measured in LFH layers could be related to substrate quality but that litter-N inputs could not account for the size of mineralization rates measured. Williams also noted that there was a close correlation between the availability of readily mineralized forms of N and the microbial biomass measured by Alexander (1986). Anaerobic laboratory incubations suggested that either N was released from the death of microbial hyphae due to waterlogging or by the increased activity of ammonifying decomposers.

### Microbial activity

Alexander (1986) investigated the microbial activity beneath pure spruce and spruce/pine stands at Culloden. If increased N mineralization rates were the result of more effective heterotrophic decomposition, then differences between the size and activity of microbial populations would be apparent beneath pure spruce and mixed stands. Total microbial biomass was estimated by amended basal respiration and revealed that populations were indeed larger in the LFH layers of spruce/pine stands ($113\,g\,m^{-2}$) compared with pure spruce ($77\,g\,m^{-2}$). No differences were apparent when basal respiration was compared between the native peat horizons from each treatment. The rate of decomposition of microbial hyphae was greater in the soil and LFH layers of mixed stands. Faster decomposition rates may have allowed N to become available more rapidly from the peat under mixed stands. Dead microbial biomass would have immobilized N for longer periods beneath spruce compared with mixtures. The results accorded well with the findings from measurements of N mineralization. There appeared to be a larger, more active microbial population beneath mixed stands which turned over more rapidly to make a larger pool of labile N available for uptake by trees.

### Field study conclusions

The studies of N cycles in field experiments have identified the processes which were most likely to account for the high availability and uptake of N in mixed stands. Atmospheric inputs and N fixation were not considered to offer any advantage to the nutrition of mixtures over pure spruce. The most probable source of increased availability of N came from the soil. Net mineralization rates of soil organic-N and LFH layers were shown to be greater beneath mixed stands than under pure spruce. More active microbial populations have been associated with the rapid turnover of organic-N in mixtures.

The implication that larger soil N mineralization rates could account for the greater N capitals and total biomass of mixed stands deserves further consideration. The field experiments reviewed were more than 15 years old when investigations were carried out. The 'mixture effect' had become apparent in these trials from approximately 10 years after planting. Pure spruce had begun to encounter N deficiency and check at about 8 years of age, when foliar N levels of spruce in mixed stands had also declined to deficiency levels (see Table 2). Spruce in mixed stands recovered from this period of N deficiency between 8 and 10 years after planting when it is assumed that processes which increase the availability of N to trees in mixtures have been initiated. Litter layers developing under stands of this age would contain small amounts of fresh material which would tend to immobilize rather than release N during decomposition. The most likely source of mineral N for trees of this age would be the native peat on which they were planted.

Only one field study has compared the biomass production and nitrogen uptake of young spruce and pine planted in pure stands at the same site. Carey, McCarthy and Hendrick (1984) investigated 7-year-old stands of Sitka spruce and lodgepole pine growing on a raised bog in the Republic of Ireland. Both total biomass and nitrogen capitals of the pine were greater than the spruce. Pine had accumulated $150\,kg\,N\,ha^{-1}$ compared with $100\,kg\,N\,ha^{-1}$ by spruce, despite no large differences in the total root weight of each species. Nitrogen mineralization rates were not measured at this site but it may be assumed that the differences between mineralization rates observed beneath pure spruce and mixed stands will show similar trends to those present under monocultures of spruce and nurse species.

## LABORATORY AND GREENHOUSE STUDIES

### *Specific differences in N uptake*

As the nurse species in the experiments examined in the field do not generally develop N deficiency symptoms the question arises as to whether these species are capable of accessing sources or forms of nitrogen that are unavailable to spruce in pure stands. To test this Morgan (1990) conducted a 2-year experiment in a glasshouse in which 140 seedlings were established in isolated square blocks $(0.75 \times 0.75 \times 0.15\,m)$ of an oligotrophic (*Calluna–Sphagnum*) peat with an added inoculum from an uncultivated peat bog. The species tested were Sitka spruce, lodgepole pine, Japanese larch and birch (*Betula pendula* Roth.). These were planted as pure (36 plants at 12.5 cm spacing) or mixed (1:1) spruce/other species plots and all received P and K fertilizers equivalent to normal field applications.

The results, after 2 years growth (Fig. 1a), indicated that the total dry weight per plant was greater in pure pine, larch and birch, than in pure spruce. The mixed stands were intermediate between the respective pure components. Mortality of some larch occurred in the spruce/larch plots and these results have been excluded from Fig. 1b which shows that the total uptake of N in the plots followed closely the amount of growth made ($r = 0.93$). This confirmed that N

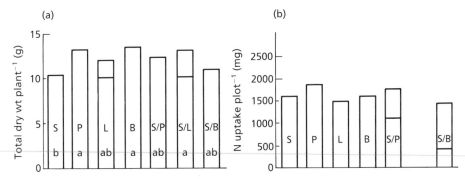

FIG. 1. (a) Mean total dry wt per plant after 2 years growth in 36-tree plots of oligotrophic peat with restricted volume. Bars in larch treatments indicate reductions to account for mortality. Bars with the same letter (a or b) are not significantly different at 5% probability. (b) Mean total uptake of N by trees (per plot) after 2 years growth on oligotrophic peat. Bars denote extent of spruce uptake. S = Sitka spruce; P = lodgepole pine; L = Japanese larch; B = birch.

was the main factor limiting growth. The proportional uptake by the spruce component in the mixtures with pine and birch demonstrated different competitive effects where mixed with pine; spruce took up 60% of the total N uptake, but only 26% when mixed with birch. On an individual plant basis this latter was less than half of the uptake in the pure spruce plots. Because of plot variability, and because the peat used in this experiment proved to be very recalcitrant with little mineralization *in situ*, Morgan (1990) concluded that there was inadequate evidence to suggest thay any one species or mixture could access greater quantities of N from a specific volume of peat.

## Microbial activity

Campbell (1988) investigated how lodgepole pine might cause more nitrogen to be released from a peat substrate than Sitka spruce. Campbell grew pine and spruce in microcosms and measured the release and uptake of N from the same volume of peat planted with each species and a 1:1 mixture of both species. Campbell demonstrated that more soluble carbohydrate was present in peat planted with pine, than when planted with spruce. Basal respiration, utilization of a glucose amendment and mineralization rates of an amino acid amendment were all greater in peat from the pine treatment, than that from spruce (Fig. 2). Soluble carbohydrate levels correlated well with tree N uptake and with mineralization rates of the amino acid amendment. It was proposed that pine may have influenced the rates of N release from peat by increasing the availability of carbon to microbial populations responsible for decomposition. The potential for labile carbon to be a driving mechanism for mineralization rates was in accordance with the study of microbial populations by Alexander (1986). In the microcosm study, pure pine had assimilated 39% more N from peat than pure spruce; however, mixed plots had only taken up the same amounts of N as pure spruce. Campbell

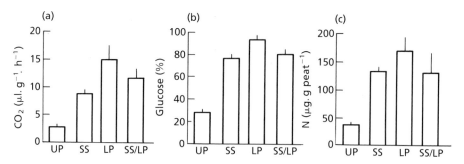

FIG. 2. Respiration (a) and utilization of amendments of glucose (b) and amino-N (c) by peats in which Sitka spruce (SS), lodgepole pine (LP) pure or mixed (SS/LP) had been grown in microcosms. Bars denote 1 S.E. UP = unplanted.

concluded that N uptake could easily have been *retarded* by the presence of spruce, rather than *enhanced* by the presence of pine.

### Specific differences in rooting habit

A *physical* mechanism which may promote the mineralization of soil organic-N remains to be described. Decomposition of oligotrophic peats is limited by poor aeration, low pH and poor substrate quality of the peat for micro-organisms. Aeration of peaty soils is influenced by the presence of trees which intercept rainfall and deplete soil moisture through transpiration. King, Smith and Pyatt (1986) reported that pure stands of pine and spruce intercepted 28% of rainfall over a peaty soil and this evaporated without reaching the soil. Low soil matric potentials under the same stands of trees were attributed to the absorption of water by roots. The potential for drying of peat in the lower soil horizons may be expected to be greater for trees with deeper root systems. King, Smith and Pyatt (1986) reported that peaty soils were better aerated and had lower summer water table levels beneath lodgepole pine compared with Sitka spruce. The roots of some nurse species may play an important role in the reduction of waterlogging and the increase of aeration of soils under mixed stands. In the Scots pine/Sitka spruce mixtures at Culloden it was shown that the pine coarse roots were distributed at a deeper level than those of spruce (Miller *et al.* 1986) and McKay and Malcolm (1988) demonstrated a similar depth increase in fine roots of Scots pine. This field experiment is on a relatively shallow organic horizon. On deeper peat it is known that lodgepole pine roots are capable of surviving anaerobic conditions (Coutts & Phillipson 1978) while those of spruce cannot.

Morgan (1990) investigated the root development and biomass of lodgepole pine and Sitka spruce seedlings in tubes of oligotrophic peat, where the roots either had free access to or were restricted from a permanent water table. The lodgepole pine distribution proved to be deeper than spruce (Fig. 3) with a greater total root length (23.1 m versus 15.6 m) and 5 cm penetration into the

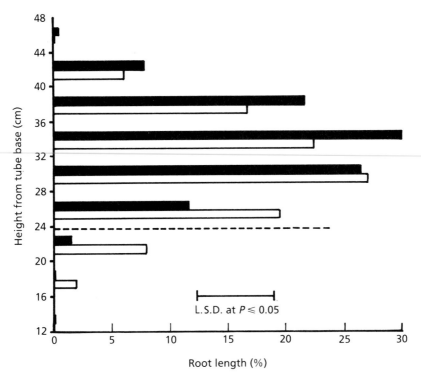

FIG. 3. Distribution of root length of Sitka spruce (solid bars) and lodgepole pine (open bars) in tubes above a fixed water table (dashed line).

anaerobic peat. Although both species had greater biomass and N accumulation in the unrestricted tubes these differences were not significant ($P \leqslant 0.05$). Nevertheless the drying effect of the greater rooting of pine in the field might enhance N mineralization as the pine grows.

## Mycorrhizal effects

A fundamental assumption for the acceptance of greater mobilization of soil organic-N as the mechanism responsible for the 'mixture effect' is that spruce has free access to N which is available for uptake. McKay (1986) has demonstrated that spruce had a greater weight of root and more root tips than pine in the soil of mixed plantations. The rooting *intensity* of spruce would appear to give it a competitive advantage for the uptake of N over pine. However, the relative uptake *efficiency* of mycorrhizal roots belonging to each species remains to be elucidated. Miller *et al.* (1986) have demonstrated that a wider range of mycorrhizal fungi were present beneath spruce/pine compared with pure spruce at Culloden.

Subsequent experimentation has been designed to clarify the role of mycorrhizal fungi in the nutrition of spruce in pure and mixed stands. At Culloden it was found that there were different populations of mycorrhizal root tips between the pure and mixed stands particularly for Sitka spruce root tips under Scots pine trees. Isolates from both spruce and pine mycorrhizas were tested for their ability to break down a simple protein *in vitro*. It was found that the pine associate *Suillus variagatus* was capable of rapidly degrading protein, leaving the bulk of the degradation products in solution. This fungus is not an associate of spruce and it may be that it can release refractory N from organic matter in such a way that spruce may benefit.

## CONCLUSIONS

On oligotrophic sites, where soil profiles are developed over mineralogically poor substrates the dominating profile types are podzols where drainage is free or gleyed mineral horizons overlain by peaty material of varying depths. A feature of the vegetation prior to afforestation is the occurrence of *Calluna vulgaris* which after cultivation and fertilization frequently becomes dominant. The observed 'check' in the growth of spruce when planted in *Calluna* swards masked the real problem for pure stands of this and similar species, that is, the inadequate supply of available nitrogen on these sites.

Comparative studies of stands of spruce, pure and mixed with larches or pines, on these site types, clearly show that in the latter cases N becomes 'available' to the spruce component. In most trials this change in the nutritional status of spruce occurs at about 8 years from planting, i.e. shortly before canopy closure and is not dependent on, although assisted by, the removal of *Calluna* competition.

The additional N becoming available does not originate from atmospheric deposition, litterfall or dinitrogen fixation. The quantities of N involved clearly must come from the native soil organic matter through enhanced mineralization rates in mixed stands. As no real differences are apparent between species in N uptake from restricted volumes of peat, the differences in rates of organic matter mineralization that are observed must be explained by differential effects of species on these organic horizons.

Large differences occur in the distribution of the roots between 'nurse' species and spruce which tends on these sites to develop superficial root concentrations. Pines in particular appear able to explore greater volumes of the organic horizons. These horizons are then drier, broken up and better aerated leading to increased microbial activity. This effect, and the consequent increase in mineralization, is likely to become quantitatively more important as the demands of the spruce component of the mixtures increase with size, particularly with the late start to litter deposition because of spruce needle longevity.

The second difference arises from the recent work on mycorrhizas which suggests that spruce roots operating in soils dominated by the nurse species may develop a different suite of associations. In addition the discovery that one, at least, of the common pine associates is capable of degrading proteinaceous material, points to a mechanism for increased availability of nitrogen for the spruce in mixtures.

It is important to note that almost all the studies have been conducted in mixture trials in which the 'mixture effect' was already clear. These are aggrading systems where conditions for the spruce component are steadily improving. Study of the development of mixtures at a younger age (4–6 years) will be necessary to confirm the suggestion that enhanced N nutrition in mixture is likely to arise from improved conditions for mineralization of organic matter in the presence of the nurse species roots and its associated mycorrhizal fungi.

## REFERENCES

**Alexander, C.E. (1986).** Microbial activity in humus and soil beneath Sitka spruce in pure stands and admixed with Scots pine at Culloden Forest. In *Maintenance and Enhancement of Forest Productivity through Manipulation of the Nitrogen Cycle*, European R & D programme in the field of wood as a raw material, 1982–85 (Ed. by J.D. Miller *et al.*) pp. 179–204. Macaulay Institute for Soil Research, Aberdeen.

**Boggie, R. (1972).** Effects of water table height on root development of *Pinus contorta* on a deep peat in Scotland. *Oikos* **23**, 304–12.

**Busby, R.J.N. (1974).** Forest site yield guide to upland Britain. *Forestry Commission Forest Record*, **97**. HMSO, London.

**Campbell, J.M. (1988).** *Release of carbon and nitrogen from acid peats as influenced by some tree species.* Ph.D. thesis, University of Edinburgh.

**Carey, M.L., McCarthy, R.G. & Hendrick, E. (1984).** Nutrient budgets in young Sitka spruce and lodgepole pine on a raised bog. *Proceedings of the 7th International Peat Congress Dublin*, **3**, 207–18.

**Carey, M.L., McCarthy, R.S. & Miller, H.G. (1988).** More on nursing mixtures. *Irish Forestry*, **45**, 7–20.

**Carey, M.L. et al. (1986).** In *Maintenance and Enhancement of Forest Productivity through Manipulation of the Nitrogen Cycle*, European R & D programme in the field of wood as a raw material, 1982–85 (Ed. by J.D. Miller *et al.*). Macaulay Institute for Soil Research, Aberdeen.

**Carlyle, J.C. (1984).** *The nitrogen economy of larch:spruce mixtures.* Ph.D. thesis, University of Edinburgh.

**Coutts, M.P. & Phillipson, J.J. (1978).** Tolerance of tree roots to waterlogging. I. Survival of Sitka spruce and lodgepole pine. *New Phytologist*, **80**, 63–9.

**Coutts, M.P. & Phillipson, J.J. (1987).** Structure and physiology of Sitka spruce roots. *Proceedings of the Royal Society of Edinburgh*, **93B**, 131–44.

**Dickson, D.A. (1971).** The effect of form, rate and position of phosphatic fertilizers on growth and nutrient uptake of Sitka spruce on deep peat. *Forestry*, **44**, 17–26.

**Dickson, D.A. & Savill, P.S. (1974).** Early growth of *Picea sitchenis* (Bong.) Carr. on deep oligotrophic peat in Northern Ireland. *Forestry*, **47**, 57–88.

**Everard, J.E. (1974).** Fertilizers in the establishment of conifers in Wales and southern England. *Forestry Commission Booklet*, **41**, HMSO, London.

**Garforth, M.F. (1979).** Mixtures of Sitka spruce and lodgepole pine in South Scotland: History and future management. *Scottish Forestry*, **33**, 15–28.

**Handley, W.R.C. (1963).** Mycorrhizal associations and *Calluna* heathland afforestation. *Forestry Commission Bulletin*, **36**. HMSO, London.

**Henderson, D.M. & Faulkner, R.** (Eds) **(1987).** Sitka spruce. *Proceedings of the Royal Society of Edinburgh*, **98B**, 234 pp.

**King, J.A., Smith, K.A. & Pyatt, D.G. (1986).** Water and oxygen regimes under conifer plantations and native vegetation on upland peaty gley soil and deep peat soils. *Journal of Soil Science*, **37**, 485–97.

Leyton L. (**1955**). The influence of artificial shading on the nutrition and growth of Sitka spruce in a heathland plantation. *Forestry*, **28**, 1–6.

Leyton, L. & Weatherell, J. (**1959**). Coniferous litter amendments and the growth of Sitka spruce. *Forestry*, **32**, 7–13.

Locke, G.M.L. (**1987**). Census of woodlands and trees. *Forestry Commission Bulletin*, **63**. HMSO, London.

Macdonald, J.A.B. (**1936**). The effect of introducing pine species among checked Sitka spruce on a dry, *Calluna*-clad slope. *Transactions of the Royal Scottish Arboricultural Society*, **50**, 83–6.

Macdonald, J.A.B. & Macdonald, A. (**1952**). The effect of inter-planting with pine on the emergence of Sitka spruce from check on heather land. *Scottish Forestry*, **6**, 77–9.

Mackenzie, J.M. (**1972**). Early effects of different types, rates and methods of application of phosphate rock on peatland. *Proceedings: 4th International Peat Congress Helsinki*, **3**, 531–46.

Mackenzie, J.M. (**1974**). Fertilizer/herbicide trials on Sitka spruce in East Scotland. *Scottish Forestry*, **28**, 211–21.

Mackenzie, J.M., Thompson, J.H. & Wallis, K.E. (**1976**). Control of heather by 2,4-D. *Forestry Commission Leaflet*, **64**. HMSO, London.

Malcolm, D.C. (**1975**). The influence of heather on silvicultural practice—an appraisal. *Scottish Forestry*, **28**, 14–24.

McIntosh, R. (**1981**). Fertilizer treatment of Sitka spruce in the establishment phase in upland Britain. *Scottish Forestry*, **35**, 3–13.

McIntosh, R. (**1983**). Nitrogen deficiency in established phase Sitka spruce in upland Britain. *Scottish Forestry*, **35**, 185–93.

McKay, J.M. (**1986**). Fine root dynamics of the pure and mixed plantations at Culloden. In *Maintenance and Enhancement of Forest Productivity through Manipulation of the Nitrogen Cycle*, European R & D programme in the field of wood as a raw material, 1982–85 (Ed. by J.D. Miller *et al.*), pp. 121–52. Macaulay Institute for Soil Research, Aberdeen.

McKay, H.M. & Malcolm, D.C. (**1988**). A comparison of the fine root component of a pure and mixed coniferous stand. *Canadian Journal of Forestry Research*, **18**, 1416–26.

Miller, H.G. & Miller, J.D. (**1987**). Nutritional requirements of Sitka spruce. *Proceedings of the Royal Society of Edinburgh*, **98B**, 75–83.

Miller, J.D., Miller, H.G., Williams, B.L., Alexander, C.E. & McKay, H.M. (Eds) (**1986**). In *Maintenance and Enhancement of Forest Productivity through Manipulation of the Nitrogen Cycle*, European R & D programme in the field of wood as a raw material, 1982–85 (Ed. by J.D. Miller *et al.*), pp. 121–52, Macaulay Institute for Soil Research, Aberdeen.

Morgan. J.L. (**1984**). A *comparison of the intensity of rooting by larch and spruce on cultivated deep peat*. Honours thesis, University of Edinburgh.

Morgan, J.L. (**1990**). *Oligotrophic peat as a nitrogen source for different tree species*. Ph.D. thesis, University of Edinburgh.

Neustein, S.A. (**1976**). A history of plough development in British forestry. II. Historical review of ploughing on wet soils. *Scottish Forestry*, **30**, 89–111.

Norhstedt, H.-Ö. (**1988**). Nitrogen fixation ($C_2H_2$ reduction) in birch litter. *Scandinavian Journal of Forest Research*, **3**, 17–23.

O'Carroll, N. (**1978**). The nursing of Sitka spruce I. Japanese larch. *Irish Forestry*, **35**, 60–5.

Robinson, R.K. (**1972**). The production by roots of *Calluna vulgaris* of a factor inhibitory to growth of some mycorrhizal fungi. *Journal of Ecology*, **60**, 219–24.

Ross, S.M. & Malcolm, D.C. (**1988**). Modelling nutrient mobilization in an intensively mixed heathland soil. *Plant and Soil*, **107**, 113–21.

Stirling-Maxwell, Sir John (**1907**). The planting of high moorlands. *Transactions of the Royal Scottish Arboricultural Society*, **20**, 1–7.

Taylor, C.M.A. (**1985**). The return of nursing mixtures. *Forestry & British Timber*, **14**, 18–19.

Taylor, C.M.A. & Tabbush, P.M. (**1990**). Nitrogen deficiency in Sitka spruce plantations. *Forestry Commission Bulletin*, **89**. HMSO, London.

Taylor, G.M.M. (1970). Ploughing practice in the Forestry Commission. *Forestry Commission Forest Record*, **70**. HMSO, London.

Thompson, J.H. & Neustein, S.A. (1973). An experiment in intensive cultivation of an upland heath. *Scottish Forestry*, **27**, 211–21.

Weatherell, J. (1953). The checking of forest trees by heather. *Forestry*, **26**, 37–41.

Weatherell, J. (1957). The use of nurse species in the afforestation of upland heaths. *Quarterly Journal of Forestry*, **51**, 298–304.

Williams, B.L. (1986). Nutrient status of humus and soil beneath Sitka spruce in pure stands and admixed with Scots pine at Culloden forest. In *Maintenance and Enhancement of Forest Productivity through Manipulation of the Nitrogen Cycle*, European R & D programme in the field of wood as a raw material, 1982–85 (Ed. by J.D. Miller *et al.*), pp. 121–52. Macaulay Institute for Soil Research, Aberdeen.

Zehetmayer, J.W.L. (1954). Experiments in tree planting on peat. *Forestry Commission Bulletin*, **22**. HMSO, London.

Zehetmayer, J.W.L. (1960). Afforestation of upland heaths. *Forestry Commission Bulletin*, **32**. HMSO, London.

# A comparison of stemwood production in monocultures and mixtures of *Pinus contorta* var. *latifolia* and *Abies lasiocarpa*

F.W. SMITH* AND J.N. LONG†

*Department of Forest Sciences, Colorado State University, Fort Collins, CO 80523, USA; and †Department of Forest Resources and Ecology Center, Utah State University, Logan, UT 84322-5215, USA

## SUMMARY

**1** We examined leaf area index, stemwood production and leaf area efficiency of mature, well-stocked, even-aged, pure and mixed stands of *Pinus contorta* and *Abies lasiocarpa* in an area of limited variation in site quality.

**2** Leaf area index of well-stocked, pure *A. lasiocarpa* stands (mean = $8.5\,\mathrm{m^2\,m^{-2}}$) was significantly greater than the leaf area index of well-stocked *P. contorta* stands (mean = $3.6\,\mathrm{m^2\,m^{-2}}$). Stemwood increment of *A. lasiocarpa* stands ($12.1\,\mathrm{m^3\,ha^{-1}\,year^{-1}}$) was significantly greater than the stemwood increment of *P. contorta* stands ($5.5\,\mathrm{m^3\,ha^{-1}\,year^{-1}}$). Pure *A. lasiocarpa* stands had higher stem densities then pure *P. contorta* stands, and hence substantially smaller mean tree size.

**3** Leaf area index (mean = $5.6\,\mathrm{m^2\,m^{-2}}$) and stemwood production (mean = $8.6\,\mathrm{m^3\,ha^{-1}\,year^{-1}}$) of well-stocked mixed-species stands was intermediate to the means for pure stands. Leaf area index and stemwood increment were linearly related to the percentage of *P. contorta* leaf area when the data were expressed as a replacement series.

**4** Mean stemwood increment was highly correlated with mean leaf area ($r^2 = 0.86$) for pure and mixed stands. This relationship was non-linear, indicating that leaf area efficiency of stands with smaller mean stem size was higher than for stands with a high mean size, regardless of species composition. The ratio of foliar biomass to total crown biomass was negatively correlated with mean stand leaf area. Increased leaf area efficiency was then associated with decreased carbon allocation to branch wood in stands with reduced mean size.

**5** No evidence was found of synergism in mixed stands of *P. contorta* and *A. lasiocarpa*. Differences in production efficiency were largely explained by the high efficiency of small stem sizes. Stemwood production will increase as the proportion of *A. lasiocarpa* increases due to an increase in total stand leaf area, but not necessarily due to an increase in the efficiency of stemwood production.

## INTRODUCTION

There are substantial differences in the production of forest stands growing on the same site but composed of different species. Stemwood production by a population of trees is largely determined by leaf area index (LAI) and the rate of production per unit leaf area (Waring & Schlesinger 1985). Species differences in productivity under similar environmental conditions should therefore be accounted for by differences in the amount and efficiency of foliage. Species differences in LAI are considerable (Jarvis & Leverenz 1983). In general, conifers appear to support higher LAIs than do hardwoods, and late successional species support higher LAIs than do early successional species. On an individual tree basis, more leaf area is supported per unit of conducting tissue in shade-tolerant than shade-intolerant tree species (Kaufmann & Troendle 1981; Parker & Long 1989). These kinds of differences in species LAI suggest fundamentally different patterns for leaf area efficiency (e.g. stemwood production per unit leaf area) of, for example, shade-tolerant and shade-intolerant species. Generalities concerning species differences in leaf area efficiency are not as clear. However, data presented by Jarvis and Leverenz (1983) suggest that the trends for LAI (i.e. conifer > hardwoods and late successional > early successional) may also apply to leaf area efficiency.

If species differences occur in either maximum LAI or leaf area efficiency, production in mixed species stands will depend on the distribution of leaf area between components of the population. For example, Horn (1974) suggests that early successional species are inherently more efficient than later successional species due to differences in canopy geometry. In this case, an increase in the proportion of total leaf area represented by late successional species would result in a decline in production compared to a stand in which a high proportion of total leaf area was contributed by early successional species.

In a previous study of mature *P. contorta* var. *latifolia* Dougl. stands of widely different site quality (Long & Smith, 1990), we concluded that the total leaf area carried by a stand was directly related to the site quality as measured by site index. However, difference in site quality did not appear to influence the efficiency with which leaf area produced stemwood. In the current study, we examine mature stands of *P. contorta*, *Abies lasiocarpa* Hook(Nutt.) and mixtures of the two species across a limited range of site quality in order to assess the role of species differences in canopy architecture on the production ecology of coniferous forests.

*P. contorta* and *A. lasiocarpa* are substantially different in their patterns of ecological behaviour. *P. contorta* is a fire-adapted, shade-intolerant, pioneering species, that forms extensive, even-aged stands following catastrophic disturbance (Clements 1910). In contrast, *A. lasiocarpa* is an exceedingly shade-tolerant species that commonly dominates late-successional stands (Peet 1988). Less commonly in the study area, *A. lasiocarpa* may act as a pioneer following disturbance, forming even-aged stands either by itself or in mixture with *P. contorta* (Schimpf, Henderson & MacMahon 1980).

The objective of this study was to examine canopy structure and the influence of these structural differences on productivity of these two very different coniferous species. Specifically, we compare and contrast *P. contorta* and *A. lasiocarpa* with respect to:

1 individual tree leaf area;
2 LAI of dense, even-aged, single-species stands;
3 species differences in stemwood production and leaf area efficiency; and
4 LAI, stemwood production, and production efficiency in mixed-species stands.

## SITE DESCRIPTION AND METHODS

This study was conducted on the Utah State University Experimental Forest in the Bear River Range of northern Utah. The Experimental Forest is at an elevation of about 2500 m with a continental climate typical of the west-central Rocky Mountains (Peet 1988). Mean temperatures for August and January are about 17 and −11°C, and the majority of the 1040 mm mean annual precipitation falls as snow outside of an approximately 3-month growing season. More detailed descriptions of the climate and vegetation are presented by Hart and Lomas (1979) and Schimpf, Henderson and MacMahon (1980).

A total of 106 plots were established in 70–120 year-old stands of pure *P. contorta*, pure *A. lasiocarpa* and mixed stands of the two species. Mixed-species stands were characterized as those in which *P. contorta* or *A. lasiocarpa* contributed no more than 80% of the total basal area ($m^2 ha^{-1}$ at a height of 1.3 m). *Picea engelmannii* Parry was present in some stands, but never exceeded 5% of the stand basal area. Stands were selected to include well-stocked plots (Table 1).

TABLE 1. Stand statistics for pure *Pinus contorta*, pure *Abies lasiocarpa*, and mixed *P. contorta/A. lasiocarpa* stands in the Utah study area. Means and (ranges) are presented

|  | P. contorta | A. lasiocarpa | Mixed |
|---|---|---|---|
| n | 38 | 27 | 29 |
| Density (ha$^{-1}$) | 1806 | 6656 | 3004 |
|  | (250–6300) | (800–17000) | (635–12000) |
| Basal area (m$^2$ ha$^{-1}$) | 50.9 | 78.4 | 60.6 |
|  | (17.1–82.1) | (35.7–136.4) | (33.7–96.4) |
| Average stand diameter (cm) | 22.9 | 14.3 | 18.9 |
|  | (12.3–39.4) | (7.9–23.8) | (8.5–27.7) |
| Leaf area index (m$^2$ m$^{-2}$) | 3.64 | 8.46 | 5.63 |
|  | (2.4–4.9) | (3.0–14.1) | (3.3–7.7) |
| Mean leaf area (m$^2$) | 39.0 | 19.0 | 30.0 |
|  | (6.6–112.0) | (4.9–37.3) | (4.9–87.4) |
| Stemwood increment (m$^3$ ha$^{-1}$ year$^{-1}$) | 5.53 | 12.07 | 8.64 |
|  | (2.2–9.0) | (7.4–18.9) | (4.5–15.2) |
| Mean stemwood increment (m$^3$ year$^{-1}$) | 0.0047 | 0.0030 | 0.0043 |
|  | (0.0013–0.0113) | (0.0006–0.0099) | (0.0009–0.0104) |
| Foliar biomass : crown biomass | 0.33 | 0.50 | 0.35 |
|  | (0.12–0.43) | (0.37–0.61) | (0.26–0.50) |

Plot sizes varied such that the number of live trees per plot ranged from eighty-two on a small high-density plot to fifteen on a large low-density plot. The average number of trees per plot was thirty-nine.

For each tree taller than 1.3 m, we measured: diameter at breast height (DBH), total height, height to the base of the live crown, sapwood cross-sectional area at breast height and 5-year radial increment. Five-year radial increment was used to backdate tree dimensions for volume increment determination. Regression equations based on thirty *P. contorta* and thirty *A. lasiocarpa* felled trees were used to estimate total stem volume, 5-year volume increment, projected leaf area, and ovendry needle and branch biomass for each tree on a plot. Tree numbers, projected leaf area, and biomass were summed for each plot and converted to a hectare basis. Stand growth was estimated by summing the estimated stem volume increment for each tree on the plot, dividing this figure by 5 years and expressing stand growth on a hectare basis. Stand growth is an estimate of gross stemwood volume increment because growth during the previous 5 years of trees that had died during this period was also estimated. Details concerning the sampling procedures and the leaf area and volume prediction equations are presented elsewhere (Dean & Long 1986; Long & Smith 1988; Dean, Long & Smith 1988; Smith & Long 1989).

Site index, an indirect measure of site quality, for each plot was determined from standard height–age relationships based on the height and age of dominant trees and expressed as the height (m) of dominant trees at 100 years of age. Site index for *P. contorta* was estimated using the curves of Alexander (1967a); site index for *A. lasiocarpa* was estimated using *Picea engelmannii* curves (Alexander 1967b) because *A. lasiocarpa* site index curves have not been published.

## RESULTS

Density, LAI, and growth of the *P. contorta* and *A. lasiocarpa* stands examined in this study vary substantially, although they represent a narrow range of stand ages and site indexes (Table 1). For both *P. contorta* and *A. lasiocarpa*, projected leaf area is a non-linear function of sapwood cross-sectional area at breast height, tree height and crown dimensions (Long & Smith 1988, 1989). While *A. lasiocarpa* supports more leaf area for a given sapwood area and crown size than does *P. contorta*, the *A. lasiocarpa* trees in our sample tended to carry less leaf area because they were, on average, smaller than trees in *P. contorta* stands (Table 1). However, on a stand basis, LAI of *A. lasiocarpa* stands (mean = $8.5\,\mathrm{m^2\,m^{-2}}$) exceeded LAIs of *P. contorta* stands (mean = $3.6\,\mathrm{m^2\,m^{-2}}$) ($P < 0.05$) since the increased number of trees per hectare in stands dominated by *A. lasiocarpa* more than compensated for their lower mean leaf area (Table 1). Total stemwood increment of the *A. lasiocarpa* stands (mean = $11.9\,\mathrm{m^3\,ha^{-1}\,year^{-1}}$) was greater than that of the *P. contorta* stands (mean = $5.5\,\mathrm{m^3\,ha^{-1}\,year^{-1}}$) ($P < 0.05$) (Table 1).

Leaf area index and stemwood increment of mixed-species stands was inter-

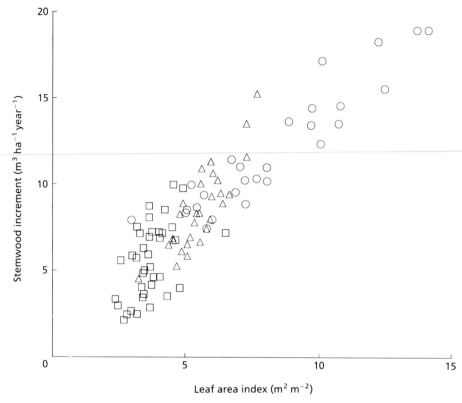

FIG. 1. Stemwood increment increases with increased leaf area index for pure *Pinus contorta* (△), pure *Abies lasiocarpa* (○) and mixed *P. contorta/A. lasiocarpa* (□) stands, but the relation is variable.

mediate between the high leaf areas of *A. lasiocarpa* and low leaf area of *P. contorta* stands (Table 1). Stemwood increment increased as LAI increased for pure *A. lasiocarpa*, pure *P. contorta* and mixed stands, but there was considerable variability in the relation (Fig. 1). In general, pure stands of *A. lasiocarpa* had high LAI and high stemwood increment, pure *P. contorta* stands had low LAI and low stemwood increment, and mixed stands were intermediate in LAI and increment. Total stand LAI declined as the percentage of lodgepole leaf area in a stand increased (Fig. 2). The relationship was significant ($P < 0.05$) and linear. The decline in stemwood production was linear and significant ($P < 0.05$) as the proportion of stand leaf area in *P. contorta* increased (Fig. 3).

Mean stemwood production (stemwood increment/stand density) increased for pure *A. lasiocarpa*, pure *P. contorta* and mixed stands as positive non-linear function of mean leaf area (stand leaf area/stand density), regardless of species composition ($r^2 = 0.86$) (Fig. 4). As mean leaf area increased, volume growth increased, but with a decreasing rate of increase. Since the increase in volume increment declined at high mean leaf area, leaf area efficiency (i.e. volume

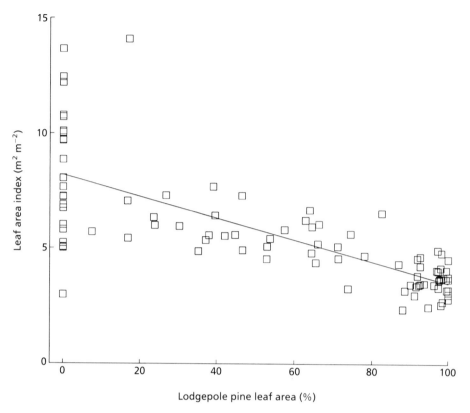

FIG. 2. Leaf area index of pure and mixed-species *Pinus contorta* and *Abies lasiocarpa* stands declines linearly as the proportion of *P. contorta* leaf area increases ($y = 8.19 - 0.047x$, $r^2 = 0.55$).

growth per unit of leaf area) declined with increasing mean leaf area. Residual analysis indicated that the relationship was unbiased with respect to species composition.

The decline in leaf area efficiency associated with increasing mean leaf area has, for *P. contorta* stands, been attributed to decreasing amounts of foliage as a percentage of total crown weight in trees with large crowns (Long & Smith, 1990). The ratios of foliar biomass and total crown biomass (i.e. the sum of foliage and branch biomass) for *A. lasiocarpa* stands in our study were, in general, higher than for *P. contorta* (Table 1). For pure *P. contorta*, pure *A. lasiocarpa* stands and mixed stands in this study, the ratio of foliar biomass to crown biomass was inversely proportional to increasing crown size, or mean leaf area (Fig. 5). This relationship differed for the two species (Fig. 5), since the ratio of foliar to crown biomass was higher for *A. lasiocarpa* than for *P. contorta* stands of the same mean leaf area.

Since mean leaf area was generally lower in *A. lasiocarpa* stands than in *P.*

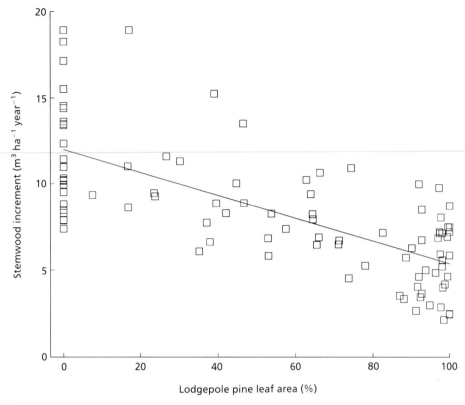

Fig. 3. Stemwood increment of pure and mixed-species *Pinus contorta* and *Abies lasiocarpa* stands declines linearly as the proportion of *P. contorta* leaf area increases ($y = 11.98 - 0.66x$, $r^2 = 0.49$).

*contorta* stands, mean growth was also lower, but leaf area efficiency was greater (Fig. 4). High efficiency combined with generally high LAIs results in greater total stemwood increment for stands dominated by *A. lasiocarpa* than for those dominated by *P. contorta* (Table 1). For pure *P. contorta*, pure *A. lasiocarpa* stands and mixed stands, stemwood increment ($I_v$) was highly correlated with LAI and ratio of foliar biomass to total crown biomass (FB:CB):

$$I_v = 3.60 * \text{LAI}^{0.71} * \text{FB:CB}^{0.40}$$
$$r^2 = 0.85$$
$$n = 106.$$

Residual analysis indicated that this relationship was unbiased with respect to species composition and mean leaf area. Thus, the relationship between stemwood increment and leaf area was independent of species composition once the typically small mean leaf area, and therefore high efficiency, of *A. lasiocarpa* stands was taken into account.

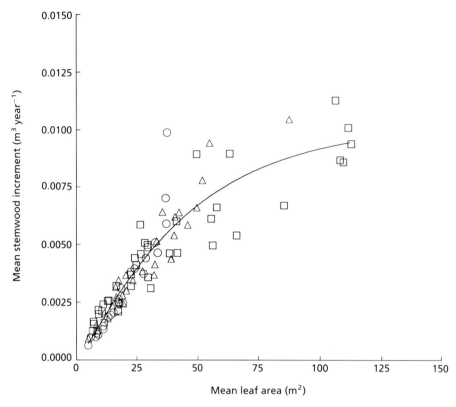

FIG. 4. Mean stemwood increment is related to mean leaf area for pure *Pinus contorta* ($\triangle$), pure *Abies lasiocarpa* ($\bigcirc$) and mixed *P. contort/A. lasiocarpa* ($\square$) stands ($y =$ 0.010$[1 - \exp(-0.024x)]^{1/(1-0.17)}$, $r^2 = 0.86$). Non-linearity in the relation indicates higher leaf area efficiency for small than for large trees.


## DISCUSSION

Differences in the way individual *A. lasiocarpa* and *P. contorta* trees carry leaf area are consistent with conventional wisdom concerning shade-tolerant and shade-intolerant species. Shade-intolerant, early successional tree species, for example, are presumed to be characterized by rapid juvenile height growth but relatively ineffective light capture (Daniel, Helms & Baker 1979). This generality is consistent with the observed patterns of leaf area development for *P. contorta* and *A. lasiocarpa*.

Contrasts between the leaf area relations of these two species also exist at the stand-level. Not only do individual *A. lasiocarpa* trees support more leaf area than do *P. contorta* trees of the same size, but stands of *A. lasiocarpa* carry more trees of a given mean size per hectare than stands of *P. contorta*. These two factors, greater leaf area per tree and greater relative stand density, result in

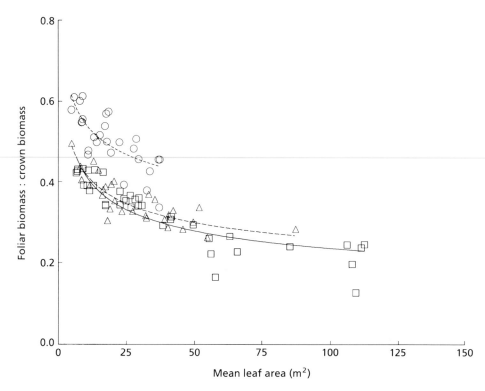

FIG. 5. The ratio of foliar biomass to total crown biomass declines with increasing mean leaf area for pure *Pinus contorta* ($\triangle$) ($y = 0.75\,x^{-0.25}$), pure *Abies lasiocarpa* ($\bigcirc$) ($y = 0.80\,x^{-0.17}$) and mixed *P. contorta*/*A. lasiocarpa* ($\square$) ($y = 0.87x^{-0.27}$) stands.

substantially greater LAIs in stands dominated by *A. lasiocarpa* than stands dominated by *P. contorta* (Table 1).

When LAI is displayed in the form of a 'replacement series' (i.e. LAI as a function of per cent *P. contorta* leaf area), there is no indication of other than a linear replacement of leaf area from 0 to 100% *P. contorta* leaf area (Fig. 2). If foliar 'packing' were more or less effective in mixed-species populations than in single-species populations, the response of LAI to changes in mixture in the replacement series should be other than linear. In fact, leaf area of *P. contorta* and *A. lasiocarpa* appears in a sense to be interchangeable. A unit of *P. contorta* leaf area is apparently interchangeable with a unit of *A. lasiocarpa* foliage. The units, however, are not equal in terms of actual leaf area since a 'unit' of *A. lasiocarpa* foliage represents more leaf area than does a 'unit' of *P. contorta* foliage.

Previous work with *P. contorta* stands (Smith & Long 1989) established that stand structure, e.g. stand density and mean crown size, has considerable impact on leaf area efficiency. Additionally, we determined that the relation between leaf area efficiency and stand structure of *P. contorta* stands is independent of site quality (Long & Smith, 1990). Accounting for differences in mean crown size is critical to a meaningful comparison of leaf area efficiency between species. The apparent species differences in leaf area efficiency between *P. contorta* and *A. lasiocarpa* are largely the result of differences in mean size (e.g. mean leaf area). Apparently, there is no substantial difference in leaf area efficiency between these two very dissimilar coniferous species, once we have accounted for differences in mean leaf area. While leaf area efficiency is, on average, higher for *A. lasiocarpa* stands than for *P. contorta* stands, this difference between species in leaf area efficiency appears to directly follow from the fact that the crowns of *A. lasiocarpa* are generally smaller, and therefore more efficient, than those of *P. contorta*.

We suspect that decreasing leaf area efficiency with increasing crown size results from an associated increase in the proportion of non-photosynthetic, and respiring, woody branches in large crowns (e.g. Fig. 4). The decreased ratio of foliar biomass to crown biomass is an indication of increased carbon allocation to branchwood, possibly at the expense of stemwood production. Further, as the ratio of foliage to branch biomass increases, a decreasing proportion of photo-synthate produced by the foliage on a branch may be available for export from the branch to other parts of the tree, resulting in a decline in stemwood produc-tion (Assmann 1970; Long & Smith, 1990). There were differences between the two species in the relationship between the ratio of foliar to crown biomass and mean leaf area (Fig. 5). However, given the fundamental differences in relative shade-tolerance between *P. contorta* and *A. lasiocarpa*, the overall simi-larity in their foliar biomass:crown biomass–LAI relations is surprising.

Production is a function of the amount and efficiency of leaf area supported by a stand. In contrast with *P. contorta* stands, *A. lasiocarpa* stands typically have greater LAI and the same or greater (since mean leaf area is usually smaller) leaf area efficiency. Total stemwood production by *A. lasiocarpa* stands is greater than production by *P. contorta* dominated stands (Table 1). This is primarily due to the high leaf area of *A. lasiocarpa* stands coupled with a stand structure characterized by small, and therefore, efficient trees.

These results have implications for the production ecology of mixed-species stands. Since there are no substantial inherent differences in the leaf area effic-iency of the two species, an increase in LAI results in greater stemwood produc-tion. A change in species composition that results in an increase in LAI also results in an increase in stemwood production. For the species studied here, an increase in the proportion of *A. lasiocarpa* at the expense of *P. contorta*, there-fore results in greater stemwood production. For stands of similar leaf area, those with small trees will have higher production than those with large trees, since small trees produce more stemwood per unit of leaf area than do large trees. In effect, young stands or stands of relatively high density will have high leaf area efficiency. The high leaf area efficiency of small trees may be related to changes in carbon allocation associated with age and density, as small trees had high foliar biomass compared with total crown biomass, while proportionally less of the total crown biomass was allocated to foliage in stands of large trees.

When total stemwood production is displayed in the form of a 'replacement series' (i.e. production as a function of per cent *P. contorta* leaf area), there is no indication of other than a linear replacement of leaf area from 0 to 100% *P. contorta* leaf area (Fig. 2). Further, there is no indication in these data of the mixed-species synergism (Fig. 3) which results in higher stemwood production in other mixed-species forests (Schuler & Smith 1988). The suggestion that crowns of two species might occupy space more effectively than the crowns of a single species is not supported by these data.

## ACKNOWLEDGMENTS

We acknowledge the contributions of P. Conklin, T. Dean, S. Jack, J.B. McCarter and S. Roberts to this project. This work was supported in part by USDA grant 85-FSTY-9-0124.

## REFERENCES

Alexander, R.R. (1967a). *Site indexes for lodgepole pine, with corrections for stand density: instructions for field use.* USDA Forest Service Research Paper RM-24.

Alexander, R.R. (1967b). *Site indexes for Engelmann spruce.* USDA Forest Service Research Paper RM-32.

Assmann, E. (1970). *The Principles of Forest Yield Study.* Pergamon Press, Oxford.

Clements, F.E. (1910). *The life history of lodgepole burn forests.* USDA Forest Service Bulletin 79.

Daniel, T.W., Helms, J.A. & Baker, F.S. (1979). *Principles of Silviculture* (2nd edn). McGraw-Hill, New York.

Dean, T.J. & Long, J.N. (1986). Variation in sapwood area–leaf area relations within two stands of lodgepole pine. *Forest Science*, **32**, 749–58.

Dean, T.J., Long, J.N. & Smith, F.W. (1988). Bias in leaf area–sapwood area ratios and its impact on growth analysis in lodgepole pine. *Trees*, **2**, 104–9.

Hart, G.E. & Lomas, D.A. (1979). Effects of clear-cutting on soil water depletion in an Engelmann spruce stand. *Water Resources Research*, **15**, 1598–602.

Horn, H.S. (1974). The ecology of secondary succession. *Annual Review of Ecology and Systematics*, **5**, 25–37.

Jarvis, P.G. & Leverenz, J.W. (1983). Productivity of temperate, deciduous and evergreen forests. In *Encyclopedia of Plant Physiology*, New Series, Vol. 12D: Physiological Plant Ecology IV (Ed. by O.L Lange, P.S. Noble, C.N. Osmond & H. Ziegler), pp. 234–80. Springer, Berlin.

Kaufmann, M.R. & Troendle, C.A. (1981). The relationship of leaf area and foliage biomass to sapwood conducting area in four subalpine forest tree species. *Forest Science*, **27**, 477–82.

Long, J.N. & Smith, F.W. (1988). Leaf area–sapwood area relations of lodgepole pine as influenced by stand density and site index. *Canadian Journal of Forest Research*, **18**, 247–50.

Long, J.N. & Smith, F.W. (1989). Estimating leaf area of *Abies lasiocarpa* across range of stand density and site quality. *Canadian Journal of Forest Research*, **19**, 930–32.

Long, J.N. & Smith, F.W. (1990). Determinants of stemwood production in *Pinus contorta* var. *latifolia* forests: the influence of site quality and stand structure. *Journal of Applied Ecology* **27**, 847–56.

Parker, J.N. & Long, J.N. (1989). Intra- and inter-specific tests of some traditional indicators of relative tolerance. *Forest Ecology and Management*, **28**, 177–89.

Peet, R.K. (1988). Forests of the Rocky Mountains. In *North American Terrestrial Vegetation* (Ed. by M.G. Barbour & W.D. Billings), pp.63–101. Cambridge University Press, Cambridge.

Schimpf, D.F., Henderson, D.F. & MacMahon, J.F. (1980). Some aspects of succession in the spruce–fir forest zone of Northern Utah. *Great Basin Naturalist*, **40**, 1–26.

**Schuler T.M. & Smith, F.W. (1988).** Effect of species mix on size/density and leaf area relations in southwest pinyon–juniper woodlands. *Forest Ecology and Management*, **25**, 211–20.

**Smith, F.W. & Long, J.N. (1989).** The influence of canopy structure on stemwood production and growth efficiency of *Pinus contorta* var. *latifolia*. *Journal of Applied Ecology*, **26**, 681–91.

**Waring, R.H. & Schlesinger, W.H. (1985).** *Forest Ecosystems: Concepts and Management*. Academic Press, Orlando.

# Mixtures of nitrogen$_2$-fixing and non-nitrogen$_2$-fixing tree species

D. BINKLEY

*Department of Forest Sciences, Colorado State University, Fort Collins, CO 80523, USA*

## SUMMARY

The inclusion of nitrogen (N)-fixing trees in stands with non-N-fixing trees has revealed a wide range of effects on ecosystem production and nutrient cycling, and a wide variation in these effects across species, locations, and stand designs. The presence of N-fixing trees appears to increase rates of N cycling in all situations, but the effects on ecosystem productivity and on the growth of associated species have been variable. Productivity on N-limited sites generally increases in the presence of N-fixing species, in some cases doubling relative to pure stands of non-N-fixing species. On N-rich sites, productivity remains unaffected or even decreases in the presence of N-fixing trees (due to increased competition and mortality). The growth of associated non-N-fixing trees is generally greater in stands mixed with N-fixing trees than in pure stands only on N-limited sites. Mixtures of N-fixing and non-N-fixing trees typically show increased rates of cycling of nutrients such as phosphorus. Increased growth in mixed stands may entail increased nutrient demands, and the increased rates of nutrient cycling may not be sufficient for maintaining maximum growth rates. Soil acidification is commonly observed in mixed stands with N-fixing trees, driven by both greater nitrate leaching and accumulation of stongly acidic soil organic matter. The components of competition between N-fixing and non-N-fixing trees have received experimental investigation in a few cases; competition for light appears most critical, but competition for water and nutrients other than N are likely to be important in some cases. Nitrogen-fixing trees provide a powerful silvicultural tool for managing forest stands on N-limited sites, especially when concerns over biodiversity or the availability of inorganic-N fertilizer shape silvicultural prescriptions.

## INTRODUCTION

The potential increased productivity of plants growing near N-fixing plants has been recognized for centuries:

> The alder whose fat shadow nourisheth
> Each plant set neere to him long flourisheth.
> (Browne 1613)

The effects of N-fixing trees on interplanted non-N-fixing trees have most often been characterized in terms of heights and diameters (see Kittredge 1948; Berntsen

1961; Remezov & Pogrebnyak 1965; Newton, El Hassan & Zavitkovski 1968; Miller & Murray 1978; Domingo 1983; Mikola, Uomala & Mälkönen 1983; Turvey & Smethurst 1983; Schlesinger & Williams 1984; Cole & Newton 1986; Walstad, Brodie & McGinley 1986; Heilman 1990a). Intensive research over the past few decades has provided a relatively solid foundation for understanding many of the major ecological interactions that occur in mixed stands of N-fixing and non-N-fixing trees. On nitrogen-deficient sites, mixed stands present an ecological opportunity for increasing both total stand growth and the growth of the non-N-fixing trees. On sites with ample supplies of N, mixed stands typically perform more poorly than pure stands of non-N-fixing species.

Mixtures of N-fixing and non-N-fixing trees differ from other sets of species by the direct and indirect effects of increased N supply. In stands where N availability would limit production in the absence of N fixation, these effects typically include increased N mineralization, nutrient uptake, leaf area, biomass production, and nutrient leaching. The key features of mixed stands depend upon the magnitude of these increases, and how some of the increases are apportioned between the N-fixing and the non-N-fixing trees. This summary focuses first on observed rates of N fixation and subsequent cycling in mixed stands relative to single-species stands, and then examines other biogeochemical aspects. The components that determine productivity in mixed stands are then described, followed by a discussion of silvicultural opportunities.

## RATES OF N FIXATION

Most estimates of rates of N fixation in mixed stands deal with red alder (*Alnus rubra* Bong.) mixed with Douglas fir (Table 1). These studies have generally found N fixation rates of about $20–85 \, kg \, ha^{-1} \, year^{-1}$. Other combinations, such as red alder with black cottonwood (*Populus trichocarpa* Torr. & Gray) and poplar and black alder [*Alnus glutinosa* (L.) Gaertn.] appear to have similar rates.

Three approaches to estimating N fixation were used in these studies. Acetylene reduction assays the current rate of N fixation by substituting acetylene ($C_2H_2$) for dinitrogen ($N_2$) as a substrate for the nitrogenase enzyme; changes in concentrations of acetylene and its product, ethylene ($C_2H_4$) are easier to detect than changes in dinitrogen or the normal products of N fixation. Estimates of N fixation rates by acetylene reduction have high uncertainty, due to errors in estimating average rates per gram of nodule, the biomass of nodules per plant (or area), and in the conversion ratio of moles of acetylene reduced to moles of dinitrogen reduced (Silvester 1983). Nitrogen accretion methods are used most commonly, where the N content of a soil horizon or all ecosystem components for the non-N-fixing stand is subtracted from the N-fixing stand. These estimates tend to be more precise than acetylene reduction, and are generally good if nitrogen leaching is minimal. The third method is by far the best, and involves adding labelled nitrogen (the heavy, stable isotope $^{15}N$) to the soil prior to planting. Subsequent dilution of N pools with $^{14}N$ provides a direct measure of the N added

TABLE 1. Rates of N fixation for mixed-species stands (kg ha$^{-1}$ year$^{-1}$)

| N-fixer | Location | Associated trees | Stand age/period | Rate | Component | Method | Reference |
|---|---|---|---|---|---|---|---|
| Red alder | Mt Benson, BC | Douglas fir | 23 | 130 | Ecosystem | C$_2$H$_2$ reduction | Binkley 1981 |
| | Washington | | 0–23 | 65 | Ecosystem | Accretion | Binkley 1983 |
| | Oregon | | 0–23 | 42 | Ecosystem | Accretion | Binkley 1983 |
| | | | 0–17 | 13 | Soil to 15 cm | Accretion | Berg & Doerksen 1975 |
| | | | 0–17 | 21 | | Accretion | |
| | Oregon | | 0–5 | 0 | Soil to 15 cm | Accretion | Cole & Newton 1986 |
| | Wind River, Washington | | 0–26 | 40 | Soil to 90 cm | Accretion | Tarrant and Miller 1963 |
| | | | 0–55 | 49 | Ecosystem | Accretion | Binkley et al. 1991 |
| | Cascade Head, Oregon | | 55 | 73 | Soil to 90 cm | Accretion | Franklin et al. 1968 |
| | | | 0–40 | 26 | Ecosystem | C$_2$H$_2$ reduction | Binkley et al. 1991 |
| | | | 0–55 | 74 | Ecosystem | Accretion | |
| | | | 55 | 85 | Ecosystem | C$_2$H$_2$ reduction | |
| | Camas, Washington | Black cottonwood | 0–4 | 40 | Soil to 15 cm | Accretion | DeBell & Radwan 1979 |
| | Mt St Helens, Washington | Hybrid poplar, Black cottonwood | 0–6 | 56 | Soil to 30 cm | Accretion | Heilman 1990a |
| | Lennox Forest, Scotland | Sitka spruce | 0–20 | 36 | Forest floor Soil to 5 cm | Accretion | Malcolm et al. 1985 |
| Sitka alder | Mt Benson, BC | Douglas fir | 23 | 20 | Ecosystem | C$_2$H$_2$ reduction | Binkley 1981 |
| | | | 0–23 | 30 | Ecosystem | Accretion | Binkley et al. 1984 |
| Black alder | Laval, Quebec | Hybrid poplar | 0–3 | 16–34 | Ecosystem | $^{15}$N dilution | Cote & Camire 1984, 1987 |
| Black locust | Coweeta, North Carolina | Hardwoods | 4 | 30 | Ecosystem | C$_2$H$_2$ reduction | Boring & Swank 1984a,b |
| | | | | 48 | Soil to 50 cm | Accretion | |
| | | | 17 | 75 | Soil to 50 cm | Accretion | |
| | | | 38 | 33 | Soil to 50 cm | Accretion | |

from the atmosphere. Cote and Camire (1984, 1987) used this method to determine that black alder mixed with poplar fixed 16–34 kg-N ha$^{-1}$ year$^{-1}$ (depending on alder density), and that alders obtained about 55% of their N from the atmosphere (and 45% from the soil).

The factors that determine the rate of N fixation in mixed species stands have generally not been studied under field conditions. Heilman and Stettler (1985) and Binkley, Lousier and Cromack (1984) noted that acetylene reduction rates by red alder and Sitka alder dropped to low levels when alders were overtopped by associated trees. Binkley (1986) noted that a bioassay of red alder seedlings grown in soils from a red alder/Douglas fir stand responded strongly to P fertilization. Fertilized seedlings demonstrated five-fold greater acetylene reduction rates than controls. Experimentation with a variety of species, relative proportions of species, and nutrition treatments are clearly warranted.

Early studies on the effects of nitrogen supply in solution culture on rates of N fixation by alder seedlings indicated that high availability of ammonium or nitrate might depress the rate of N fixation. The rates used in these experiments greatly exceeded rates found in soil solutions from even very fertile soils, and many studies now show that fixation of nitrogen by alders is not inhibited by high availability of ammonium or nitrate (Ingestad 1980; Mackay, Simon & Lalonde 1987) or by high levels of total soil N (Franklin *et al.* 1968; Luken & Fonda 1983; Heilman & Stettler 1985; Binkley *et al.* 1991). In fact, red alder stands commonly show rates of nitrate leaching of 50 kg-N ha$^{-1}$ year$^{-1}$ (see later).

Why do alders continue expending energy on N fixation rather than making greater use of abundantly available soil N (as agricultural leguminous crops do)? The carbon 'costs' of assimilating nitrogen from $NH_4^+$, $NO_3^-$, and $N_2$ appear to strongly favour use of both ammonium and nitrate (0.1 mol glucose mol$^{-1}$ ammonium, and 0.33 mol glucose mol$^{-1}$ nitrate; Schubert 1982) rather than dinitrogen (0.38–0.50 mol glucose mol$^{-1}$ dinitrogen-N fixed, depending on hydrogen evolution and recovery). The total costs of each source may also depend on the investment in biomass of fine roots and mycorrhizae to obtain the molecules, and a potential reduction in biomass investment to obtain each type of N may partly offset the costs of assimilation (Gutschick 1981). The need for further research on mechanisms controlling the responses of actinorhizal plants to increasing supplies of ammonium and nitrate (at realistic levels) is clear.

## BIOGEOCHEMISTRY

### *Litterfall*

The cycling of nutrients in above-ground litterfall is much greater in mixed stands with N-fixing species than in stands containing only non-N-fixing trees (Table 2). Reported rates of litterfall cycling of N are three to eight times greater in mixed stands. Perhaps more surprising is the two- to seven-fold greater litterfall cycling of phosphorus, sulphur, calcium, magnesium and potassium. The mechanisms

TABLE 2. Annual litterfall biomass and nutrient content (kg ha⁻¹) in single-species and mixed stands. * Approximated from graph

| Location | Vegetation | Stand ages | Biomass | N | P | S | Ca | Mg | K | Reference |
|---|---|---|---|---|---|---|---|---|---|---|
| Wind River, Washington | Conifer | 55 | 4030 | 10 | 5.3 | 1.6 | 18.4 | 2.0 | 4.0 | Binkley et al. 1991 |
|  | Alder/conifer | 55 | 9700 | 74 | 10.2 | 6.4 | 59.6 | 7.7 | 16.8 |  |
| Cascade Head, Oregon | Conifer | 28 | 4740 | 36 |  |  |  |  |  | Tarrant et al. 1969 |
|  | Alder/conifer | 28 | 6640 | 116 |  |  |  |  |  |  |
|  | Conifer | 55 | 6400 | 29 | 6.5 | 3.8 | 19.2 | 4.1 | 7.4 | Binkley et al. 1991 |
|  | Alder/conifer | 55 | 21 000 | 140 | 16.1 | 8.6 | 48.0 | 13.2 | 28.5 |  |
| Mt Benson, BC | Douglas fir | 23 | 1420 | 16 | 2.0 |  | 14.0 | 1.9 | 2.0 | Binkley et al. 1984 |
|  | Douglas fir/ Sitka alder | 23 | 5210 | 112 | 7.1 |  | 44.0 | 13.0 | 14.8 |  |
|  | Douglas fir/ red alder | 23 | 6400 | 136 | 5.6 |  | 54.0 | 10.3 | 16.0 | Binkley unpubl. |
| Laval, Quebec* | Hybrid poplar | 3 | 2000 | 25 | 1.7 |  | 19 | 3.7 | 5.0 | Cote & Camire 1987 |
|  | + 33% black alder |  | 2900 | 58 | 2.8 |  | 32 | 5.5 | 5.5 |  |
|  | + 67% black alder |  | 3000 | 70 | 3.0 |  | 30 | 6.5 | 9.0 |  |
| Hilo, Hawaii | Eucalyptus | 6 | 8000 | 25 | 1.6 |  |  |  |  | Dunkin & Binkley unpubl. |
|  | + 14% Albizia |  | 7300 | 71 | 3.3 |  |  |  |  |  |
|  | + 25% Albizia |  | 7300 | 85 | 4.9 |  |  |  |  |  |
|  | + 34% Albizia |  | 8000 | 84 | 4.8 |  |  |  |  |  |
|  | + 50% Albizia |  | 7200 | 84 | 6.4 |  |  |  |  |  |
|  | + 75% Albizia |  | 8500 | 95 | 6.1 |  |  |  |  |  |

underlying greater N content of litter include the contribution of N fixation to the litter from the N-fixing species, and the rapid decomposition of N-rich litter allowing greater N uptake by all trees. The mechanisms underlying greater cycling rates of the other nutrients are less clear (Binkley *et al.* 1991). For P and S, increased turnover of organic matter may increase the role of organic pools of these elements relative to inorganic pools. The mechanisms determining the greater cycling rate of cation nutrients probably varies among sites, but would include various combinations of greater depletion from the exchange complex (Van Miegroet *et al.* 1989), greater weathering of primary minerals, and perhaps greater turnover of cations bound in soil organic matter.

## Phosphorus

The cycling of P in mixed stands is particularly intriguing, although few examples are available from mixed stands of N-fixing trees and non-N-fixing trees. Mechanisms that have been demonstrated in some cases include: greater rooting depth (Malcolm, Hooker & Wheeler 1985; Heilman 1990b), rhizosphere acidification (Gillespie & Pope 1990a,b), production of low-molecular-weight organic acids and chelates (Reid, Reid & Szaniszlo 1985; Jayachandran, Schwab & Hetrick 1989; Ae *et al.* 1990) and increased phosphatase activity (Ho 1979). For example, B. Caldwell (pers. comm.) found that phosphatase activity (both phosphomonesterase and phosphodiesterase) in replicated plots of N-fixing albizia was double that found for plots with eucalyptus.

Rates of nutrient cycling in litterfall may not correspond with nutritional status of trees for several reasons. A low rate of nutrient cycling could represent either a low supply of nutrients, or a strong constraint of non-nutrient resources on productivity (Chapin, Van Cleve & Vitousek 1986). In addition, a high rate of cycling in litterfall may not be well allocated between the N-fixing and the non-N-fixing trees. For example, the concentrations of P in Douglas fir foliage on the Mt. Benson site in British Columbia averaged $220 \mu g \, g^{-1}$ without alder, compared with $110 \mu g \, g^{-1}$ in stands mixed with Sitka alder [*Alnus sinuata* (Regel.) Rydb.] and $90 \mu g \, g^{-1}$ in stands mixed with red alder (Binkley 1983; Binkley, Lousier & Cromack 1984). The P concentration in Sitka alder leaves was $300 \mu g \, g^{-1}$, and $230 \mu g \, g^{-1}$ in red alder leaves. Bioassays with Douglas fir seedlings (Binkley 1986) indicated the soil from the Sitka alder/Douglas fir stand had the greatest P availability (1.4 mg/seedling), followed by the red alder/Douglas fir soil (0.7 mg/ seedling) and the Douglas fir soil (0.3 mg/seedling). This pattern is consistent with the ranking of the 3 stands with respect to the P content of litterfall, but almost opposite the ranking based on the P content of Douglas fir foliage collected in the field. Both alder species appeared to be better competitors for P than Douglas fir; the P nutrition of Douglas fir was impaired in the mixed stands, despite an apparently greater supply in the soil under alder. Studies on some other sites have found no effect of alder on the P nutrition of Douglas fir (Binkley 1983; Binkley *et al.* 1991), illustrating the site-specific nature of interactions between species.

One case has been reported where red alder appeared to deplete the pools of phosphorus available in the soil at the University of Washington's Thompson Research Center. Brozek (1990) reported that foliage of Douglas fir seedlings planted on a site that formerly contained Douglas fir had 210 μg g$^{-1}$ of P, compared with 125 μg g$^{-1}$ P in seedlings on a former red alder site. Available P in the soil was about four times greater in the soils from the former Douglas fir stand. The soil from the latter contained twice the total P found in the rockier alder soil (Johnson & Lindberg 1991), and the difference between sites of over 1800 kg ha$^{-1}$ is too great to have developed from one rotation of alder. Some of the apparent alder effect may reflect pre-existing differences between the two sites.

## Nutrient cations

Information is sparse on the effects of mixed stands on pools of exchangeable cations and mineral weathering. At the Cascade Head site in Oregon, the pool of exchangeable base cations (0–90 cm) was 600 kmol$_c$ ha$^{-1}$ under conifers, compared to 320 kmol$_c$ ha$^{-1}$ under an adjacent stand of conifers and red alder (Binkley & Sollins 1990). If the soils were equivalent at the time the stands were established, the rate of depletion in the mixed stand would be about 5.5 kmol$_c$ ha$^{-1}$ year$^{-1}$ for 55 years. This rate would roughly match the rate of accumulation of cations in biomass plus depletion through leaching, suggesting no increase in mineral weathering under the influence of alder. In contrast, the pools of exchangeable cations in a similar pair of stands at the Wind River site in Washington were virtually identical (1250 kmol$_c$ ha$^{-1}$) in a Douglas fir stand and a red alder/Douglas fir stand of the same age (Binkley & Sollins 1990), indicating a slight increase in mineral weathering to supply the greater accumulation of cations in biomass and the cations leached from the soil in the mixed stand. In a comparison of pure red alder and Douglas fir stands, Van Miegroet and Cole (1988) found no depletion of base cations from exchangeable pools under alder, despite 3.0 kmol$_c$ ha$^{-1}$ greater leaching losses of base cations from the alder stand. Their results suggest that mineral weathering was much greater under the influence of red alder. In a study of mixtures of black alder and hybrid poplar, pools of exchange base cations in the 0–5 cm soil depth tended to increase during the first 3 years of plantation development, probably due to transfer from lower depths via plant uptake and litterfall (Cote & Camire 1987).

## Decomposition

The litter from N-fixing species generally decomposes faster than litter of non-N-fixing species, and the addition of N-fixer litter may accelerate the decomposition of other litter types. For example, Taylor, Parsons and Parkinson (1989) found that leaf litter of green alder [*Alnus crispa* (Ait.) Pursh] in litter bags would require about 11.5 years to reach 95% decomposition, compared with 14.5 years for leaf litter from aspen (*Populus tremuloides* Michx.). Aspen litter mixed with alder litter in the same bag decomposed as rapidly as the alder litter. The decomposition of most N-poor litter generally involves a period where the N content of the decomposing material actually increases (from throughfall, asymbiotic N fixation, or import from the soil through fungal hyphae). However, Edmonds (1987) found that decomposition of small-diameter (6–10 cm) branches from red alder trees proceeded relatively rapidly with no net immobilization of N from outside the branch material.

Decomposition rates can also be estimated from the rate of litterfall and the biomass of the forest floor, assuming the forest floor is at steady state and the measured rate of litterfall represents a long-term average. This approach with several combinations of conifers and alders illustrates that in all cases the rate of decomposition in terms of $kg\,ha^{-1}$ of material respired is always greater in mixed stands (because litterfall inputs are always greater), but that the mean residence time does not differ consistently between single-species stands and mixed stands (Table 3). For example, the greater rates of litterfall in the alder/Douglas fir stands at the Mt. Benson site were matched by proportional increases in the biomass of the forest floor. In fact, the residence time for N in the forest floor appeared greater in the mixed stands, despite the greater annual rate of N release in these stands. At the Wind River site, the mixed stand showed only a marginally lower residence time for biomass, but a 50% reduction in residence time for N. At Cascade Head, the residence time for both biomass and N were much lower in the mixed stand. These few studies all demonstrated different patterns, and underscore the need for further work.

## *Nitrogen acquisition by the non-N fixer*

A variety of mechanisms has been suggested for the transfer of N from the N-fixing trees to associated plants. Red'ko (1958, cited in Remezov & Pogrebnyak 1965) claimed that in many cases the root hairs of poplar (probably *Populus deltoides* Bartr. ex Marsh) trees would invade the N-fixing nodules of black alder and absorb N compounds. Zavitkovski and Newton (1968) claimed that 60% of the N accretion observed under red alder derived from either secretion of N from the nodules and roots of alder, or from N fixation by free-living microbes. Tarrant and Trappe (1971) thought that decomposition of alder nodules could be a great source of N; however, they reported a nodule biomass of only $244\,kg\,ha^{-1}$ and a 3% N concentration would amount to only $7\,kg\,ha^{-1}$ (compared to above-ground litterfall N of over $100\,kg\,ha^{-1}$ in the same stand). Direct sharing of a common mycorrhizal network could also be a mechanism for transfer of N (M. Schoeneberger, pers. comm.; Malajczuk & Grove 1980).

In the absence of good evidence for the quantitative importance of any of these proposed mechanisms, the simplest explanation for the transfer of N to non-N-fixers is of course the decomposition of tissues that senesce from the N fixer.

The length of time required for newly established N fixers to benefit associated species has not been well characterized. Under tropical environmental conditions

TABLE 3. Decomposition of litter in stands with assumed steady-state forest floor biomass

| Location | Stand | Litterfall Biomass (Mg ha$^{-1}$) | Litterfall N (kg ha$^{-1}$) | Forest floor Biomass (Mg ha$^{-1}$) | Forest floor N (kg ha$^{-1}$) | Residence time Biomass (years) | N | Reference |
|---|---|---|---|---|---|---|---|---|
| Mt Benson, BC | Douglas fir | 1.4 | 16 | 7.0 | 36 | 4.9 | 2.3 | Binkley 1983 |
| | Douglas fir/Sitka alder | 5.2 | 112 | 22.1 | 282 | 4.3 | 2.5 | Binkley et al. 1984 |
| | Douglas fir/Red alder | 6.4 | 136 | 33.6 | 502 | 5.3 | 3.7 | |
| Wind River, Washington | Douglas fir | 4.0 | 10 | 15.6 | 104 | 3.9 | 10.4 | Binkley et al. 1991 |
| | Douglas fir/red alder | 9.7 | 74 | 29.3 | 350 | 3.0 | 4.7 | |
| Cascade Head, Oregon | Conifer | 6.4 | 29 | 25.6 | 234 | 4.0 | 8.1 | Binkley et al. 1991 |
| | Conifer/red alder | 21.0 | 140 | 20.5 | 329 | 1.0 | 2.4 | |

in Hawaii, eucalyptus trees interplanted with N-fixing albizia exhibited improved N status in less than 4 years. Under more temperate conditions, the elegant use of $^{15}$N techniques by Cote and Camire (1987) showed that the cumulative effects of 3 years of N fixation by alder apparently stimulated the mineralization of soil N in mixed stands of black alder and hybrid poplar. The biomass accumulation of N derived from soil pools (rather than from N fixation) was 101 kg ha$^{-1}$ over 3 years in the poplar stand, compared with 140 kg ha$^{-1}$ for the 33% alder/67% poplar stand and 126 kg ha$^{-1}$ for the 67% alder/33% poplar stand. Cole and Newton (1986) found no evidence of increased levels of total or available N or needle concentrations of N in 5-year-old Nelder-design plots of red alder mixed with Douglas fir versus pure Douglas fir. In mixtures of red alder and Douglas fir that are older than 20 years, N enrichment of the soil and conifer needles are apparent (Tarrant 1961; Binkley 1983; Binkley, Lousier & Cromack 1984). The dynamics between ages 5 and 20 remain unexamined for alder/conifer mixtures. The decomposition patterns in Table 3 indicate that a lag time of 2–5 years may be required for the processing of N-rich alder litter into pools of available N.

## Soil leaching losses

Only one study has characterized leaching losses from stands of non-N-fixers and from mixed stands. Binkley *et al.* (1991) examined nitrogen leaching from two pairs of 55-year-old Douglas fir and red alder/Douglas fir stands. At the relatively poor site at Wind River, leaching of inorganic-N past 80 cm depth was about 1 kg ha$^{-1}$ year$^{-1}$, compared to about 4 kg ha$^{-1}$ of organic N. The adjacent mixed stand had about 25 kg ha$^{-1}$ of annual N leaching, again dominated primarily by organic N (Fig. 1). At the more fertile Cascade Head site, N leaching exceeded 20 kg ha$^{-1}$ year$^{-1}$ from the conifer stand, with roughly equal components of ammonium, nitrate, and organic-N. The loss from the alder/conifer stand was about 50 kg ha$^{-1}$, with nitrate comprising the major fraction. Greater rates of leaching of inorganic-N have been found in all pure stands of red alder that have been examined to date (Bigger & Cole 1983; Miller & Newton 1983; Van Miegroet & Cole 1984; Van Miegroet, Cole & Homann 1990). Indeed, harvesting of red alder stands tends to lower nitrate leaching rates to near zero within 2–3 years (Bigger & Cole 1983; Van Miegroet & Cole 1988). Greater leaching of nitrate from stands containing N-fixing trees is important for four reasons:

1   loss of N slows the rate of N accretion from N fixation;

2   production of nitrate generates H$^+$; leaching of nitrate from the ecosystem allows the H$^+$ generation to remain unbalanced by consumption during reduction of nitrate;

3   increased concentrations of nitrate in soil solutions increase ionic strength, tending to lower solution pH and increase concentrations of potentially toxic aluminium (Reuss 1989); and

4   increased leaching losses of nutrient cations, and export of aluminium to aquatic ecosystems.

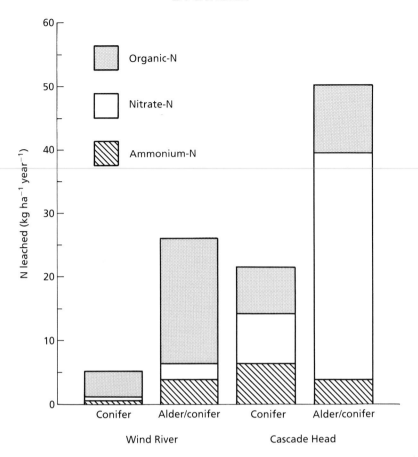

FIG. 1. Nitrogen leaching from stands at Cascade Head, estimated with tension-free lysimeters (from data of Binkley *et al.* 1991).

These processes have been examined only in pure stands of red alder or mixed stands of red alder with Douglas fir on only a few types of sites. The degree to which these ecosystems might represent the responses of other combinations of species on other sites is probably limited, and needs to be investigated.

### Soil acidification

Mixed stands with N-fixing trees commonly have lower soil pH than non-N-fixing stands (Franklin *et al.* 1968; DeBell, Whitesell & Schubert 1989; Van Miegroet *et al.* 1989; Binkley & Sollins 1990). Five mechanisms could be involved in any observed difference in the pH of soils and of soil solutions (Binkley *et al.* 1989a,b). The simplest deals with the ionic strength (total concentration of ions; Richter *et al.* 1988). For example, the pH of 0–15 cm depth soil samples from the Douglas fir stand at Wind River was 5.4 in deionized water, compared with 5.1

for soils from the adjacent red alder/Douglas fir stand (Binkley & Sollins 1990). The pH measured in 0.01 M $CaCl_2$ was 4.3 for both stands. The quantity of water soluble anions in the Douglas fir soils was about $175\,\mu mol_c\,l^{-1}$, compared with $330\,\mu mol_c\,l^{-1}$ for the mixed stand; the difference was due to the greater concentration of nitrate in soils from the mixed stand. The extra $155\,\mu mol_c\,l^{-1}$ of anions in the soils from the mixed stand was of course balanced by equivalent concentrations of cations, and about $3.6\,\mu mol_c\,l^{-1}$ of the greater cation charge came from $H^+$. Although $H^+$ balanced relatively little of the anion charge in the alder/conifer soil, it was enough to lower pH by 0.3 units. The pH of both soils dropped substantially with the addition of 0.01 M $CaCl_2$, which again was due to increased quantity of anion charge allowing a greater concentration of $H^+$ to be maintained in the solution. This effect of changes in ionic strength is especially important in mixed stands with N-fixing trees for two reasons. The higher concentrations of nitrate can lead to lower pH of soil solutions, with consequences for leaching of nutrient cations and other processes. In addition, trivalent cations such as $Al^{3+}$ respond more dramatically to changes in ionic strength than do divalent or monovalent cations (Reuss & Johnson 1986; Reuss 1989). Van Miegroet and Cole (1988) reported that seasonal variations in nitrate concentrations in lysimeters in a red alder stand were closely matched by variations in solution aluminium, with aluminium balancing roughly 1/3 of the nitrate charge.

The second mechanism controlling soil pH is the degree of dissociation of soil acids. Both soil organic matter and clay particles (collectively referred to as the cation exchange complex) act as stabilized weak acids, accepting or releasing $H^+$ in response to pH. An exchange complex saturated with $H^+$ (and acidic $Al^{3+}$) represents a largely undissociated acid, and maintains a low pH of the soil solution. Saturation of the complex with so-called base cations indicates dissociated acid conditions, and therefore a high pH is maintained in the soil solution. Soil scientists typically index the degree of saturation of soil acid as base saturation, which is the proportion of the exchange complex occupied by so-called base cations ( = degree of dissociation of the acid). The inclusion of N-fixing trees in mixed stands could lower soil pH if the accumulation of base cations in biomass were greater than in non-N-fixing stands, or if nitrate leaching led to removal of base cations from the soil. For example, Binkley and Sollins (1990) estimated that the lower base saturation of 0–15 cm depth soil in the red alder/conifer stand (9%) at Cascade Head relative to the conifer stand (29%) accounted for perhaps 1/3 of the lower pH under the mixed stand.

The third mechanism is the quantity of acid stabilized in the soil (cation exchange capacity). The inclusion of N-fixing trees in an ecosystem often stimulates production and may lead to an increase in soil organic matter; this increase can lower soil pH. The best illustration of this mechanism comes from a comparison of adjacent pure stands of red alder and Douglas fir (Van Miegroet & Cole 1988). They found that the content of exchangeable base cations was similar between the 50-year-old stands at the Thompson Research Forest in Washington, but that exchangeable acidity (a measure of the undissociated weak acids stored in the soil) was 40% greater under alder. The base saturation was therefore lower under alder because of the increase in the size of the exchange complex, not because of any leaching losses of cations that might have been driven by nitrate leaching.

The fourth mechanism is the strength of the acids present in the soil. Some acids are stronger than others; stronger acids (characterized by low $pK_a$s) maintain lower solution pH than weaker acids with the same quantity of acid at the same degree of dissociation. At both the Wind River and Cascade Head sites, the greater strength of organic acids accumulated under the mixed stands of red alder and Douglas fir appeared to be the most important factor in determining differences in soil pH (Binkley & Sollins 1990).

The final mechanism involves the redox potential of a soil. In a thermodynamic system, the sum of pe + pH remains constant (pe is a scale of redox potential; Bartlett 1986). Therefore, greater soil aeration in one stand could lead to higher pe, which would simultaneously lower pH proportionately with no change in the acid/base status of the soil. If pe were lowered at a later point, pH would recover to the earlier level. This mechanism might be important on poorly drained sites if transpiration rates differed between single-species and mixed stands.


## PRODUCTIVITY

A variety of studies have characterized above-ground net primary productivity (ANPP) and biomass accumulation for single-species stands and for mixed stands (Table 4). The results fall into two distinct categories. On sites that appear to be N deficient, ANPP and biomass accumulation are much greater in mixed stands, although any benefits to the non-N-fixing trees may take decades to develop. On more fertile sites, ANPP and biomass accumulation are greater in single-species stands of non-N-fixers. For the N-poor site at Wind River, ANPP at age 28 in the Douglas fir stand was 15% greater in the pure stand of Douglas fir. The greater ANPP in the mixed stand resulted solely from the growth of the alder. By age 55, however, the accumulated stemwood of Douglas fir was about 20% greater in the mixed stand (and it was distributed on larger individual stems, Miller & Murray 1978). The nitrogen enrichment of the soil at Wind River was a long-term process, and the benefits to conifer growth took more than 30 years to become apparent due to competitive interactions between the species. A somewhat analogous pattern is evident in the fast-growing plantations of eucalyptus and albizia in Hawaii (Table 4, DeBell, Whitesell & Schubert 1989). Low proportions of albizia in the stands appeared to compete with the eucalyptus without the offsetting benefits of substantially improved nitrogen nutrition of the eucalyptus. Higher proportions of albizia apparently improved the N nutrition of eucalyptus to the point that competitive interactions were more than offset.

Other studies that have documented parameters of growth other than ANPP and biomass include: Berntsen (1961), Newton, El Hassan and Zavitkovski (1968),

*Mixtures with N$_2$-fixing species*

TABLE 4. Productivity (Mg ha$^{-1}$ year$^{-1}$) and accumulated stem biomass (Mg ha$^{-1}$) of non-N-fixing stands and mixed stands. * 2 year coppice. †Calculated from volume data, assuming wood density of 0.58 for oak, and 0.37 for other species. ‡By age 6, productivity in some mixtures exceeded that of either species in pure stands (D. DeBell pers. comm.)

| Location | Stand | Species | Age | ANPP | Stem biomass | Reference |
|---|---|---|---|---|---|---|
| Mt Benson, BC | Douglas fir, with Sitka alder | Douglas fir | 23 | 6.9 | 35.0 | Binkley 1983 |
| | | Douglas fir | | 9.0 | 69.7 | |
| | | Sitka alder | | 4.9 | 23.8 | |
| | with red alder | Douglas fir | | 6.4 | 41.2 | |
| | | Red alder | | 9.3 | 53.5 | |
| Skykomish, Washington | Douglas fir, with red alder | Douglas fir | 23 | 23.2 | 218.2 | Binkley 1983 |
| | | Douglas fir | | 15.5 | 140.9 | |
| | | Red alder | | 7.0 | 38.3 | |
| Wind River, Washington | Douglas fir, with red alder | Douglas fir | 28 | 6.4 | 44.4 | TARRANT 1961, Binkley 1982 |
| | | Douglas fir | | 5.4 | 38.6 | |
| | | Red alder | | 10.5 | 50.7 | |
| | Douglas fir, with red alder | Douglas fir | 55 | 3.7 | 108.0 | Binkley et al. 1991 |
| | | Douglas fir | | 6.4 | 132.0 | |
| | | Red alder | | 3.8 | 69.7 | |
| Cascade Head, Oregon | Conifers, with red alder | Conifers | 55 | 19.2 | 418 | Binkley et al. 1991 |
| | | Conifers | | 6.1 | 135 | |
| | | Red alder | | 4.2 | 120 | |
| Laval, Quebec | Poplar, 1/3 black alder | Poplar | 3 | | 4.2 | Cote & Camire 1987 |
| | | Poplar | | | 8.6 | |
| | | Black alder | | | 3.7 | |
| | 2/3 black alder | Poplar | | | 4.9 | |
| | | Black alder | | | 8.5 | |

| Location | Treatment | n | Species | Value | Reference |
|---|---|---|---|---|---|
| Camas, Washington | Black cottonwood, with red alder | 2* | Cottonwood | 11.4 | DeBell & Radwan 1979 |
| | | | both species | 19.5 | |
| Oregon, Washington, British Columbia | Hybrid poplar with red alder | 4 | Poplar | 66.8 | Heilman & Stettler 1985 |
| | | | Poplar+alder | 63.6 | |
| Gisburn Forest, England | Scots pine† | 26 | Scots pine | 60.4 | Harrison & Brown pers. comm. |
| | Scots pine/alder | | Scots pine | 41.4 | |
| | | | Black alder | 6.7 | |
| | Norway spruce | | Norway spruce | 35.3 | |
| | Norway spruce/alder | | Norway spruce | 30.6 | |
| | | | Black alder | 2.6 | |
| | Oak | | Oak | 15.7 | |
| | Oak/alder | | Oak | 10.4 | |
| | | | Black alder | 3.3 | |
| Hilo, Hawaii‡ | Eucalyptus saligna, with N fertilization | 4 | Eucalyptus | 93.7 | DeBell et al. 1989 |
| | with 11% Albizia | | Eucalyptus | 58.1 | |
| | | | Albizia | 8.9 | |
| | with 25% Albizia | | Eucalyptus | 48.1 | |
| | | | Albizia | 28.9 | |
| | with 34% Albizia | | Eucalyptus | 57.8 | |
| | | | Albizia | 44.8 | |
| | with 50% Albizia | | Eucalyptus | 65.8 | |
| | | | Albizia | 39.0 | |
| | with 75% Albizia | | Eucalyptus | 53.4 | |
| | | | Albizia | 50.8 | |

Miller and Murray (1978), Schlesinger and Williams (1984), Malcolm, Hooker and Wheeler (1985), Campbell and Dawson (1989) and Heilman (1990a).

## COMPONENTS OF COMPETITION

On relatively infertile sites, competition for limiting resources between species in mixed stands determines the relative success of each tree species. On fertile sites, experience with red alder in the Pacific Northwest clearly indicates that mixed stands of alder and conifers tend to become pure alder stands due to conifer mortality (Newton, El Hassan & Zavitkovski 1968; Walstad, Brodie & McGinley 1986). These empirical observations may seem unexpected; an N-fixing tree would seem to be competitively most superior on N-poor sites, with any advantage over non-N-fixing trees diminishing as site N supply increased. In reality, the competition between N fixers and non-N fixers depends upon complex ecophysiological processes. The success of alders in mixed stands on N-rich sites probably derives from more rapid deployment of leaf area (reaching canopy closure in about 5 years, compared with decades for conifers; Zavitkovski & Newton 1971), greater leaf area per kilogram of leaf (about $20\,m^2\,kg^{-1}$ for red alder, versus $13\,m^2\,kg^{-1}$ for Douglas fir (Krueger & Ruth 1969)), and greater rates of photosynthesis per unit of leaf area (about twice that of Douglas fir (Krueger & Ruth 1969)). For these reasons, early growth of red alder on fertile sites easily surpasses that of conifers. On less fertile sites, however, alder growth is retarded, and the potential benefits of N-enrichment of the site provide an opportunity for competition pressures to be at least partially offset by improved N nutrition.

These patterns from alder/conifer stands are typically reversed in mixtures of alder and poplar on N-rich sites. Under fertile conditions, poplars can grow faster than alder, and the productivity of pure poplar stands exceeds that of mixed alder/poplar stands (Heilman & Stettler 1985). On N-poor sites, the alder/poplar patterns of production are analogous to the alder/conifer patterns, where the poplar benefit more from increased N supply than they suffer from increased competition (Heilman 1990a,b).

### Light

Competition for light between trees of each species in mixed stands depends upon the leaf area index (LAI) of each species, the duration of LAI (evergreen or deciduous), the light intercepted per unit of leaf area, and the relative heights of leaf area between the species. Surprisingly few measurements from mixed stands are available. Cole and Newton (1986) measured quantities of light penetrating canopies of 5-year-old plots of Douglas fir and red alder/Douglas fir. At 2.5 m height above ground, 25% of incoming light was intercepted by the Douglas fir canopy, compared with 70% for the mixed plots. Leaf morphology of the Douglas

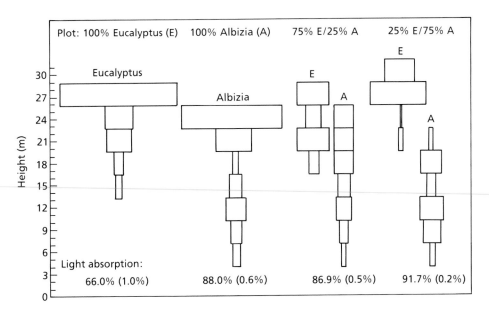

FIG. 2. Leaf area profiles (proportion of total leaf area in 3 m height intervals; determined with a zoom lens rangefinder), and total light interception (mean of 36 measurements with a Decagon Ceptometer, with standard error of the mean in parentheses (Binkley unpubl.)). Total leaf area data are not available.

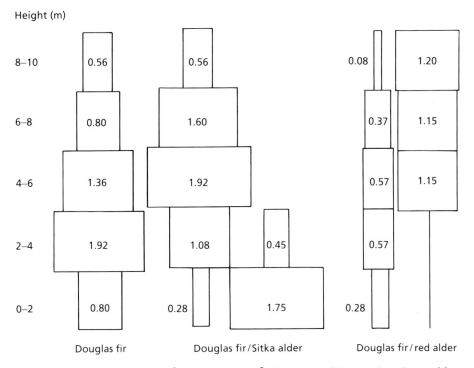

FIG. 3. Distribution of leaf area (m² of leaf area per m² of ground area) for the plantations on Mt. Benson (from data of Binkley 1982, 1984).

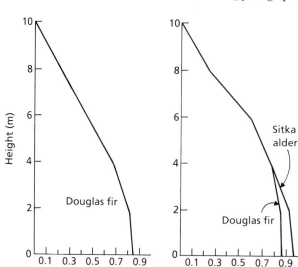
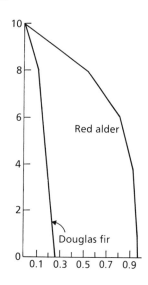

Height (m)

Proportion of light extinction

FIG. 4. Approximate profile of light extinction (using Beer/Lambert equation) for the stands in Fig. 3 based on a light extinction coefficient of 0.4 for Douglas fir and 0.6 for alders.

fir adapted to the shaded conditions in the mixed plots by producing leaves with greater surface area per kilogram ($13\,m^2\,kg^{-1}$ in the mixed plot, $12\,m^2\,kg^{-1}$ in the Douglas fir plots).

In the replicated plantations of eucalyptus and albizia in Hawaii (DeBell, Whitesell & Schubert 1989), the pure eucalyptus plots intercepted only 66% of incoming radiation, compared with 88% for the pure albizia plots (Fig. 2). The proportional display of leaf area with canopy height was similar between species. Light interception was marginally (and significantly) greater in the plots containing 25% eucalyptus/75% albizia, with the leaf area shifted upward for eucalyptus and downward for albizia.

Binkley (1984 and unpubl. data) examined leaf area profiles in the Mt. Benson plantations of Douglas fir with Sitka alder and red alder. Leaf area indexes for Douglas fir were $5.4\,m^2$ of leaf area per $m^2$ of ground area, in both the pure stand and in the mixture with Sitka alder (Fig. 3). Douglas fir LAI with red alder was only 1.9. The LAI of the alders were 2.2 for Sitka alder, and 3.5 for red alder. The cumulative light intercepted by these canopies (Fig. 4) revealed that the short canopy of the Sitka alder had little effect on light interception by Douglas fir. In fact, shading of Sitka alder by Douglas fir appeared to be driving the high rate of mortality of alder (Binkley, Lousier & Cromack 1984). In the red alder/Douglas fir stand, most of the light was captured by the canopy of red alder, due to high leaf area of alder in the upper crown. The persistence and relatively good growth of Douglas fir mixed with red alder probably depends on substantial photo-

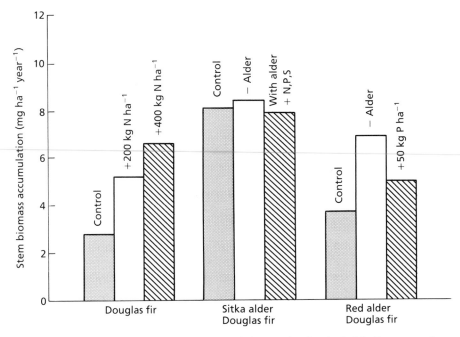

FIG. 5. Five-year stem growth response of Douglas fir in 0.125 ha plots in the Mt. Benson stands to: fertilization in the Douglas fir stand; removal of the Sitka alder or leaving the alder and adding 200 kg ha$^{-1}$ N, 50 kg ha$^{-1}$ P, and 50 kg ha$^{-1}$ S; removal of red alder, or leaving the alder and adding 50 kg ha$^{-1}$ P. The Douglas fir was strongly N-limited in the pure stand; no limitations or competition with alder was apparent in the mixed stand with Sitka alder; and strong competition and P limitation were apparent in the mixed stand with red alder (Binkley, unpubl.).

synthesis before alder budbreak and after alder leaf-fall (Waring & Franklin 1979). Thinning and fertilization studies in these stands documented N limitation on Douglas fir growth in the Douglas fir stand, no apparent limitations in the Sitka alder/Douglas fir stand, and strong limitation by P and by competition with red alder in the red alder/Douglas fir stand (Fig. 5).

## Moisture

Surprisingly little work has focused on water relations in mixed stands; none appears to be available for trees older than 5 years. Cole and Newton (1986) documented lower xylem pressure potentials in 5-year-old Douglas fir trees mixed with red alder (about −1.5 MPa) than in pure plots (about −1.3 MPa) in September on a relatively dry site, but no differences on more moist sites. Red alder trees generally had higher potentials (by about 0.4 MPa) than Douglas fir seedlings, indicating either greater competitive ability for water (which might include deeper rooting; Heilman 1990b) or lower transpiration rates.

*Allelopathy*

Allelopathic interactions between species are almost impossible to prove under field conditions (Harper 1977), though some possibilities of allelopathy in mixed stands have been suggested. Neave and Dawson (1989) demonstrated in solution cultures that $2 \times 10^{-6}$ M juglone (an allelochemical from black walnut) reduced nitrogen fixation by black alder seedlings. Sustained effects of juglone in soils were not apparent even at concentrations of $10^{-3}$ M. Younger and Kapustka (1983) found that nitrogenase activity and nodule biomass of speckled alder (*Alnus incana* spp. *rugosa* Clausen) was greater in a single stand of alder than in a single stand of alder mixed with aspen. In addition to differences in site and plant vigour, they speculated that allelochemic inhibition of the alder by the aspen may be occurring (based on inhibition of alder seed germination by extracts from leaves of balsam poplar; Jobidon & Thibault 1981). Heilman and Stettler (1985) also found that ground poplar leaves and litter reduced growth and N fixation by red alder, but that water extracts of these materials did not. These authors concluded that the lack of any reduction in growth and N fixation by alders when planted in soils from poplar stands clearly indicated that normal decomposition processes inactivated whatever allelopathic chemicals were present in fresh litter. Given the complex competitive interactions between species in mixed stands, inferences of allelopathic interactions under field conditions are very difficult to demonstrate and no evidence is currently strong enough to support a role for allelopathy in such stands.

## BIODIVERSITY

Mixed-species stands by definition have greater diversity of tree species than pure stands, but the effects of mixtures on other species is also substantial. Franklin and Pechanec (1968) examined the composition of the understorey in the Cascade Head stands. A 375 m² plot in the conifer stand contained nineteen understorey species, compared with twenty-three in the red alder/conifer stand (a pure alder stand contained thirty-six understorey species). At the Wind River site (Binkley *et al.* 1991), the understorey of the Douglas fir stand was dominated by a well-developed stand of the ericaceous shrub salal (*Gaultheria shallon* Pursh), which is noted for inducing water stress in associated trees (Price, Black & Kelliher 1986). The alder/Douglas fir stand had few salal shrubs, and the understorey was dominated instead by widely spaced swordfern (*Polystichum munitum* (Kaulf.) Presl.) Indeed, the leaf biomass of salal in the Douglas fir stand roughly matched that of the alder in the mixed stand.

Differences in animal populations between N-fixing and non-N-fixing stands have not been characterized, but are probably substantial. For example, many of the dominant Douglas fir trees in the mixed stand at Wind River were damaged (and many killed) by black bears (*Ursus americanus* Pallas) that removed bark to eat the cambium in early spring. No trees in the Douglas fir were damaged (R.

Miller, pers. comm.). Apparently the alder greatly improved the food quality of Douglas fir for black bears.

With growing interests in non-timber aspects of forestry, the role of the biodiversity implications of mixed stands will likely increase in importance.

## SILVICULTURAL IMPLICATIONS

The inclusion of N-fixing species into stands can result in greater productivity of associated species, or it can result in the mortality of the associated species. Harper (1977) noted that mixtures of species almost never exceed the productivity of pure stands of the more-productive member of the mixture, and available evidence supports extension of this generalization to mixtures with N-fixing trees. At Cascade Head, the pure alder forest was more productive than the mixed stand for the first 20 years (Binkley & Greene 1983), whereas the pure conifer stand was more productive than the mixed stand during later periods. In the Gisburn Forest, the biomass of black alder in mixed stands was too low to compensate for the reduction in biomass of Scots pine and Norway spruce relative to the pure-conifer plots. At several fertile sites across the Pacific Northwest, productivity in pure poplar stands exceeded productivity in poplar/alder stands (Heilman & Stettler 1985). In Hawaii, the productivity of pure stands of N-fixing albizia was greater than stands mixed with eucalyptus through age 4 years (DeBell, Whitesell & Schubert 1989), though the mixed stands began to out-perform the pure-albizia stand at age 6 years. Differences in the value of biomass between species may still result in economic superiority of mixed stands over single-species stands, but economic gains are likely to be realized only if competition between species is well controlled. The key feature of mixtures is the relative dominance of the species; dominance by the N fixer may result in suppressed growth or mortality of associated trees, whereas dominance by associated species may result in low vigour and low rates of N fixation by the N fixer. Empirical examples are available to illustrate the gamut of possible interactions. The most successful case studies involve either N fixers with low stature (such as Sitka alder or shrubs and herbaceous plants not discussed here; see Gadgil (1979) and Dyck et al. (1988) for excellent examples), early removal of the N-fixing trees to allow canopy closure of the crop trees, or systems that provide the non-N-fixing trees several years of advanced growth before establishment of the N-fixing trees. Silvicultural pre-scriptions that take advantage of the opportunities presented by N-fixing species need to account for the ecophysiological aspects of each species, and for changes over time in the availability of site resources. The development of a realistic computer simulation model would be a productive step.

## REFERENCES

Ae, N., Arihara, J., Okada, K., Yoshihara T. & Johansen. C. (1990). Phosphorus uptake by pigeon pea and its role in cropping systems of the Indian subcontinent. *Science*, **248**, 477–80.

**Bartlett, R.J. (1986).** Soil redox behavior. *Soil Physical Chemistry* (Ed. by D.L. Sparks), pp. 179–201. CRC Press, Boca Raton, Florida.

**Berg, A. & Doerksen, A. (1975).** *Natural fertilization of a heavily thinned Douglas-fir stand by understory red alder.* Oregon State University Forest Research Laboratory Research Note 56, Corvallis.

**Berntsen, C.M. (1961).** *Growth and development of red alder compared with conifers in 30-year-old stands.* USDA Forest Service Research Paper PNW-38, Portland, Oregon.

**Bigger, C.M. & Cole, D.W. (1983).** Effects of harvesting intensity on nutrient losses and future productivity in high and low productivity red alder and Douglas-fir stands. *IUFRO Symposium on Forest Site and Continuous Productivity* (Ed. by R. Ballard & S. Gessel), pp. 167–78. USDA Forest Service General Technical Report PNW-163, Portland.

**Binkley, D. (1981).** Nodule biomass and acetylene reduction rates of red alder and Sitka alder on Vancouver Island, BC *Canadian Journal of Forest Research*, **11**, 281–6.

**Binkley, D. (1982).** *Case studies of red alder and Sitka alder in Douglas-fir plantations.* Ph.D. thesis, Oregon State University.

**Binkley, D. (1983).** Interaction of site fertility and red alder on ecosystem production in Douglas-fir plantations. *Forest Ecology and Management*, **5**, 215–27.

**Binkley, D. (1984).** Douglas-fir stem growth per unit of leaf area increased by interplanted Sitka alder and red alder. *Forest Science*, **30**, 259–63.

**Binkley, D. (1986).** *Forest Nutrition Management.* Wiley, New York.

**Binkley, D. & Greene, S. (1983).** Production in mixtures of conifers and red alder: the importance of site fertility and stand age. *IUFRO Symposium on Forest Site and Continuous Productivity* (Ed. by R. Ballard & S.P. Gessel), pp. 112–17. USDA Forest Service General Technical Report PNW-163, Portland, OR.

**Binkley, D. & Sollins, P. (1990).** Soil acidification in adjacent conifer and alder/conifer stands. *Soil Science Society of America Journal*, **54**, 1427–33.

**Binkley, D., Driscoll, C., Allen, H.L., Schoeneberger, P. & McAvoy, D. (1989a).** *Acidic Deposition and Forest Soils.* Springer, New York.

**Binkley, D., Lousier, J.D. & Cromack, K. Jr. (1984).** Ecosystem effects of Sitka alder in a Douglas-fir plantation. *Forest Science*, **30**, 26–35.

**Binkley, D., Valentine, D., Wells, C. & Valentine, U. (1989b).** An empirical analysis of the factors contributing to 20-year decrease in soil pH in an old-field plantation of loblolly pine. *Biogeochemistry*, **7**, 39–54.

**Binkley, D., Sollins, P., Bell, R., Sachs, D. & Myrold, D. (1991).** Biogeochemistry of adjacent conifer and alder/conifer ecosystems. Submitted to *Ecology*.

**Boring, L & Swank, W. (1984a).** Symbiotic nitrogen fixation in regenerating black locust (*Robinia pseudoacacia* L.) stands. *Forest Science*, **30**, 528–37.

**Boring, L. & Swank, W. (1984b).** The role of black locust (*Robinia pseudoacacia* L.) in forest succession. *Journal of Ecology*, **72**, 749–66.

**Browne, W. (1613).** *Britannia's Pastorals*, Book 1, Song 2, Lines 357–8, London.

**Brozek, S. (1990).** Effect of soil changes caused by red alder (*Alnus rubra*) on biomass and nutrient status of Douglas-fir (*Pseudotsuga menziesii*) seedlings. *Canadian Journal of Forest Research*, **20**, 1320–5.

**Campbell, G.E. & Dawson, J.O. (1989).** Growth, yield, and value projections for black walnut interplantings with black alder or autumn Olive. *Northern Journal of Applied Forestry*, **7**, 289–99.

**Chapin, F.S.T. III, Van Cleve, K. & Vitousek, P. (1986).** The nature of nutrient limitation in plant communities. *American Naturalist*, **127**, 148–58.

**Cole, E.C. & Newton, M. (1986).** Nutrient, moisture, and light relations in 5-year-old Douglas-fir plantations under variable competition. *Canadian Journal of Forest Research*, **16**, 727–32.

**Cote, B. & Camire, C. (1984).** Growth, nitrogen accumulation, and symbiotic dinitrogen fixation in pure and mixed plantings of hybrid poplar and black alder. *Plant and Soil*, **78**, 209–20.

**Cote, B. & Camire, C. (1987).** Tree growth and nutrient cycling in dense planting of hybrid poplar and black alder. *Canadian Journal of Forest Research*, **17**, 516–23.

**DeBell, D.S. & Radwan, M.A. (1979).** Growth and nitrogen relations of coppiced black cottonwood and red alder in pure and mixed plantings. *Botanical Gazette*, **140**(Suppl.), S97–S101.

**DeBell, D.S., Whitesell, C.D. & Schubert, T.S. (1989).** Using $N_2$-fixing Albizia to increase growth of eucalyptus plantations in Hawaii. *Forest Science*, **35**, 64–75.

**Domingo, I.L. (1983).** Nitrogen fixation in Southeast Asian forestry: research and practice. *Biological Nitrogen Fixation in Forest Ecosystems: Foundations and Applications* (Ed. by J.C. Gordon & C.T. Wheeler), pp. 295–315. Martinus-Nijhoff/Dr W. Junk, The Hague.

**Dunkin, K. (1989).** *Nitrogen-fixing Albizia increases nitrogen cycling in Eucalyptus plantations.* M.S. thesis, Colorado State University.

**Dyck, W., Beets, P., Will, G. & Messina, M. (1988).** Nitrogen buildup in a sand-dune pine ecosystem. *Forest Site Evaluation and Long-term Productivity* (Ed. by D. Cole & S. Gessel), pp. 107–12. University of Washington Press, Seattle.

**Edmonds, R.L. (1987).** Decomposition rates and nutrient dynamics in small-diameter woody litter in four forest ecosystems in Washington, U.S.A. *Canadian Journal of Forest Research*, **17**, 499–509.

**Franklin, J. & Pechanec, A.A. (1968).** Comparison of vegetation in adjacent alder, conifer, and mixed alder–conifer communities. I. Understory vegetation and stand structure. *Biology of Alder* (Ed. by J. Trappe *et al.*), pp. 37–44. USDA Forest Service Portland, OR.

**Franklin, J., Dyrness, C., Moore, D. & Tarrant, R. (1968).** Chemical properties under coastal Oregon stands of alder and conifers. *Biology of Alder* (Ed. by J. Trappe *et al.*), pp. 157–72. USDA Forest Service, Portland, OR.

**Gadgil, R. (1979).** The nutritional role of *Lupinus arboreus* in sand dune forestry. *New Zealand Journal of Forestry Science*, **9**, 324–36.

**Gillespie, A.R. & Pope, P.E. (1990a).** Rhizosphere acidification increases phosphorus recovery of black locust: I. Induced acidification and soil response. *Soil Science Society of America Journal*, **54**, 533–7.

**Gillespie, A.R. & Pope, P.E. (1990b).** Rhizosphere acidification increases phosphorus recovery of black locust: II. Model predictions and measured recovery. *Soil Science Society of America Journal*, **54**, 538–41.

**Gutschick, V. (1981).** Evolved strategies in nitrogen acquisition by plants. *American Naturalist*, **118**, 607–37.

**Harper, J. (1977).** *Population Biology of Plants.* Academic Press, New York.

**Heilman, P. (1990a).** Growth and N status of *Populus* in mixture with red alder on recent volcanic mudlfow from Mount Saint Helens. *Canadian Journal of Forest Research*, **20**, 84–90.

**Heilman, P. (1990b).** Growth of Douglas-fir and red alder on coal spoils in western Washington. *Soil Science Society of America Journal*, **54**, 522–7.

**Heilman, P. & Stettler, R.F. (1985).** Mixed, short-rotation culture of red alder and black cottonwood: growth, coppicing, nitrogen fixation, and allelopathy. *Forest Science*, **31**, 607–16.

**Ho, I. (1979).** Acid phosphatase activity in forest soil. *Forest Science*, **25**, 567–8.

**Ingestad, T. (1980).** Growth, nutrition, and nitrogen fixation in grey alder at varied rate of nitrogen addition. *Physiologia Plantarum*, **50**, 353–64.

**Jayachandran, K., Schwab, A.P. & Hetrick, B.A.D. (1989).** Mycorrhizal mediation of phoshorus availability: synthetic iron chelate effects on phosphorus solubilization. *Soil Science Society of America Journal*, **53**, 1701–6.

**Jobidon, R. & Thibault, J.R. (1981).** Allelopathic effects of balsam poplar on green alder germination. *Bulletin of the Torrey Botanical Club*, **108**, 413–18.

**Johnson, D.W. & Lindberg, S.E. (1991).** *Atmospheric Deposition and Nutrient Cycling in Forest Ecosystems.* Springer, New York.

**Kittredge, J. (1948).** *Forest Influences.* McGraw-Hill, New York.

**Krueger, K.W. & Ruth, R.H. (1969).** Comparative photosynthesis of red alder, Douglas-fir, Sitka spruce, and western hemlock seedlings. *Canadian Journal of Botany*, **47**, 519–27.

**Luken, J.O. & Fonda, R.W. (1983).** Nitrogen accumulation in a chronosequence of red alder communities along the Hoh River, Olympic National Park, Washington. *Canadian Journal of Forest Research*, **13**, 1228–37.

**Mackay, J., Simon, L. & LaLonde, M. (1987).** Effect of substrate nitrogen on the performance of *in vitro* propagated *Alnus glutinosa* clones inoculated with $Sp^+$ and $Sp^-$ *Frankia* strains. *Plant and Soil*, **103**, 21–31.

**Malajczuk, N. & Grove, T.S. (1980).** Vesicular arbuscular mycorrhizae—a possible link in the transfer of nutrients between legumes and eucalypts. *Managing Nitrogen Economies of Natural and Man Made Forest Ecosystems* (Ed. by R.A. Rummery & F.J. Hingston), pp. 195–201. CSIRO, Canberrra.

Malcolm, D.C., Hooker, J.E. & Wheeler, C.T. (1985). *Frankia* symbiosis as a source of nitrogen in forestry: a case study of symbiotic nitrogen-fixation in a mixed *Alnus–Picea* plantation in Scotland. *Proceedings of the Royal Society of Edinburgh*, **85B**, 263–82.

Mikola, P., Uomala, P. & Mälkönen, E. (1983). Application of biological nitrogen fixation in European silviculture. *Biological Nitrogen Fixation in Forest Ecosystems: Foundations and Applications* (Ed. by J.C. Gordon & C.T. Wheeler), pp. 279–94. Martinus-Nijhoff/Dr W. Junk, The Hague.

Miller, J. & Newton, M. (1983). Nutrient loss from disturbed forest watersheds in Oregon's Coast Range. *Agro-ecosystems*, **29**, 127–37.

Miller, R.E. & Murray, M.D. (1978). The effects of red alder on growth of Douglas-fir. *Utilization and Management of Alder* (compiled by D.G. Briggs, D.S. DeBell & W.A. Atkinson), pp. 283–306. USDA Forest Service General Technical Report PNW-70, Portland, OR.

Neave, I.A. & Dawson, J.O. (1989). Juglone reduces growth, nitrogenase activity, and root respiration of actinorhizal black alder seedlings. *Journal of Chemical Ecology*, **15**, 1823–36.

Newton, M., El Hassan, B.A. & Zavitkovski, J. (1968). Role of red alder in western Oregon forest succession. *Biology of Alder* (Ed. by J. Trappe *et al.*), pp. 73–84. USDA Forest Service, Portland, OR.

Price, D.T., Black, T.A. & Kelliher, F.M. (1986). Effects of salal understory removal on photosynthetic rate and stomatal conductance of young Douglas-fir trees. *Canadian Journal of Forest Research*, **16**, 90–7.

Red'ko, G.L. (1958). *Influence of common alder on the productivity of Carolina poplar*. Doklady AN Ukranian Soviet Socialist Republic, Vol. 3, (in Ukranian).

Reid, R.K., Reid, C.P.P. & Szaniszlo, P.J. (1985). Effects of synthetic and microbially produced chelates on the diffusion of iron and phosphorus to a simulated root in soil. *Biology and Fertility of Soils*, **1**, 45–52.

Remezov, N.P. & Pogrebnyak, P.S. (1965). *Forest Soil Science*. English translation, Israel Program for Scientific Translations, Jerusalem.

Reuss, J. (1989). Soil-solution equilibria in lysimeter leachates under red alder. *Effects of Air Pollution on Western Forests* (Ed. by R.K. Olson & A.S. LeFohn), pp. 547–59. Air and Waste Management Association, Pittsburgh.

Reuss, J. & Johnson, D. (1986). *Acid Deposition and Acidification of Soils and Waters*. Springer, New York.

Richter, D., Comer, P., King, K., Sawin, D. & Wright, D. (1988). Effects of low ionic strength solution on pH of acid forested soils. *Soil Science Society of America Journal*, **52**, 261–4.

Schlesinger, R.C. & Williams, R.D. (1984). Growth response of black walnut to interplanted trees. *Forest Ecology and Management*, **9**, 235–43.

Schubert, K. (1982). *The Energetics of Biological Nitrogen Fixation*. American Society of Plant Physiologists, Rockville, Maryland.

Silvester, W. (1983). Analysis of nitrogen fixation. *Biological Nitrogen Fixation in Forest Ecosystems: Foundations and Applications* (Ed. by J.C. Gordon & C.T. Wheeler), pp. 173–212. Martinus-Nijhoff/Dr W. Junk, The Hague.

Tarrant, R. (1961). Stand development and soil fertility in a Douglas-fir/red alder plantation. *Forest Science*, **7**, 238–47.

Tarrant, R.F. & Miller, R.D. (1963). Soil nitrogen accumulation beneath a red alder/Douglas fir plantation. *Soil Science Society of America Proceedings*, **27**, 231–4.

Tarrant, R. & Trappe, J. (1971). The role of *Alnus* in improving the forest environment. *Plant and Soil*, **Special Volume 1971**, 335–48.

Tarrant, R., Lu, K.C., Bollen, W.B. & Franklin, J. (1969). *Nitrogen enrichment of two forest ecosystems by red alder*. USDA Forest Service Research Paper PNW-76, Portland, OR.

Taylor, B.R., Parsons, W. & Parkinson, D. (1989). Decomposition of *Populus tremuloides* leaf litter accelerated by addition of *Alnus crispa* litter. *Canadian Journal of Forest Research*, **19**, 674–9.

Turvey, N. & Smethurst, P. (1983). Nitrogen fixing plants in forest plantation management. *Biological*

*Nitrogen Fixation in Forest Ecosystems*: *Foundations and Applications* (Ed. by J.C. Gordon & C.T. Wheeler), pp. 233–60. Martinus-Nijhoff/Dr. W. Junk, The Hague.

**Van Miegroet, H. & Cole, D. (1984).** The impact of nitrification on soil acidification and cation leaching in a red alder forest. *Journal of Environmental Quality*, **13**, 585–90.

**Van Miegroet, H. & Cole, D. (1988).** Influence of nitrogen-fixing alder on acidification and cation leaching in a forested soil. *Forest Site Evaluation and Long-term Productivity* (Ed. by D. Cole & S. Gessel), pp. 113–24. University of Washington Press, Seattle.

**Van Miegroet, H., Cole, D., Binkley, D. & Sollins, P. (1989).** The effect of nitrogen accumulation and nitrification on soil chemical properties in alder forests. *Effects of Air Pollution on Western Forests* (Ed. by R.K. Olson & A.S. LeFohn), pp. 515–28. Air and Waste Management Association, Pittsburgh.

**Van Miegroet, H., Cole, D. & Homann, P. (1990).** *The effect of alder forest cover and alder forest conversion on site fertility and productivity.* Proceedings of the 7th North American Forest Soils Conference. Vancouver, BC, Canada, 25–28 July 1988.

**Walstad, J.D., Brodie, J.D. & McGinley, B.C. (1986).** Silvicultural value of chemical brush control in the management of Douglas-fir. *Western Journal of Applied Forestry*, **1**, 69–73.

**Waring, R. & Franklin, J. (1979).** Evergreen coniferous forests of the Pacific Northwest. *Science*, **204**, 1380–6.

**Younger, P.D. & Kapustka, L.A. (1983).** $N_2(C_2H_2)$ase activity by *Alnus incana* ssp. *rugosa* (Betulaceae) in the northern hardwood forest. *American Journal of Botany*, **70**, 30–9.

**Zavitkovski, J. & Newton, M. (1968).** Effect of organic matter and combined nitrogen on nodulation and nitrogen fixation by red alder. *Biology of Alder* (Ed. by J. Trappe *et al.*), pp. 209–24. USDA Forest Service, Portland, OR.

**Zavitkovski, J. & Newton, M. (1971).** Litterfall and litter accumulation in red alder stands in western Oregon. *Plant and Soil*, **35**, 257–68.

# Functioning of mixed-species stands at Gisburn, N.W. England

A.H.F. BROWN

*Institute of Terrestrial Ecology, Merlewood Research Station, Grange-over-Sands,
Cumbria LA11 6JU, UK*

## SUMMARY

This paper describes a study of replicated stands of all six two-species mixtures of four tree species (Norway spruce, Scots pine, alder and oak), together with their monocultures.

The mensurational data indicate the existence of three types of mixture effect: mutual 'cooperation', compensation and mutual inhibition. Investigations of the mechanisms involved have concentrated on the three spruce mixtures, which encompass these three types of interaction. Foliar analysis results indicate that N nutrition is implicated and the root bioassay technique suggests that P uptake may also be important. These two elements tend to become 'locked up' in organic matter and, at Gisburn, their release in the field from this source is shown to relate well to the tree growth mixture effects, both positive and negative; this is paralleled by comparable interactions in decomposer activity, with the role of enchytraeids apparently being especially relevant.

The evidence suggests that underlying mechanisms include microbial activity interactions, brought about when spruce and pine litters are mixed, and the possible involvement of inhibitory substances leached from the foliage of spruce and oak. The complementary rooting patterns found in spruce and pine, and possibly present in other combinations, provide yet another mechanism for differential resource capture.

The marked positive interaction occurring within the spruce/pine mixed stands also appears to be reflected in the soil organic matter *quality* of that mixture.

## BACKGROUND

### Introduction

On poor sites, the beneficial 'nursing' of a 'main-crop' species such as spruce by admixed pines or larch, relative to pure stands of the main-crop species, has been demonstrated in a number of studies (Weatherell 1957; Zehetmayr 1960; O'Carroll 1978; Brown & Harrison 1983; McIntosh 1983; Gabriel 1986). However, in assessing the performance of a mixture *as a whole*, comparison with both components in monoculture is necessary. In most of these studies, such a comparison was precluded by the lack of a monoculture of the nurse species.

The investigation described contains all six possible two-species combinations of four species, together with their monocultures. The mensurational data for these replicated pure and mixed stands indicate the existence of three types of mixture effect comparable to those defined (e.g. Willey 1979) in agronomy, viz. mutual cooperation, mutual inhibition and compensation. In this paper, these three types of interaction are described, and the mechanisms which appear to be involved are explored.

## The experiment

The experiment, situated in the Gisburn block of the Forestry Commission's Bowland Forest (N.G.R. (34) 750 580), was established jointly by the former Nature Conservancy and the Forestry Commission. The four pure stands—Scots pine (*Pinus sylvestris* L.), Norway spruce (*Picea abies* (L.) Karst), alder (*Alnus glutinosa* (L.) Gaertn.) and oak (*Quercus petraea* (Mattuschka) Liebl.)—and all six possible combinations of two-species mixtures were planted in 1955 as half-acre (0.2 ha) plots, together with an unplanted control. The resulting 11 treatments were replicated as three blocks (Fig. 1) on poorly drained rough grazings. To improve drainage and to provide an inverted ribbon of turf in which to plant, the area was shallow (20–25 cm) ploughed at 5 ft (*c.* 1.5 m) spacing; trees were planted at the same spacing within the rows. No fertilizers were applied either at planting or subsequently. Because of an increasingly serious wind-throw problem, the experiment was felled in 1988/89.

Advantages of the mixtures trial at Gisburn are that not only was the full range of pure and mixed stands of the species under study included, but because no phosphatic fertilizers were applied, it has enabled *inter alia* the effect of mixtures on the availability of this nutrient to be studied.

## Site

The site, some 35 km from the coast of north-west England, slopes gently (*c.* 3°) to the south-west and is very exposed to prevailing winds. Elevation ranges from 260 to 290 m and mean annual rainfall for the period 1984–89 was 1606 mm. The bedrock is Bowland grits and slates of the Carboniferous, overlain with clayey glacial till derived from the underlying strata. At this elevation and rainfall, the site is predisposed to peat formation, although its development can be prevented by good husbandry (Pearsall 1950; Hall & Folland 1970). Consistent with this trend, the soils were transitional between cambic stagnogleys and stagnohumic gleys (Avery 1980). The soil of block I, with its higher organic content, differs from that of the other blocks. Recent information on soils is provided by Moffat and Boswell (1990).

Prior to afforestation the vegetation was a species-poor *Festuca–Agrostis* rough pasture with *Nardus stricta*, *Deschampsia cespitosa* and *Juncus* spp. The vegeta-

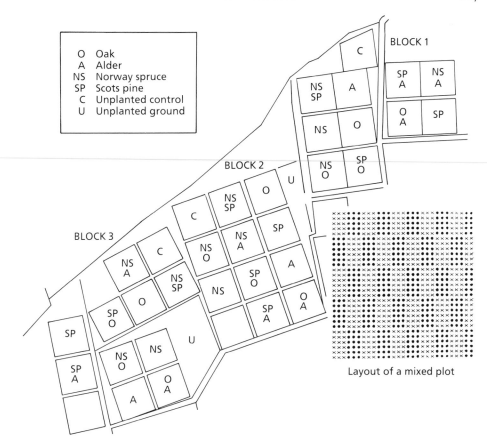

FIG. 1. Layout of experiment, and planting pattern in the mixtures. (Blank plots represent a discontinued sheep-grazed comparison.)

tion of block I contained appreciably more *Holcus lanatus* and *Rumex acetosa* but less *Nardus* reflecting its different soil.

## Methods

It is not feasible to give here all the sampling details for the relatively wide range of variables discussed in this paper. Two general points are made, however.

1  The mixed stands consist of a chequerboard pattern of groups (3 × 6 trees) of each species (Fig. 1). Such group mixtures were used to reduce the risk of a more vigorous species suppressing a less vigorous one, in which it has been entirely successful. It also enables that portion of the stand beneath the canopy of each component species to be separately sampled and compared with the appropriate pure stand; or it can be combined with the other component to provide whole

stand information. Separate sampling of each mixture component has been done for all parameters discussed in this paper.

2   The restriction of the experimental design to only three replicate blocks is statistically far from ideal. As noted above, one of these (block I) has soil which is different from the other two; more seriously, it also exhibits marked within-block heterogeneity. Some studies have, therefore, been confined to the two other relatively uniform blocks. Yet other investigations, for reasons of practicality and logistics, have been done in a single block only.

Hence, although it may often be possible to show statistically significant differences between *stands*, based on subsample replication and variability, these differences cannot necessarily be rigorously ascribed to *treatment* effects on the basis of only one-, two- or at best three-fold replication. Nevertheless, in blocks II and III there is no evidence of any marked variability in soils, topography or otherwise, and many of the results presented in this paper conform with comparable findings elsewhere. At the very least, therefore, they provide pointers to mechanisms which could be tested more rigorously in other ways if desired.

TABLE 1. Top heights (m) of four tree species, pure and mixed, at given ages, at Gisburn, all three blocks combined. Heights in *italics* are for the pure stands. Data for a given species (i.e. within horizontal rows) with different suffix letters are significantly different, statistically. N.S.D. = no significant differences. Note: some of these *stand* differences, although statistically significant as indicated, cannot strictly be ascribed to *treatment* effects because of the small number of replicate blocks, as noted in the text. Except for year 4, height data kindly made available by the Forestry Commission Research Branch. *4 year figures are mean height, not top heights. †Ignoring the soil heterogeneity-induced effects in Block I

| Species | Age | NS | When admixed with: Oak | Alder | SP | |
|---|---|---|---|---|---|---|
| Norway spruce | 4* | *0.48* | 0.42 | 0.45 | 0.46 | N.S.D. |
| (NS) | 7 | *1.20*$^a$ | 1.04$^b$ | 1.13$^{abc}$ | 1.18$^{ac}$ | |
| | 10 | *1.86*$^a$ | 1.52$^b$ | 1.73$^{ac}$ | 2.19$^d$ | |
| | 20 | *5.87*$^a$ | 5.29$^b$ | 6.22$^{ac}$ | 7.11$^d$ | |
| | 26 | *8.80*$^a$ | 8.76$^{ab}$ | 9.84$^c$ | 10.62$^d$ | |
| Sessile oak | 4* | 0.33 | *0.39* | 0.31 | 0.37 | N.S.D. |
| | 7 | 1.17$^a$ | *1.48*$^b$ | 1.18$^a$ | 1.53$^b$ | |
| | 10 | 1.73$^a$ | *2.05*$^b$ | 1.88$^{ab}$ | 2.53$^c$ | |
| | 20 | 3.63$^a$ | *4.21*$^b$ | 4.47$^b$ | 6.19$^c$ | |
| | 26 | 5.67$^a$ | *6.58*$^b$ | 7.29$^c$ | 8.82$^d$ | |
| Common alder | 4* | 0.63 | 0.63 | *0.63* | 0.62 | N.S.D. |
| | 7 | 1.56 | 1.56 | *1.67* | 1.68 | |
| | 10 | 2.29$^a$ | 2.35$^{ab}$ | *2.46*$^{bc}$ | 2.58$^c$ | |
| | 20 | 4.90$^a$ | 4.99$^{ab}$ | *5.77*$^c$ | 6.54$^d$ | |
| | 26 | 7.57$^a$ | 7.72$^{ab}$ | *8.24*$^c$ | 9.31$^d$ | |
| Scots pine | 4* | 0.94 | 0.92 | 0.84 | *0.85* | N.S.D.$^†$ |
| (SP) | 7 | 2.16 | 2.20 | 2.04 | *1.97* | |
| | 10 | 3.49 | 3.59 | 3.38 | *3.26* | |
| | 20 | 8.30 | 8.05 | 7.94 | *8.03* | |
| | 26 | 11.56 | 11.34 | 11.12 | *11.12* | |

## MENSURATIONAL DATA AND MIXTURE EFFECTS

Top height, based on a sample of the trees with largest diameter at breast height (DBH), has been measured at Gisburn at the intervals indicated in Table 1 by the Forestry Commission Research Branch. Additionally, in 1982, they also made basal area (BA) assessments. The effects of wind-throw preclude the use of BA and height measurements made shortly before clear-felling in 1987, and even in 1982 a few corrections for wind-throw were considered necessary for the BAs, though heights measured in 1981 were considered acceptable.

Basal areas may be combined with form heights, i.e. top heights corrected for taper (or 'form') of the trees, to calculate volumes. Form heights were derived from Forestry Commission tables (Hamilton 1975), using the sycamore/ash/birch table for alder (T.J.D. Rollinson, pers. comm.). In some of the poorer-grown broad-leaf stands, this involved extrapolation of the top height/form height relationship. Whilst the absolute values estimated in this way may not be valid, they provide a first approximation for present comparisons, in which possible errors would not alter the conclusions reached. The volumes based on the product of the 1982 BAs (corrected for wind-throw where necessary) and 1981 form heights are presented in Table 2.

### Heights

The effect of one tree species on the height growth of another was, in general, consistent between the three replicate blocks. Accordingly, results from the three blocks have been combined for statistical analysis and for presentation in Table 1.

Significant height differences ($P < 0.01$) became apparent from age 7 years, even before canopy closure, and have in general persisted although, as noted above, they cannot always be ascribed to treatment effects. In summary:

1 Pine provided much the earliest, and the largest, 'nursing' effect on all three other species; and did so with no detriment to its own height growth, which, in general, was unaffected by mixing. A positive interaction in pine mixtures is therefore suggested. The effect was especially marked in oak, where tree form was also improved; a comparable finding was reported by Gabriel (1986) for hardwood/pine mixtures on the North Yorkshire Moors. The earlier concern (Lines & Nimmo 1966) that oak might be suppressed by the pine at Gisburn has proved unfounded.

2 Admixed alder has improved recent heights of oak and spruce, though to a smaller extent than pine, and only at the expense of its own growth, i.e. in a compensatory fashion. The alder nursing effect was also very slow to manifest itself, only becoming significant at age 26 years.

3 Spruce and oak not only reduce the height of alder in mixture, but until very recently, at least, have reduced each other's heights as well. The oak/spruce mixture therefore results in an apparently negative interaction.

TABLE 2. Volumes (m³ ha⁻¹) of mixed stands relative to their components in monoculture. *The two components in a mixture are each the volume of half a stand, their sum giving the whole mixture volume. †For equal area comparability, the component monoculture comparisons are therefore each given as half the pure stand volume, their sum thus providing the mean monoculture yield (MMY). ‡For each mixture, the more productive of the two component monocultures is also given (as the whole stand volume) for easier comparison. §For each component, its volume in mixture relative to its volume pure (for an equal half plot area) is given in the final column. The mean of these two relative volumes gives the relative yield total (RYT) for each mixture—see text

| Mixed stands* | | Component monocultures† | | More productive monoculture‡ | | Relative yields§ | | |
|---|---|---|---|---|---|---|---|---|
| SP (NS) | 120 | SP pure | 79 | | | | 1.51 | |
| NS (SP) | 67 | NS pure | 49 | | | | 1.37 | |
| Mixture | 187 | MMY | 128 | 159 | RYT | 1.44 | | |
| SP (Al) | 109 | SP pure | 79 | | | | 1.38 | Pine |
| Al (SP) | 18 | Al pure | 12 | | | | 1.52 | mixtures |
| Mixture | 127 | MMY | 91 | 159 | RYT | 1.45 | | RYT > 1 |
| SP (Oak) | 118 | SP pure | 79 | | | | 1.49 | |
| Oak (SP) | 22 | Oak pure | 13 | | | | 1.65 | |
| Mixture | 140 | MMY | 93 | 159 | RYT | 1.57 | | |
| Al (NS) | 7 | Al pure | 12 | | | | 0.59 | |
| NS (Al) | 85 | NS pure | 49 | | | | 1.75 | Other two |
| Mixture | 92 | MMY | 61 | 98 | RYT | 1.17 | | alder |
| Al (Oak) | 9 | Al pure | 12 | | | | 0.77 | mixtures |
| Oak (Al) | 18 | Oak pure | 13 | | | | 1.36 | RYT ≈ 1 |
| Mixture | 27 | MMY | 25 | 27 | RYT | 1.06 | | |
| NS (Oak) | 63 | NS pure | 49 | | | | 1.29 | Spruce/oak |
| Oak (NS) | 4 | Oak pure | 13 | | | | 0.32 | RYT < 1 |
| Mixture | 68 | MMY | 62 | 98 | RYT | 0.81 | | |

## Volumes

In comparing the performance of the mixture as a whole with that of the pure stands, the most practicable measure of yield is an estimate of volume. The yield of the mixed stand can be compared with the mean yield of the component species (on an equal area basis), termed the mean monoculture yield (MMY) by Trenbath (1974). However, the MMY may commonly be exceeded by the performance of the mixture through competition between the components, particularly where the productivities of the component species are very disparate, so that the more efficient species in resource utilization increases its yield to a greater extent than that foregone by the less efficient species. That is, there is a redeployment rather than an increase in the use of site resources. Where biomass productivity is the sole measure, as in some agricultural crops, no benefit occurs unless the mixture outperforms the pure stand of the more productive component. In tree mixtures, on the other hand, other considerations of value, often non-quantifiable benefits, may outweigh higher productivity obtained from a pure stand.

In addition to the competitive reallocation of resources envisaged in the de Wit (1960) model there is the possibility of a mixture stimulating enhanced growth of one or both components through accessing more resources, or their more efficient utilization. A further possibility is an overall inhibition of growth in mixtures.

A convenient measure of these possible outcomes is based on the yield of each species in mixture as a proportion of its yield in monoculture. The mean of these two relative yields has been termed the *relative yield total* (RYT) by de Wit and van den Bergh (1965).

$$RYT = \frac{1}{2}\left(\frac{\text{yield sp. A in mixture with sp. B}}{\text{yield sp. A pure}} + \frac{\text{yield sp. B in mixture with sp. A}}{\text{yield sp. B pure}}\right)$$

In cases where only competition for resources occurs, the RYT is theoretically unity, even though the mixture yield may exceed the MMY. Where the use of extra resources appears to apply, e.g. where *both* components are more productive in mixture than when pure, or where the mixture as a whole outyields the more productive component in monoculture, then RYT > 1. With overall growth reduction, and RYT < 1, some inhibition of the availability of resources may be occurring. This index, or modified versions, have commonly been applied to mixtures of arable and grassland crops (e.g. Francis 1989), and although its use under certain circumstances has been criticized by Connolly (1987), such criticism does not apply in the present case of fixed density, 1:1 mixtures.

Calculating RYTs from the volume data it is evident from Table 2 (which also provides the MMY calculations) that at Gisburn each of these mixture effects occurs:

1  In all three pine mixtures, the volume of both components is enhanced relative to their pure stands, thus creating an unambiguous overall mixture benefit, reflected in RYTs all >1. In only one case, however, (pine/spruce) does the mixture volume exceed the better of the two component monoculture volumes (i.e. that of pine).

2  In the case of the two other alder mixtures, RYTs are little different from unity, even though in the case of the alder/spruce mixture, at least, its volume is appreciably greater than its MMY. Much, if not all, of the effect can be ascribed to a compensatory redeployment of resources.

3  In the case of spruce/oak, although the mixture volume is very similar to the MMY, the RYT of appreciably <1 (0.8) provides further evidence (as suggested by the heights) that growth is probably inhibited in this combination.

These relative volume yields therefore confirm in a more definitive manner the tentative interpretations based on heights. Research on the mechanisms of these different mixture effects at Gisburn has concentrated on spruce mixtures; both because of the practical importance of this species and because its mixtures encompass all three types of effect (i.e. RYTs < 1, c.1 and > 1). These different spruce mixture effects are presented separately in Fig. 2 to provide comparability with the next section.

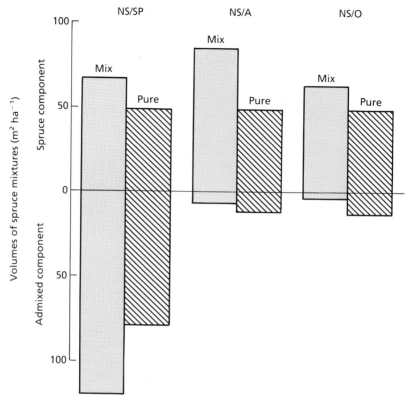

FIG. 2. Volumes of spruce mixtures (m² ha⁻¹). For each mixture, the bars above the zero line represent the spruce: on the left when in mixture, on the right the monoculture comparison on an equal (half plot) basis. Bars below the line represent the comparable data for the admixed species. The combined left-hand bar thus gives the volume of the mixture as a whole, the right hand bar, overall, provides the MMY (mean monoculture yield). A, alder; O, oak; NS, Norway spruce; and SP, Scots pine.

TABLE 3. Relative yield totals* for different variables estimated in spruce mixtures at Gisburn. Separate enchytraeid data for the two parts of the mixtures were not presented by Chapman (1986) who gave results for the mixtures as a whole. This precludes calculation of RYTs for this variable, although that for spruce/pine must be >1. *See text for explanation

|  | Tree volumes | Forest floor flux | | Soil NO₃ | Earthworms | Cotton cloth decomposition |
|---|---|---|---|---|---|---|
|  |  | Inorg-N | P |  |  |  |
| Spruce/pine | 1.44 | 1.22 | 1.50 | 3.79 | 3.70 | 1.18 |
| Spruce/alder | 1.17 | 0.77 | 0.59 | 1.59 | 1.52 | 1.02 |
| Spruce/oak | 0.81 | 0.59 | 0.18 | 0.71 | <1 | — |

# STUDIES OF POSSIBLE MECHANISMS FOR MIXTURE EFFECTS IN SPRUCE AT GISBURN

In these studies, the mean value of a variable in the two monocultures (the 'expected' value of Chapman (1986) and Chapman, Whittaker & Heal (1988)) has been used as the primary basis of comparison with that variable estimated in a mixed stand. In most cases, however, an index equivalent to the RYT calculation is also presented in Table 3.

## Foliar analysis

To test for a nutritional involvement in the differential growth of spruce at Gisburn, spruce foliage was sampled in November 1982 using standard techniques based on Everard (1973) and chemically analysed for N, P and K. Although tree heights and foliar K were uncorrelated, there was a significant correlation ($P <$ 0.001) with needle N concentrations (Fig. 3). The lowest, near deficient (Binns, Mayhead & MacKenzie 1980) N levels were associated with the low spruce top heights found in pure stands and those in mixture with oak; intermediate levels of both variables occurred in spruce when mixed with alder; and highest, near optimal (Binns, Mayhead & MacKenzie 1980) foliar N concentrations were found in the mixture with pine—the best grown spruce trees. The relationship with P was not so close, though still significant ($P < 0.05$). Both the calculated shape of this latter curve, and comparison of the foliar P concentrations with those given by Binns, Mayhead and MacKenzie (1980) indicate that P levels appear to be mostly near optimal. This does not necessarily mean that availability of P is much less important than that of N, but that foliar analysis used to diagnose deficiencies in older trees may be less reliable for P than for N (McIntosh 1983). This is supported by the application of the root bioassay test for P to these trees (Brown & Harrison 1983). The test measures the uptake of $^{32}$P by roots from a solution *in vitro* to estimate the extent of deficiency (Harrison & Helliwell 1979). Results for samples of spruce roots from pure and mixed stands at Gisburn are shown in Fig. 4, which suggest that, within each of the two blocks studied, pure spruce and spruce with oak are appreciably deficient in P relative to spruce with pine, with the alder mixture again being intermediate. There is also a clear relationship between root assays of P deficiency and tree height.

Uptake of both N and P by spruce is apparently enhanced more in the pine mixture than that with alder. Although improved N nutrition has been demonstrated in studies of mixture effects elsewhere, e.g. Sitka spruce when grown with lodgepole pine or larch (e.g. Carlyle & Malcolm 1986; Carey, McCarthy & Miller 1988), in most such cases the routine application of long-lasting phosphatic fertilizers has precluded any determination of the possible role of P nutrition as a further factor.

Increased uptake of N and P by spruce in mixture clearly points to an increase in the availability or 'capture' of these two nutrients in the soil or forest floor. What evidence is there for this?

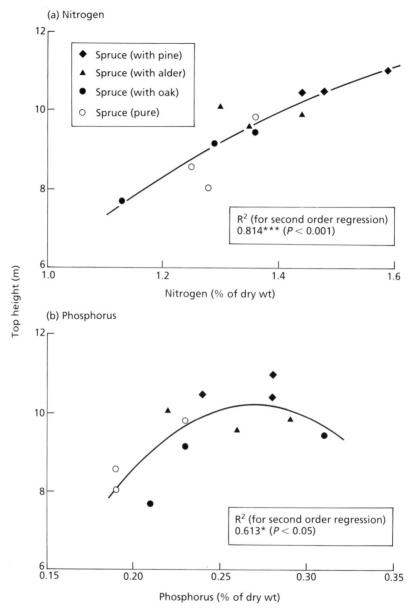

FIG. 3. Regressions of spruce top height on foliar concentrations of (a) nitrogen and (b) phosphorus.

## Fluxes from forest-floor lysimeters

The source of additional N involved in spruce mixture effects has been widely assumed to be the organic matter; that both N and P may be implicated as suggested by the Gisburn data, strongly supports this view. The forest-floor

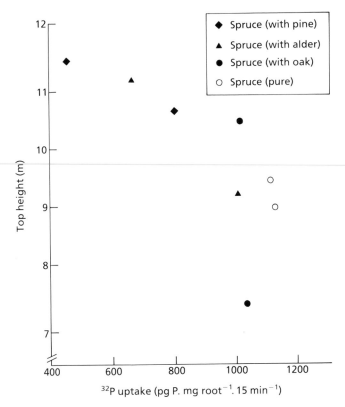

FIG. 4. Relationship between top height of spruce, pure and mixed, and uptake of $^{32}$P by excised roots at Gisburn, 1981.

represents a major store of organic matter within the forest ecosystem, and release of N and P in a soluble form from this source may be estimated by lysimeter studies. Using this approach, Chapman (1986) sampled forest-floor leachates at fortnightly intervals, from April 1983 to April 1984, in a single replicate (block II) at Gisburn. His results (Fig. 5) for total inorganic-N and for PO$_4$–P show:

1  enhanced release of N and P in *both* components of the spruce/pine mixture, i.e. good evidence of a positive interaction;

2  reduced mineralization in the alder part of the spruce/alder stands; the spruce appears to 'switch off' much of the activity associated with N and P release within the alder; and

3  reduced mobilization—mostly very considerable—in both portions of spruce/oak, i.e. a negative interaction.

   If the N and P are both derived from the same source, i.e. by mineralization of organic materials, then the quantities of the two elements should parallel each other within the different stands. In fact, a significant correlation ($r = 0.778$;

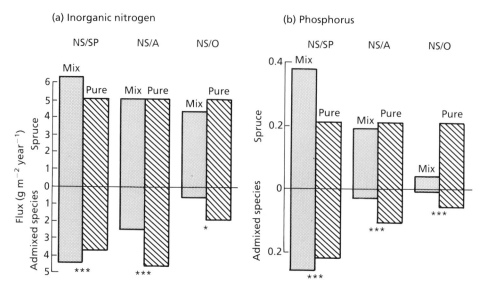

FIG. 5. Flux of (a) inorganic nitrogen and of (b) phosphorus from forest-floor lysimeters in pure and mixed spruce (g m$^{-2}$ year$^{-1}$). Data from Chapman (1986). For each mixture, the bars above the zero line are the spruce data: on the left when in mixture, on the right the monoculture comparison. Bars below the line are the comparable data for the admixed species. The combined left-hand bar thus represents the average ($\times 2$) for the mixture as a whole; the right-hand bar, overall, provides the mean monoculture figure ($\times 2$). The significance of the difference between the two is shown: *$P < 0.05$, ***$P < 0.001$. A, alder; O, oak; NS, Norway spruce; and SP, Scots pine.

$P < 0.01$) between inorganic-N and PO$_4$–P, calculated for Chapman's forest-floor lysimeter data, strongly supports this supposition. The presence of this correlation does raise the question of whether tree growth is directly influenced by both N and P, or whether one of these nutrients is only indirectly related via this interrelationship.

### Soil chemistry

To check the extent to which these interactions in the fluxes of mineralized N and P were reflected in soil chemistry, samples of the uppermost 5 cm of mineral soil from pure and mixed spruce stands in all three blocks were analysed for N and P. Results for both total N, and P extractable by 2.5% acetic acid were inconsistent between replicate blocks; and there was no interpretable relationship with the spruce growth mixture effects. Conversely, KCl extractable NH$_4$–N and NO$_3$–N in a single 'spot check' in September 1986 did reflect the performance of the spruce mixtures, especially in the case of NO$_3$ including the clear compensation effect in spruce/alder (Fig. 6).

In an earlier study (Brown & Harrison 1983) which included various measures of the N status of the soils in the spruce mixtures, extractable NO$_3$ provided the

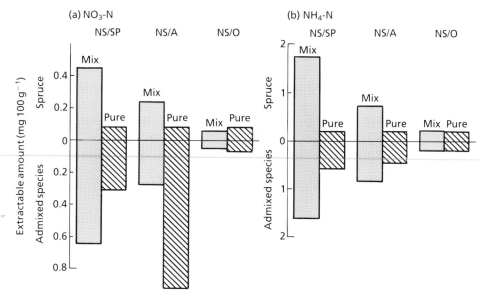

FIG. 6.   KCl extractable $NO_3$–N and $NH_4$–N from 0 to 5 cm mineral soil in pure and mixed stands at Gisburn, September 1986 (mg 100 g$^{-1}$). A, alder; O, oak; NS, Norway spruce; and SP, Scots pine.

best relationship with other parameters; and in the present data the correlation between spruce heights and extractable $NO_3$–N in 0–5 cm mineral soils is highly significant ($r = 0.838$; $P < 0.01$) whereas that with $NH_4$–N is non-significant. At Gisburn, the evidence suggests that spruce growth varies more in relation to the level of mineralized $NO_3$ than to $NH_4$; and that in turn $NO_3$ levels are enhanced by admixed alder and pine, but not by oak.

### Microbial and faunal decomposer activity

Evidence that enhanced turnover of the organic matter in the forest floor and/or surface soil, in the mixed stands, is a concomitant of the increased levels of N and P, and of improved spruce growth, is provided by studies of decomposer activity and of the organisms known to be involved.

### Cotton strip decomposition

A field index of the potential for microbial cellulolytic decomposer activity in the soils of the mixed and pure spruce was obtained using the buried cotton cloth method (Latter & Howson 1977). Breakdown of a standard cellulose-based substrate (cotton cloth) was estimated by measuring the loss in the cloth's tensile strength following 9 weeks' burial in the soils of both components of the spruce mixtures, and in the pure stands of those species.

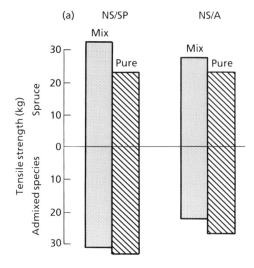

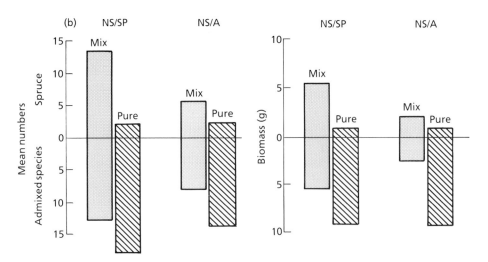

FIG. 7. Indices of microbial and faunal decomposer activity in some pure and mixed stands at Gisburn. (a) Tensile strength loss of cotton cloth (after burial for 9 weeks, summer 1980). (b) Earthworms (two blocks and two sampling periods in 1981 combined: mean numbers, and preserved fresh weight, full gut per trap). A, alder; NS, Norway spruce; and SP, Scots pine.

In this study oak and spruce/oak were not included (Brown 1988). Results are shown in Fig. 7a. They parallel those for spruce growth: the Scots pine mixture shows, overall, an increased loss in tensile strength; and in the alder mixture, although the spruce component indicates some increase in rotting—though less than with pine—this is balanced by less rotting in the alder portions, i.e. again confirming an apparently compensatory mechanism for the spruce/alder mixture.

*Earthworms*

It is generally accepted that microbial decomposition of plant materials is helped by the comminution and mixing brought about by soil animals; and that earthworms can play an important role in this respect.

Earthworm populations were studied in 1981 (Brown & Harrison 1983) again excluding the spruce/oak mixture. The results (Fig. 7b) show marked effects within the spruce portions of the stands comparable with the patterns provided by other variables; in the mixtures as a whole earthworm numbers show a slightly better correspondence with other mixture effects than earthworm weights. Subsequent investigations of earthworms (numbers only) by Chapman (1986) confirmed the earlier findings for spruce/pine and spruce/alder; and also indicated that the small numbers in pure spruce were reduced even further in spruce/oak where no earthworms were found in the samples, i.e. supporting the apparently negative interaction suggested by the tree performance and nutrient release data for this mixture.

*Enchytraeids and other soil fauna*

Other groups of soil animals have also been investigated in the mixed and pure spruce plots by sampling in April, July, October and January (1983/84) (Chapman 1986). Averaging the results from the four sampling occasions showed that Collembola densities were significantly higher in the spruce/pine mixture than would be expected from the mean monoculture populations. Numbers of enchytraeid worms (Fig. 8) showed a similar positive interaction in spruce/pine, no significant overall change in spruce/alder but a significant negative interaction in spruce/oak. In contrast to the earthworm and cotton strip results, the positive interaction of enchytraeid numbers in the spruce/pine mixture considerably exceeded both the monoculture mean and the larger population in pure pine, thus paralleling tree volumes. Indeed, the populations of enchytraeids reflect the tree growth responses in mixture more closely than any other index or measure of decomposer activity. How important a component are they of the soil fauna populations?

Petersen and Luxton (1982) concluded that lumbricid earthworms and enchytraeids are the two most important groups in terms of metabolically respired energy—and hence in aiding decomposition—and that enchytraeids tended to play the major role, except on sites particularly favourable to lumbricids. Although metabolic activity is proportional to biomass the only data for Gisburn comparing both groups in the same season (1983) and on the same sampling occasions, are for numbers only (Chapman 1986). The earthworm biomass and numbers data for 1981 (Fig. 7b) enable a mean weight per individual worm to be calculated for the different tree stands; and using a factor of 0.1 (Petersen & Luxton 1982) for conversion of fresh weight (full gut) to dry weight (empty gut), dry weights can be estimated for the 1983 earthworm numbers (means of four

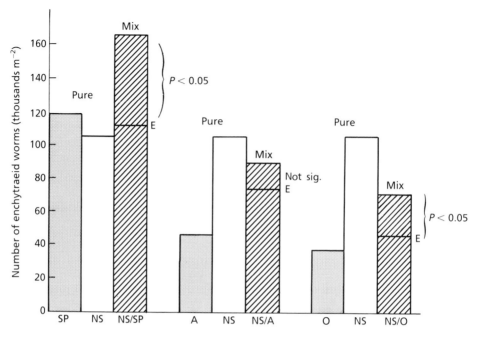

FIG. 8. Numbers of enchytraeid worms in pure and mixed spruce at Gisburn. Means of four sampling occasions, 1983/84 (thousands $m^{-2}$). The significance of differences between observed mixture numbers and 'expected' (E) mean monoculture numbers is indicated. A, alder; O, oak; NS, Norway spruce; and SP, Scots pine.

sampling occasions). Enchytraeid numbers can similarly be converted to approximate dry weight, assuming a general mean dry weight per individual of 15.4 μg derived from Edwards (1967), O'Connor (1967) and Petersen and Luxton (1982), for mainly temperate evergreen conifers. The comparisons (Table 4) confirm that, for these stands, enchytraeids are much more important than earthworms, except in the pure alder (where enchytraeid weight may be underestimated through the use of a general conifer-based conversion factor).

The positive interaction in enchytraeid numbers found in spruce/pine is consistent with that noted by Williams (1988) for *in vitro* mixtures of Sitka spruce and Scots pine litter, in which proliferation of enchytraeid worms showed a greater positive mixture effect than microbial respiration, N mineralization or nitrification. Williams and Griffiths (1989) have also shown from similar leaching column experiments the importance of enchytraeids in enhancing N and P mineralization from coniferous litters.

Enchytraeids may therefore play a key role in mixture effect mechanisms on sites such as Gisburn where they are the dominant faunal component, are known to be important in N and P release and their population densities interact in mixed stands in a way which best reflects tree growth interactions.

TABLE 4. Comparison of dry weights (mg m$^{-2}$) of lumbricid earthworms and enchytraeid worms in pure and mixed spruce stands at Gisburn (calculated as described in text). >, significantly greater than ( $P < 0.05$ ); <, significantly less than ( $P < 0.05$ ); ≈, not significantly different from; O = estimates based on observations; and E = mid-monoculture 'expected'

|  | Enchytraeids | | Lumbricids | |
|---|---|---|---|---|
|  | O | E | O | E |
| Spruce | 1610 | | 25 | |
| Oak | 560 | | 0 | |
| Alder | 710 | | 1040 | |
| Pine | 1820 | | 490 | |
| Spruce/oak | 710 < 1085 | | 0 | 12.5 |
| Spruce/alder | 1370 ≈ 1160 | | 230 ≈ 533 | |
| Spruce/pine | 2550 > 1715 | | 400 > 258 | |

## RYT calculations for the above variables

Although the relative yield total (RYT), described earlier, is normally used to assess relative responses in crop yield to resource availability, it can be argued that nutrient release or decomposer activity are also responses to the availability of other resources. Comparable calculations are therefore given in Table 3.

It is evident that all values for the spruce/pine mixture are >1, and those derived from spruce/oak are all <1, thus fully confirming the positive and negative interactions in those mixtures, respectively. The indices calculated for spruce/alder vary but average out at about 1, possibly indicating that the compensatory alder mixture effect is the net result of some positive and some negative interactions.

## Possible mechanisms influencing availability of N and P

Although the proximate cause of differential nutrient release from organic matter can be assumed to be differences in microbial activity, aided by variations in soil fauna, other underlying mechanisms may bring these differences about; or may lead to other differences in resource capture. Three possible mechanisms are (i) the mixing of litters, (ii) the role of inhibitory substances, or (iii) complementarity of rooting patterns.

### Decomposition and nutrient release in mixed litters

Laboratory studies of the effects of mixing forest-floor materials enable other environmental variables such as temperature, moisture and abundance of roots and ground vegetation to be controlled; they also facilitate exploring the role of the stage of organic matter breakdown at which mixture interactions might occur.

Chapman (1986) used laboratory microcosms (Anderson & Ineson 1982) in which decomposing forest-floor materials were leached at intervals to follow nutrient release; and from which $CO_2$ was monitored as a measure of concomitant

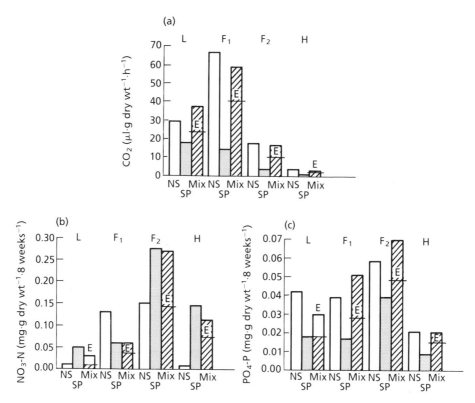

FIG. 9. (a) Respiration and mineralization of (b) $NO_3$–N and (c) $PO_4$–P, in microcosm, from different forest-floor materials in pure and mixed spruce/pine. L, litter; $F_1$, upper F-layer; $F_2$, lower F-layer; and H, humus. E indicates 'expected' mean monoculture levels. NS, Norway spruce; and SP, Scots pine.

respiration, which was assumed to be largely from microbial decomposer activity. The microcosms were used to compare pure spruce, pure pine and a 50:50 mixture of the two for four separate forest-floor horizons (i.e. stages of decomposition), namely, the fresh litter (L), the upper F-layer ($F_1$), the lower F-layer ($F_2$) and the humus (H). Results are shown in Fig. 9.

The release of $NO_3$–N and $PO_4$–P per unit of respired $CO_2$ was initially very low, rising with age of material, a result consistent with the view that nitrogen initially is retained because of the high C/N ratio. Significant mixture interactions occurred, in which the observed values differed from the mean of the two separate components. These effects were all positive, except for N and P release from L-layer material, despite a positive interaction with respect to $CO_2$ evolution. However, the negative effect of mixed L materials was considerably outweighed by the positive interactions of the other horizons, with the $F_2$ layer in particular dominating in the forest-floor interaction as a whole.

*The possible role of inhibitory substances in mixture effects*

If a tree produces substances which, in terms of some process such as organic matter turnover and nutrient release, are either self-inhibitory or inhibitory to other admixed species, this might provide a mechanism to further explain some mixture effects. In the case of self-inhibition, mixture with other trees would lead to dilution of the inhibitory material and hence to possible diminution of any adverse effect, i.e. a second possible mechanism for beneficial mixture effects. Inhibition of processes associated with other species in mixture would provide a possible explanation of negative mixture effects.

It has been commonly reported that aqueous extracts of tree foliage can inhibit microbial processes *in vitro* (e.g. Benoit & Starkey 1968; Beck, Dommergues & van den Driessche 1969). Evidence for the presence of apparently inhibitory substances in canopy leachates for the Gisburn species is available from studies of throughfall. Using microcosms again, Chapman (1986) studied the effects of additions of freshly collected throughfall waters on the mobilization of $NO_3-N$, $NH_4-N$ and $PO_4-P$, and on $CO_2$ evolution from rehydrated air-dried litter. Throughfall collected beneath spruce, pine and alder was compared (i.e. excluding oak) with controls of rainwater and distilled water. Only spruce litter was studied.

In comparing chemical analyses of microcosm leachates, allowance was made for the inputs of N and P in the throughfall additions. Although the effects of pine and alder throughfall were not significantly different from those of rainwater or distilled water on the mobilization of $NO_3-N$, $NH_4-N$ or $PO_4-P$, the addition of spruce throughfall significantly reduced levels of all three nutrients by 97%, 40% and 55%, respectively, relative to the distilled water control. $CO_2$ evolution was also reduced by 42%. Although the effects of these throughfalls on other species of litter have not been tested, it seems likely that if spruce throughfall is markedly inhibitory to the breakdown and mineralization of its own litter, it could well have at least a comparable inhibitory effect on the turnover of the litter of other species. Earlier studies by A.H.F.Brown and Desirée Masson (unpubl.) indicate a general inhibition of microbial activity by Norway spruce leachates; and also indicate that oak leachates (excluded in Chapman's study) can have a similar inhibitory effect. Aqueous foliar extracts of all four species, prepared in the laboratory, were compared in their effect on the development of a liquid culture of mixed soil organisms present in remaining farmland soil at Gisburn (i.e. roughly equivalent to the experimental stands prior to planting). Procedures followed those of Beck, Dommergues & van den Driessche (1969) and the results (Fig. 10) indicate no inhibition by alder, little or no inhibition by pine, but marked inhibition by both spruce and oak.

It is likely therefore that the influence of such inhibitors in pure stands of spruce and oak would be modified—if only by dilution—by admixture with those species having little or no such inhibitory materials, i.e. alder and pine. It might

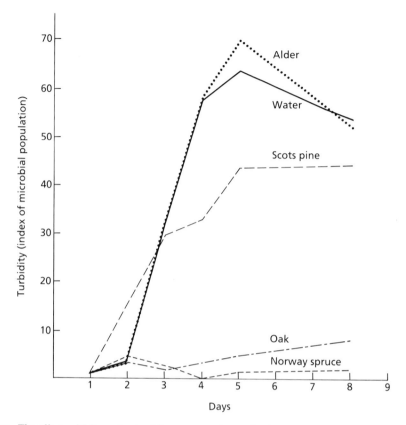

FIG. 10. The effects of foliar extracts of four tree species on the development of a mixed soil culture in liquid medium.

also explain another aspect of the spruce/alder effect in which, as suggested above, mineralization of N and P within the alder parts of the mixture appeared to have been to a large extent 'switched off' by the presence of the spruce. Similar mechanisms may help to explain the negative interactions in the spruce/oak mixture, both components of which were shown to produce inhibitors of microbial activity. In turn, such inhibition of nutrient release could be expected adversely to influence tree growth.

*Rooting patterns in mixture*

A long-supposed possible advantage of mixtures is that they might be able to exploit different resources in a complementary way by differential rooting patterns. At Gisburn, only the spruce/pine mixture has been studied quantitatively in this respect (although general observation suggests what is likely to be the case with the other mixtures).

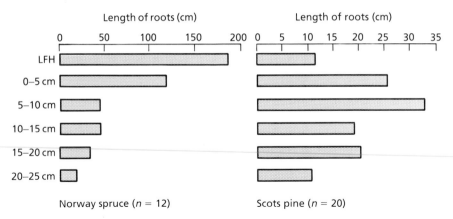

FIG. 11. Distribution of live fine roots (<1 mm) of spruce and pine with depth at Gisburn, 1983. Mean lengths of roots (cm) per 24 mm diameter core. *Note* different scales for the two species. LFH, forest floor horizon of litter (L), partially decomposed litter (F) and humus (H). (Data of J.M. Sykes & S.M.C. Robertson.)

In 1983, sample soil cores, from pure spruce and pure pine, were hand sorted to provide estimates of the distribution of live fine (<1 mm) roots down the profile (J.M. Sykes & S.M.C. Robertston unpubl.). The results (Fig. 11) show that peak quantities of fine spruce roots occur in the forest floor. In contrast, relatively few pine roots were found in this horizon, the median value occurring well into the mineral soil.

Chapman (1986) similarly hand sorted roots from cores taken in the forest floor only, but including the mixed spruce/pine stand in addition to the two separate monocultures. His results (Table 5) confirm for the monocultures the disparity in the amounts of roots found in the forest-floor layer of the earlier study. In a 50:50 mixture, the pro rata 'expected' weight of roots per species would be half that found in pure stands. In fact, although the weight of pine roots was virtually as expected, the weight of spruce roots per unit area in the mixture, was at least as much as that in the pure stand: they had proliferated throughout the whole stand to take advantage of a forest-floor resource apparently scarcely exploited by the admixed pine. Chapman (1986) then analysed the leachates from a set of forest-floor lysimeters into which tree roots had been introduced via

TABLE 5. Density of fine (<2 mm) roots (g dry wt m$^{-2}$) in the forest floor at Gisburn. Data from Chapman (1986)

|  |  | Observed | 'Expected' for 50:50 mixture |
|---|---|---|---|
|  | NS pure | 108.6 |  |
|  | SP pure | 10.2 |  |
| NS/SP mixture { | NS (SP) | 131.8 | 54.3 |
|  | SP (NS) | 5.4 | 5.1 |

watertight seals. These permitted him to infer nutrient uptake, by comparison with data from non-rooted lysimeters. These uptake results demonstrated that, in mixture with pine, the spruce not only responded by increasing its density of rooting but also took up the extra N and P released in the mixed forest floor. Its total requirements (calculated from the literature) were virtually satisfied from this horizon. In contrast, the pine obtained only a minor portion of its calculated needs of both N and P from the forest floor. By inference therefore, the pine acquired much of its N and P from the mineral soil where the bulk of its feeding roots were situated.

The complementary nature of the root distribution pattern for pine and spruce is therefore also reflected in the nutrient uptake of these species when mixed. This provides not only a further mechanism for the mixture effect in these two species at Gisburn but also an explanation of how the spruce in mixture can benefit from the extra release of N and P without detriment to the pine. Differential rooting depths of two species of the cereal *Avena*, when in mixture, have similarly been shown to produce a RYT much greater than unity (Trenbath & Harper 1973).

A similar explanation is probably applicable at least to the pine/oak mixture, because oak—from frequent field observation at Gisburn—also has many fine feeding roots in the forest floor. Whether it also applies to the mix of pine and alder is less clear since, in general, alder is thought to be the most deeply rooted of all the Gisburn species. On the other hand, deep rooting of alder may well provide one possible mechanism for the beneficial influence of alder when mixed with the shallow-rooting spruce or oak.

## OTHER INDICES OF MIXTURE EFFECTS

### *Organic matter quality*

As the organic matter of the forest floors in the various stands differs in rates of metabolic activity, and N and P mineralization, these organic materials may have developed differences in quality. In particular, if the mixed spruce/pine stand, as an example of a positive interaction, has qualitatively different organic matter from that of its components when pure, it adds weight to the view that the organic matter is the site of the interactions and that the mixture is more than the 'sum of its parts'.

Ogden (1986) made some qualitative comparisons of the organic matter in the unplanted grassland, the pure spruce, pine, oak and alder stands, and the mixed spruce/pine stand at Gisburn. These comparisons were made *inter alia* on the borate soluble fraction, representing the most recent soil organic matter. They included:

1  A determination of the ratio of low molecular weight to high molecular weight materials, using gel filtration with Sephadex columns. This ratio provides an index

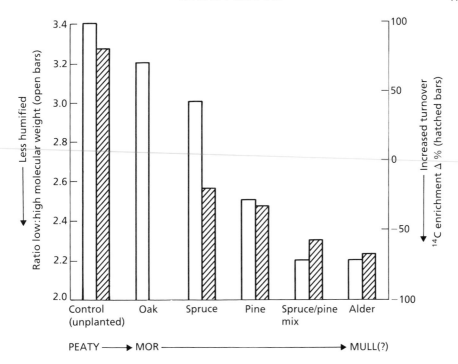

FIG. 12. Qualitative analyses of borate-soluble soil organic matter (0–5 cm) at Gisburn 1984. (Based on data of Ogden 1986.)

of the degree of humification of the organic matter (Swift, Thornton & Posner 1970), with a high ratio representing high humification.

2   The degree of enrichment with excess $^{14}$C ('bomb' carbon), providing a measure of the accumulation within the surface (0–5 cm) soils of 'recent' organic matter derived from the present vegetation.

These indices of soil organic matter quality are presented in Fig. 12. They show that the unplanted control, having retained the pre-planting tendency to incipient peat formation, is at one extreme, and alder at the other. Alder has the most biologically active forest floor of the experimental stands at Gisburn, the most rapid turnover of litter, and has developed somewhat mull-like conditions at the soil surface in contrast to the mor-humus of the other species. It may also be inferred that relative to the unplanted grassland control, all the tree stands can be regarded as site-improving in the sense of reducing both the degree of humification and the accumulation of recently fixed carbon in soil which otherwise showed a tendency towards peat-formation. Oak and spruce have least effect in these respects, alder most, with pine intermediate. On both scales of organic matter quality, it is evident that the mixed stand of spruce and pine is very similar to alder and not intermediate between its two pure components. This reconfirms the interaction occurring in its forest floor and also suggests that this mixture has site-improving qualities relative to planting either component pure.

## CONCLUSIONS

The mensurational data from the six mixed combinations of Norway spruce, Scots pine, sessile oak and common alder indicate that three types of mixture effect are present: a positive interaction in all three pine mixtures; a mainly compensatory mixture effect in the other two alder mixtures; and a negative interaction in the case of the spruce/oak combination.

As far as the three spruce mixtures are concerned, the evidence strongly indicates that $NO_3$-N, and possibly $PO_4$-P availability, are key factors. In turn the release of these nutrients is related to organic matter turnover brought about by microbial and faunal activity, with enchytraeids apparently particularly important. Interestingly, in his studies of declining productivity of the spruce monocultures which had replaced the more natural mixed stands in Saxony during the nineteenth century, Wiedemann (1923, 1924 quoted by Troup 1952) reached a similar conclusion: that $NO_3$ mobilization was reduced in monoculture relative to mixtures through the 'unfavourable decomposition of humus'; and that such reduction in nitrate was one important cause of the problem.

In the positive interactions at Gisburn, decomposer activity may be enhanced by an apparent synergism of unknown mechanism, occurring through mixing of the litters. The negative interaction possibly results from reduced decomposition brought about by microbial inhibitors present in oak and spruce foliage. It is postulated that a further mechanism for enhancing nutrient release and tree response may result from the dilution of self-inhibitory effects in oak or spruce by admixing with species lacking inhibitors (pine and alder). Yet a further mechanism seems highly probable in spruce/pine (and possibly in some other combinations), where complementary resource capture can be inferred via demonstrable differences in rooting patterns.

In spruce/pine, all these positive mechanisms may well occur in an additive fashion, resulting in the clear benefit to both mixture components. In spruce/alder circumstantial evidence suggests that spruce inhibits activity and nutrient release in the alder, whether from organic matter turnover or N fixation; but that in compensation the spruce possibly gains through dilution of its self-inhibitory effects, by differential rooting or a generally greater competitive ability to acquire the mineralized N. As oak and spruce both apparently produce many fine feeding roots competing within the same surface layers, and probably because both also produce inhibitory substances, there appears to be a net negative interaction between them in mixture.

## ACKNOWLEDGMENTS

Only through the foresight of those who established this experiment involving long-term mixtures, and their belief in the provision of research opportunities for their successors, has it been possible to make these studies: the roles of J.D. Ovington, encouraged by W.H. Pearsall and supported by E.M. Nicholson (the then Director-General of the Nature Conservancy); M.V. Edwards, Forestry Commission Silviculturist (North) in consultation with his Chief Research Officer, M.V. Laurie, and Director of Research, J. Macdonald; and J.S.R. Chard, Forestry Commission Conservator, in whose area the experiment was planted, must all be acknowledged.

In addition to establishing the experiment and providing all subsequent care and management, the Forestry Commission kindly made available the mensurational data from these stands, and also carried out the field sampling for the foliar analyses.

Many temporary members of staff, in addition to A. Carlisle, C.L. Gardener, G. Howson and S.M.C. Robertson, have helped in these studies; I thank them all.

## REFERENCES

Anderson, J.M. & Ineson, P. (1982). A soil microcosm system and its application to measurements of respiration and nutrient leaching. *Soil Biology and Biochemistry* 14, 415–16.

Avery, B.W. (1980). *Soil classification for England and Wales (Higher categories), Soil survey of England and Wales*. Technical Monograph No. 14. Harpenden.

Beck, G., Dommergues, Y. & van den Driessche, R. (1969). L'effet litière II étude expérimentale du pouvoir inhibiteur des composés hydrosolubles des feuilles et des litières forestières vis-à-vis de la microflore tellurique. *Oecologia Plantarum*, 4, 237–66.

Benoit, R.E. & Starkey, R.L. (1968). Inhibition of decomposition of cellulose and some other carbohydrates by tannin. *Soil Science*, 105, 291–6.

Binns, W.O., Mayhead, G.J. & MacKenzie, J.M. (1980). *Nutrient deficiencies of conifers in British forests*. Forestry Commission Leaflet No. 76. HMSO, London.

Brown, A.H.F. (1988). Discrimination between the effects on soils of 4 tree species in pure and mixed stands using cotton strip assay. In *Cotton Strip Assay: an Index of Decomposition in Soils* (Ed. by A.F. Harrison, P.M. Latter & D.W.H. Walton), pp. 80–3 (ITE Symposium No. 24). Institute of Terrestrial Ecology, Grange-over-Sands.

Brown, A.H.F. & Harrison, A.F. (1983). Effects of tree mixtures on earthworm populations and nitrogen and phosphorus status in Norway spruce (*Picea abies*) stands. In *New Trends in Soil Biology* (Ed. by P. Lebrun, H.M. André, A. de Medts, C. Grégoire-Wibo & G. Wauthy), pp. 101–8. Ottignies-Louvain-la-Neuve, Dieu-Brichart.

Carey, M.L., McCarthy, R.G. & Miller H.G. (1988). More on nursing mixtures. *Irish Forestry*, 45, 7–20.

Carlyle, J.C. & Malcolm, D.C. (1986). Nitrogen availability beneath pure spruce and mixed larch + spruce stands growing on deep peat. I. Net N mineralization measured by field and laboratory incubations. *Plant and Soil*, 93, 95–113.

Chapman, K. (1986). *Interaction between tree species: decomposition and nutrient release from litters*. Ph.D. thesis, University of Lancaster.

Chapman, K., Whittaker, J.B. & Heal, O.W. (1988). Metabolic and faunal activity in litters of tree mixtures compared with pure stands. *Agriculture, Ecosystems and Environment*, 24, 33–40.

Connolly, J. (1987). On the use of response models in mixture experiments. *Oecologia (Berlin)*, 72, 95–103.

Edwards, C.A. (1967). Relationships between weights, volumes and numbers of soil animals. In *Progress in Soil Biology* (Ed. by O. Graff & J.E. Satchell), 585–91.

Everard, J. (1973). Foliar analysis: sampling methods, interpretation and application of the results. *Quarterly Journal of Forestry*, 67, 51–66.

Francis, C.A. (1989). Biological efficiencies in multiple cropping systems. In *Advances in Agronomy* (Ed. by N.C. Brady), 42, 1–42.

**Gabriel, K.A.S. (1986).** Growing broadleaved trees on the North Yorks moors. *Quarterly Journal of Forestry*, **80**, 27–32.

**Hall, B.R. & Folland, C.J. (1970)** *Soils of Lancashire. Soil Survey of Great Britain*. England & Wales Bull. No. 5. Harpenden.

**Hamilton, G.J. (1975).** *Forest Mensuration Handbook*. Forestry Commission Booklet No. 39. HMSO, London.

**Harrison, A.F. & Helliwell, D.R. (1979).** A bioassay for comparing phosphorus availability in soils. *Journal of Applied Ecology*, **16**, 497–505.

**Latter, P.M. & Howson, G. (1977).** The use of cotton strips to indicate cellulose decomposition in the field. *Pedobiologia*, **17**, 145–55.

**Lines, R. & Nimmo, M. (1966).** Long-term mixtures. *Report on Forest Research, London*, 1965, 35–8.

**McIntosh, R. (1983).** Nitrogen deficiency in establishment phase Sitka spruce in upland Britain. *Scottish Forestry*, **37**, 185–93.

**Moffat, A.J. & Boswell, R.C. (1990).** Effect of tree species and species mixtures on soil properties at Gisburn Forest, Yorkshire. *Soil Use and Management*, **6**, 46–51.

**O'Carroll, N. (1978).** The nursing of Sitka spruce. 1. Japanese larch. *Irish Forestry*, **35**, 60–5.

**O'Connor. F.B. (1967).** The Enchytraeidae. In *Soil Biology* (Ed. by A. Burges & F. Raw), 213–57. Academic Press, London.

**Ogden, J.M. (1986).** *Some effects of afforestation on soil organic matter*. Ph.D. thesis, University of Stirling.

**Pearsall, W.H. (1950).** *Mountains and Moorlands*. Collins, London.

**Petersen, H. & Luxton, M. (1982).** A comparative analysis of soil fauna populations and their role in decomposition processes. *Oikos*, **39**, 291–422.

**Swift, R.S., Thornton, B.K. & Posner, A.M. (1970).** Spectral characteristics of a humic acid fractionated with respect to molecular weight using an agar gel. *Soil Science*, **110**, 93–9.

**Trenbath, B.R. (1974).** Biomass productivity of mixtures. *Advances in Agronomy*, **26**, 177–210.

**Trenbath, B.R. & Harper, J.L. (1973).** Neighbour effects in the genus *Avena* I. Comparison of crop species. *Journal of Applied Ecology* **10**, 379–400.

**Troup, R.S. (1952).** *Silvicultural Systems* (Ed. by E.W. Jones), 2nd edn. Clarendon Press, Oxford. (References to Wiedemann, p. 19.)

**Weatherell, J. (1957).** The use of nurse species in the afforestation of upland heaths. *Quarterly Journal of Forestry*, **51**, 298–304.

**Willey, R.W. (1979).** Intercropping—its importance and research needs. Part 1. Competition and yield advantages. *Field Crop Abstracts*, **32**, 1–10.

**Williams, B.L. (1988)** *Mixtures of tree species enhance nitrogen availability*. Annual Report Macaulay Land Use Research Institute for 1987, p. 76.

**Williams, B.L. & Griffiths, B.S. (1989).** Enhanced nutrient mineralization and leaching from decomposing Sitka spruce litter by Enchytraeid worms. *Soil Biology and Biochemistry*, **21**, 183–8.

**Wit, C.T. de (1960).** On competition. *Verslagen van het landbouwkunig onderzoek in Nederland*, **66**, 1–82.

**Wit, C.T. de & Bergh, J.P. van den (1965).** Competition between herbage plants. *Netherlands Journal of Agricultural Science* **13**, 212–21.

**Zehetmayr, J.W.L. (1960).** Afforestation of upland heaths. *Bulletin of the Forestry Commission London*, No. 32, 145 pp.

# Mycorrhizal fungi in mixed-species forests and other tales of positive feedback, redundancy and stability

D.A. PERRY,* T. BELL* AND M.P. AMARANTHUS†

*Department of Forest Science, Oregon State University, Corvallis, OR 97331, USA; and †US Forest Service, Siskiyou National Forest, Grants Pass, OR 97526, USA*

## SUMMARY

Individual trees, either of the same or of different species, can be physically linked by the hyphae of mycorrhizal fungi that allow carbon and nutrients to pass between them. Carbon transfer is enhanced when recipient plants are shaded and donors are in full sunlight. The few studies on the effect of mycorrhizal fungi on competition between plants of different species consistently show that mycorrhizal fungi convert a negative interaction between plants of different species into one that is either neutral or positive. Evidence suggests that this phenomenon is due, at least in part, to a more even distribution of nutrients, and perhaps carbon, among mycorrhizal than among non-mycorrhizal plants. This effect is not exhibited by all species of mycorrhizal fungi that have been studied.

Mycorrhizal fungi also link plant species in time. In the mixed conifer/hardwood forests of southern Oregon, various hardwood species that recover quickly from disturbance by sprouting from roots stabilize the mycorrhizal fungi required by conifers. The hardwoods may also stabilize bacteria that live in conifer rhizospheres, including associative nitrogen fixers. This apparent co-operation between potentially competing tree species may have evolved because the unpredictable behaviour of fire produces a large degree of uncertainty as to which species have the adaptations needed for survival or quick recovery from a particular event. In such a milieu, there would be selective pressure for mycorrhizal fungi to evolve generality in order to ensure continuity in hosts and for plant species to tolerate this generality because it ensures continuity in their fungal partner. The indirect result of such selection would be groups, or guilds, of plants that are interlinked through a common interest in mycorrhizal fungi. Similar patterns of reciprocal altruism may evolve in any system that is characterized by strong positive feedback links among components and high levels of uncertainty in the environment. Much more knowledge is needed about this phenomenon because it has significant implications for the stabilizing role of biodiversity during climate change.

## INTRODUCTION

Mycorrhizae—literally 'fungus-roots'—are mutualistic symbioses between plants and fungi. Mycorrhizal fungi benefit plants by gathering water and nutrients (especially immobile nutrients, such as phosphorus) and by extending the life of feeder roots. As the primary interface between plant roots and the below-ground environment, mycorrhizal fungi also perform various ecological services, such as protecting plants against pathogens and toxic heavy metals (Hayman 1978; Marx & Krupa 1978; Dueck *et al.* 1986; Jones, Dainty & Hutchinson 1988), aggregating soils (Sutton & Sheppard 1976; Lynch & Bragg 1985), weathering minerals (Cromack *et al.* 1979) and mediating interactions among plants (Perry *et al.* 1989b). The diversity of mycorrhizal fungi formed by a given plant may increase its ability to occupy diverse below-ground niches. Through this symbiotic relationship, the microbe's rapid evolutionary potential is available to the plant, possibly enhancing the plant's capacity to track the environment genetically. Reviews of various aspects of mycorrhizal ecology include: Janos (1980, 1983, 1987, 1988), Malloch, Pirozynski and Raven (1980), Pirozynski (1981), Harley and Smith (1983), St John and Coleman (1982), Perry, Molina and Amaranthus (1987), Newman (1988) and Trappe (1987). Allen and Allen (1990) discussed mycorrhizal mediation of plant competition, with an emphasis on grass- and shrublands.

The objective of this paper is to review what is known about how mycorrhizal fungi mediate interactions among tree species, distinguishing two types of mediation: (i) those concerned with gathering and allocation of resources at any one time; and (ii) those having to do with maintaining a continuity of the plant–fungus partnership through time. We set the stage by briefly reviewing the systematic aspects—who forms what with whom. In the final section we speculate on how random and unpredictable disturbance regimes might catalyse the evolution of cooperation among guilds of plant species that depend on the same mycorrhizal fungi. As used here, guilds are defined as a set of plant species that host at least one species of mycorrhizal fungus in common (Perry *et al.* 1989a).

## A BRIEF SURVEY OF THE MYCORRHIZAL SYMBIOSIS

To date, all the conifers that have been studied are consistently mycorrhizal in the wild. Of the 6507 species of angiosperms studied, 70% are always mycorrhizal and 12% are apparently facultatively mycorrhizal, sometimes forming mycorrhizae and sometimes not (Trappe 1987). With the possible exception of a few tropical pioneers, angiosperm trees are consistently mycorrhizal, while a high proportion of the plants classed by Holm *et al.* (1977) as the world's worst weeds are facultatively mycorrhizal. In general, pioneering herbs and grasses are believed to be the least reliant on mycorrhizae and late successional species the most reliant (Allen & Allen 1990).

The most common mycorrhizal types are formed by two quite distinct groups of fungi. Ectomycorrhizae are formed by several thousand fungal species in the subdivisions Basidiomycotina and Ascomycotina. They are characterized by extensive hyphal development external to the root, which results in readily visible modification of root colour and shape, and usually, but not always, by lack of penetration of host cells. Vesicular-arbuscular (VA) mycorrhizae penetrate host cells and do not modify the external appearance of the root. They are formed by several hundred species in the family Endogonaceae, subdivision Zygomycotina. Many ericaceous plants form a morphologically distinct mycorrhizal type called ericoid that may involve fungi which form ectomycorrhizae with plants in other families.

The majority of plants are VA mycorrhizal, but a large number of trees and shrubs are ectomycorrhizal, including species in the Pinaceae, Fagaceae, Ericaceae, etc. (see Harley & Smith (1983) for an exhaustive list). Some genera, e.g. poplars, eucalypts and alders, form both types of mycorrhizae. Although most ectomycorrhizal fungi have several to many hosts, a few are host-specific. Less is known about the specificity of VA mycorrhizal fungi. Common wisdom has it that they show little host specificity (e.g. Harley & Smith 1983); however recent work suggests that, like ectomycorrhizal fungi, some VA fungi are highly general while others are more specific in their host preferences (Rosendahl, Rosendahl & Sochting 1989; Sieverding 1989). Determining the specificity of species within both fungal types is made more difficult by the fact that the particular mix of fungi that form mycorrhizae with a given plant may vary widely depending on environmental conditions and physiological status or age of the host (see below).

Montane, boreal and temperate conifer forests are dominated by ectomycorrhizal trees (Pinaceae), though conifer forests of the Pacific Northwest also contain VA mycorrhizal cedars and maples. In the mixed conifer/hardwood forests that characterize mediterranean climates, both conifers and hardwoods are mostly ectomycorrhizal, but in temperate mixed-species forests with summer rains the two mycorrhizal groups tend to be evenly represented. In Britain, 36% of native woody species are VA mycorrhizal, 24% ectomycorrhizal, 17% form both types, and 20% form ericoid mycorrhizae (Fitter 1989).

Trees in moist tropical forests are predominantly VA mycorrhizal (Janos 1980), although some, including the Dipterocarpaceae and Caesalpinoideae, form ectomycorrhizae. Trees with ectomycorrhizae tend to dominate the most phosphorus-poor soils in the tropics (Gartlan et al. 1986; Hogberg 1986; Ashton 1988; Newbery et al. 1988). It is not clear why ectomycorrhizal trees should be more successful than VA mycorrhizal trees on P-infertile soils, since both types of fungi are quite efficient nutrient gatherers; Harley & Smith (1983) suggested that ectomycorrhizae have an advantage on infertile soils in climates with distinct seasons (wet–dry in the tropics) due to their ability to store nutrients in the hyphal 'mantles' that envelop the root surface—something VA mycorrhizae do not have.

On a particular site, the species of fungi that form mycorrhizae with a given plant species vary widely over time and with numerous environmental factors. Succession of ectomycorrhizal types on the same root system is well established

(Mason *et al.* 1983), and the proportions of different ectomycorrhizal fungi formed by a given tree species typically differ depending on soil type, successional stage, and, on disturbed sites, the nature of the disturbance (e.g. Slankis 1974; Schoenberger & Perry 1982; Pilz & Perry 1984; Amaranthus & Perry 1989). Proximity to an already infected root system can, in some cases, dramatically affect the types of mycorrhizae formed by plants (Fleming 1984; Borchers & Perry 1990), while in other cases it seems to have little effect (Pilz & Perry 1984).

## HOW MYCORRHIZAL FUNGI MEDIATE COMPETITION FOR RESOURCES

Mycorrhizal fungi might influence competition for nutrients and water within plant communities in at least three distinct ways.

**1** They may enable a sparsely rooted plant such as a tree to compete with a densely rooted plant such as a grass or sedge (Hall 1978; Bowen 1980). It has been argued that mycorrhizae could become detrimental to their hosts on fertile sites if the cost–benefit ratio of maintaining the fungus reduced the competitive advantage of the host (Fitter 1977). On the other hand, plants are by no means prisoners of the fungi, and are well known to readily adjust the numbers of mycorrhizae that they form in accordance with nutrient availability (at least phosphorus; Harley & Smith 1983). For example, tree seedlings growing in a greenhouse where they are well watered and fertilized seldom form mycorrhizae.

**2** Mycorrhizal fungi are known to link plants of the same and different species and to mediate the transfer of nutrients and carbon from one host plant to another.

**3** Mycorrhizal fungi have been shown to decrease competition between seedlings of different tree species, perhaps due to interlinking, or by increasing niche separation within the rooting zone.

The first of the above pertains to interactions between trees and herbaceous plants; as this conference deals with tree mixtures, we will focus on the other two mechanisms.

### Hyphal links and transfer of materials between hosts

Newman (1988) concluded that mycorrhizal links between plants are probably common; however few studies have verified this in the field. Given that many fungal species form mycorrhizae with more than one plant species within a community and that new mycorrhizae are often initiated from established mycorrhizae (Read 1988), one would expect plants to be linked, at least to some degree. Figure 1 shows a mycorrhizal hypha linking seedlings of lodgepole pine (left) and Scots pine (right).

The idea that guilds of plants within communities might be interlinked, like users on a computer network, raises numerous questions for the ecologist. We will focus on three. The first—and the subject of this particular section—is

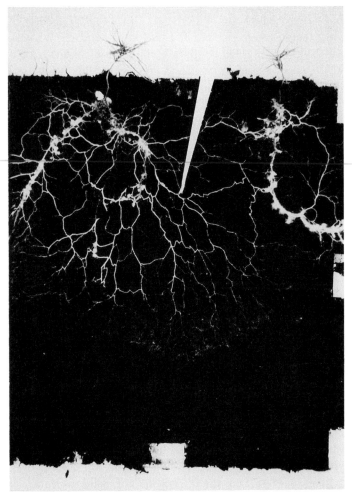

FiG. 1. The arrow shows a mycorrhizal rhizomorph linking lodgepole and Scots pine seedlings.
(Photo courtesy of David Read, University of Sheffield.)

whether or not fungi enhance material transfers from one plant to another. The
second question, dealt with in the following section, concerns how these linkages
affect community structure and ecosystem processes such as productivity. The
third question addresses the evolutionary ecology of mycorrhiza-mediated plant
guilds—how they evolved, and how to explain them within a neo-Darwinian
framework.

Numerous studies have demonstrated that carbon and nutrients move between plants of the same and different species. Bjorkman (1960) showed that $^{14}$C-
labelled sugars injected into the phloem stream of conifers appeared in nearby
*Monotropa*, a plant that is both achlorophyllous and, like conifers, ectomycorrhizal. He suggested that the transfer was through mycorrhizal hyphae, but did

not attempt to demonstrate this. Woods and Brock (1964) applied $^{45}$Ca and $^{32}$P to stumps of red maple (*Acer rubrum* L.) that were recently cut in a mixed hard-wood forest. (Maples are believed to be predominantly VA mycorrhizal, but have been reported to host ectomycorrhizae as well (Harley & Smith 1983).) Forty per cent of the trees and shrubs surrounding the treated stumps had taken up $^{32}$P within 2 days of application. Within 8 days, 80% (nineteen species) within 8 m of the stumps had $^{32}$P in their foliage, and 40% had $^{45}$Ca. Most of the recipient species were VA mycorrhizal, but ectomycorrhizal oaks were also recipients. Of the three possible transfer mechanisms—root grafts, root exudations and mycor-rhizal hyphae—Woods and Brock felt that the patterns they saw were best accounted for by some combination of root exudations and mycorrhizal hyphae. In a similar study, Read, Francis and Finlay (1985) fed $^{14}$C to a lodgepole pine sapling surrounded by seedlings of various ectomycorrhizal and VA mycorrhizal tree species. Large amounts of the tracer were transferred to nearby lodgepole pine (*Pinus contorta* Dougl. ex Loud.) seedlings—especially under shaded condi-tions; seedlings of the ectomycorrhizal hosts Sitka spruce (*Picea sitchensis* (Bong.) Carr) and birch (*Betula pubescens*) also accumulated tracer, but at lower levels than did the pine, whereas seedlings of the VA mycorrhizal species accumulated only trace amounts or none at all. Other studies showing material transfers among plants in the field include Rahteenko (1958), Crafts and Yamaguchi (1958), Reid and Woods (1969) and Chiariello, Hickman and Mooney (1982).

With one exception, studies in which mycorrhiza formation by seedlings was controlled have shown that the fungi do enhance transfer of C and nutrients (N and P) between plant species within the same guild (Whittingham & Read 1982; Francis & Read 1984; Read, Francis & Finlay 1985; van Kessel, Singleton & Hoben 1985; Finlay & Read 1986a,b; Newman & Ritz 1986). More C is trans-ferred when recipient plants are shaded while donors remain in full light (Read, Francis & Finlay 1985; Finlay & Read 1986a). The single exception is a study by Finlay and Read (1986b), who found that $^{32}$P fed to the root of a given plant did not move to other plants, but if fed to mycorrhizal hyphae, it was distributed among host-plants connected to the hyphal network. The reason for the dis-crepancy is unclear.

Evidence strongly supports the following conclusions. At least some species within plant communities are interlinked through common mycorrhizal fungi, like users on a computer network. Mycorrhizae facilitate transfer of material from one plant to another. However, more work is needed on both of these points, par-ticularly in the field, to clarify the ecological significance of networking.

## STUDIES ON MYCORRHIZAL FUNGI AND PLANT COMPETITION

We know of only three experimental studies that address how mycorrhizal fungi affect competition between plants. Two of them were conducted in our laboratory at Oregon State and the third, which deals with maize and the tropical shrub

*Solanum ocraceo*, was conducted by Puga and Janos (Puga 1985). All three studies concluded that mycorrhizal fungi decrease competition between plants and increase yields in mixture.

Perry *et al.* (1989b) grew Douglas fir (*Pseudotsuga menziesii* (Mirb.) Franco) and ponderosa pine (*Pinus ponderosa* Dougl. ex Laws.) in standard replacement series (Harper 1977), but with the added twist of varying the levels of mycorrhizal fungi. Pasteurized forest soils were reconstructed by horizon in pots. Four replacement series were run:

**1** Seedlings were inoculated with no mycorrhizal fungi. At some point during the experiment these seedlings were colonized by *Thelephora terrestris*, an opportunistic mycorrhiza-former that is a common greenhouse contaminant.

**2** Seedlings of each of the tree species were inoculated with a fungus believed to be species-specific; the two fungal species were both in the genera *Rhizopogon*, which contains many common mycobionts of the Pinaceae.

**3** Seedlings were inoculated as in **2**, but also with two common fungal species that are host-generalists (*Laccaria laccata* and *Hebeloma crustiliniforme*).

**4** Uninoculated seedlings were grown in non-pasteurized forest soil.

Figure 2 shows total pot biomass at the end of 12 months in pots. The two tree species clearly interacted negatively when uninoculated (but with the greenhouse contaminant), leading to significant underyielding (Fig. 2a). In both of the inoculation treatments, however, the relative yield total (RYT; Harper 1977) did not differ from 1 (Fig. 2b,c). In non-pasteurized soil the species again underyielded (Fig. 2d); neither species grew as well in non-pasteurized as in pasteurized soil, especially Douglas fir (Fig. 2d). Puga (1985) found similar patterns in his study on two VA mycorrhizal plants: without mycorrhizae the two species underyielded, with mycorrhizae they overyielded.

Average individual seedling biomass (Fig. 3) provides a clearer picture of how mycorrhizae influenced interactions between the two tree species in the Perry *et al.* (1989b) study. Non-inoculated Douglas fir seedlings responded negatively to the addition of any ponderosa pine; however this was not counterbalanced by increased ponderosa pine growth. In the 9/3 mixture of Douglas fir to pine, seedlings of both species were smaller than those in their respective monocultures. Although biomass was not reduced in non-inoculated pine seedlings that were growing with three or six Douglas fir seedlings, relative allocation to stem and foliage altered rather dramatically. In monoculture, non-inoculated pine seedlings averaged 18% of total biomass in foliage and 33% in stems. The addition of three Douglas fir seedlings reversed this pattern (27% foliage, 18% stem), but with six Douglas fir, the allocation was similar to that in monoculture. These patterns were consistent among all replicates.

In pots with the two species-specific fungi, the increased presence of ponderosa pine did not reduce growth in Douglas fir until the latter was in the minority (3/9). Ponderosa pine seedlings averaged greater biomass when in the minority than when in monoculture (N.S. at $P = 0.05$). Where the generalist fungi were present, Douglas fir's response to ponderosa pine was quite different from that in

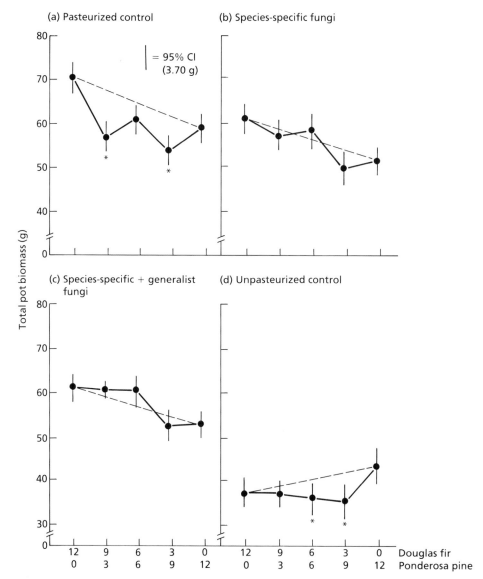

FIG. 2. Influence of ectomycorrhizal fungi on total pot biomass of Douglas fir and ponderosa pine in mixture and monoculture. (From Perry *et al.* 1989b.) (a) Seedlings grown in pasteurized soil with no added mycorrhizal fungi; (b) seedlings grown in pasteurized soil with two species-specific mycorrhizal fungi added; (c) seedlings grown in pasteurized soil with four mycorrhizal fungi added: the two species-specific ones that were also added to (b), plus two generalist species; and (d) seedlings grown in unpasteurized forest soil. *$P \leqslant 0.05$ (that the point does not differ from the line).

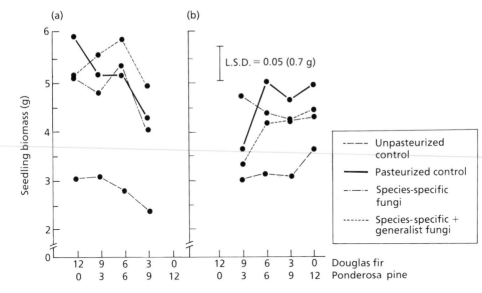

FIG. 3. Individual seedling biomasses from the experiment shown in Fig. 2 (from Perry *et al.* 1989b).

non-inoculated pots. Douglas fir biomass increased up to the 6/6 mixture, then dropped back to monoculture size when Douglas fir were in the minority (3/9). Increased Douglas fir growth in mixture was due solely to seedlings inoculated with the host-generalist fungi, *L. laccata*; in the 9/3 combination the increased growth was accompanied by lower pine growth, but this was not the case in the 6/6 mixture. In contrast to the situation in non-inoculated pots, allocation by ponderosa pine to foliage and stems did not fluctuate with species mix, but remained about 31% foliage and 16% stems for both inoculation treatments.

Because control seedlings were colonized by the greenhouse contaminant, it is not clear how competition is affected by presence versus absence of mycorrhizae, but the study does indicate that at least some types of mycorrhizae reduced competition between the two tree species. There are at least three ways that this might happen.

1 Mycorrhizae could allow plants to exploit different niches, either within the soil (e.g. humus versus mineral soil), or over time (e.g. changing temperature or moisture), and thus reduce direct competition.

2 Nutrients could be transferred through hyphal linkages, mediating an even distribution among individual plants. This would be a particularly effective mechanism where luxury consumption by one plant species (i.e. taking up nutrients in excess of growth requirements) reduces the ability of another plant species to meet its growth requirements.

3 Mycorrhizae could promote growth by detoxifying allelochemicals that are produced by one plant against the other.

Foliar analyses of seedlings in the Perry *et al.* (1989b) study indicate that inoculating with only the two species-specific mycorrhizae improved the uptake of N and P in ponderosa pine (relative to non-inoculated seedlings), and also increased the foliar P content in Douglas fir. Since (presumably) hyphal connections could not occur between tree species inoculated only with species-specific fungi, direct transfer of nutrients can be ruled out. In pots with the generalist fungi, P uptake by Douglas fir was improved, possibly at the expense of luxury consumption by the pine. Hyphal connections were possible in pots with generalist fungi, and *L. laccata* mycorrhizae were found on seedlings that were not inoculated with that fungus. This suggests that *L. laccata* may have linked seedlings at least long enough to initiate mycorrhizae. Improved P uptake was also implicated in Puga's (1985) study showing that VA mycorrhizae reduced competition between maize and *Solanum ocraceo*. Mycorrhizal fungi are especially important for plant uptake of P, an element that tends to occur in insoluble, hence immobile, chemical combinations within soils (Harley & Smith 1983).

In a second experiment (not yet submitted for publication), Douglas fir and western hemlock (*Tsuga heterophylla* (Raf.) Sarg.) were grown in soils with different levels of structural heterogeneity. We hypothesized that if mycorrhizal fungi reduce niche overlap between tree species (in this case soil niches) their effect on seedling interactions will be greater in heterogeneous than in homogeneous soils. The planting medium was a loamy mineral soil amended with three types of organic matter: (i) litter from a Douglas fir/western hemlock stand; (ii) bracken fern litter; and (iii) highly decayed wood from an old Douglas fir log. The different types of organic matter were either thoroughly homogenized within pots or systematically separated so that mixtures of the mineral soil and each type of organic matter occurred in distinct patches (heterogeneous soil). The replacement series in this case consisted of Douglas fir and western hemlock grown in pure stands of ten seedlings per pot or in 5/5 mixtures.

Seedlings were inoculated by growing them in soil from either a Douglas fir/western hemlock forest or from a recent clear-cut and then transplanted into the experimental pots. From past work we knew that both Douglas fir and hemlock seedlings form the same dominant mycorrhizal types in forest and clear-cut soils, but in different proportions; a greater number of minor types (i.e. those occurring in relatively small numbers on roots) are consistently formed in forest soils (Schoenberger & Perry 1982; Pilz & Perry 1984). Controls were not inoculated, and the soil media used in the replacement series was pasteurized to eliminate inocula.

Results of this experiment are shown in Figs 4 and 5. Non-inoculated seedlings of the two tree species interacted negatively when growing in heterogeneous soils (RYT = 0.86), but not when growing in homogeneous soils (RYT = 1). When seedlings were inoculated with soil from a clear-cut, RYT was 1.16 in heterogeneous soils and 1.25 in homogeneous soils (Fig. 4c,d). When inoculated with forest soils, RYTs averaged slightly greater than 1 but the difference was probably not real.

Response of the individual tree species was highly dependent on inocula

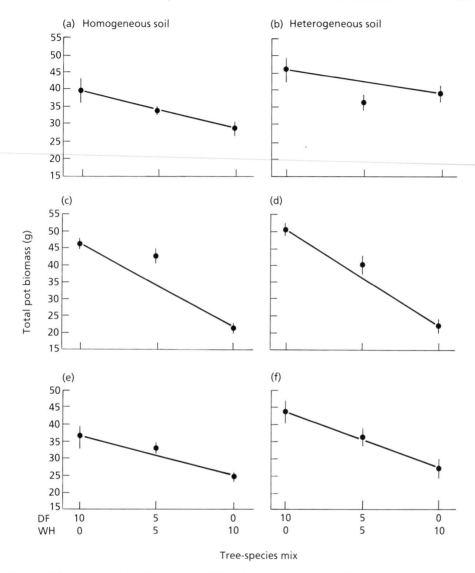

FIG. 4. Relative yield totals of Douglas fir (DF) and western hemlock (WH) replacement series growing in homogeneous or heterogeneous soils. (a, b) Seedlings are uninoculated; (c, d) inoculated with clear-cut soil; or (e, f) inoculated with forest soil.

source and soil type (Fig. 5). In homogeneous soils, non-inoculated seedlings were neutral to one another. Forest inocula may have benefited Douglas fir in mixture, but the effect was small. Clear-cut inocula, however, greatly increased Douglas fir growth in mixture, partly at the expense of hemlock seedlings, which were smaller in mixture than in monoculture by about one-third. The species of mycorrhizae occurring in clear-cut soils may benefit early successional Douglas fir

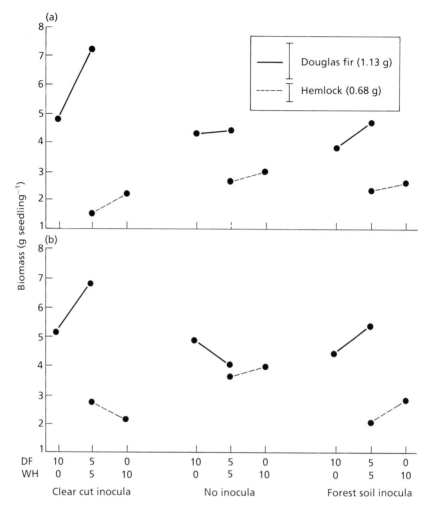

FIG. 5. Individual seedling biomasses from replacement series shown in Fig. 4. (a) Homogeneous soils; and (b) heterogeneous soils. Bar = LSD (0.05).

more than they do later successional, shade-tolerant hemlock, although hemlock readily colonizes disturbed areas in relatively moist habitats.

Patterns were quite different in heterogeneous soils. The low RYT value for non-inoculated seedlings was due primarily to lower growth of Douglas fir in mixture than in monoculture, with no compensating increase in hemlock growth. Both clear-cut and forest inocula benefited Douglas fir in mixture and, in contrast to homogeneous soils, the clear-cut inocula also benefited hemlock in mixture.

It is not clear why non-inoculated Douglas fir should be disadvantaged in heterogeneous soils—perhaps allelopathy prevented non-mycorrhizal roots from exploiting some of the heavily organic patches. The interaction between inocula source and soil type could also be related to allelopathy. Water soluble leachates from forest litter sometimes inhibit formation of mycorrhizae, but the effect is complex and varies both with type of litter and species of mycorrhizal fungi (Schoenberger & Perry 1982; Rose et al. 1983).

$^{14}$C was used to investigate possible carbon transfer between Douglas fir and hemlock. Three pots with five Douglas fir and five hemlock seedlings inoculated with forest soil were incubated with $^{14}$C. In one pot, the hemlock seedlings were covered with a plastic bag that was secured tightly around the stem just above the ground line, while the Douglas fir seedlings were uncovered. In a second pot, the Douglas fir seedlings were covered and hemlock uncovered. The third pot was a control in which all seedlings were covered. Figure 6 shows $^{14}$C counts in foliage. Seedlings in pots with all seedlings covered had small amounts of tracer, presumably because the seal at the stem base was not completely airtight. In the pot with hemlock covered and Douglas fir uncovered, hemlock seedlings averaged greater than three times more $^{14}$C than did hemlock seedlings from the control pot ($P < 0.001$)—with all five hemlock seedlings gaining more tracer than did the controls. Only one covered Douglas fir seedling growing with uncovered hemlock had more tracer than did control seedlings. These results indicate that C was probably transferred from Douglas fir to hemlock; but transfer from hemlock to Douglas fir either did not occur or was much more restricted. Transfer of $^{14}$C

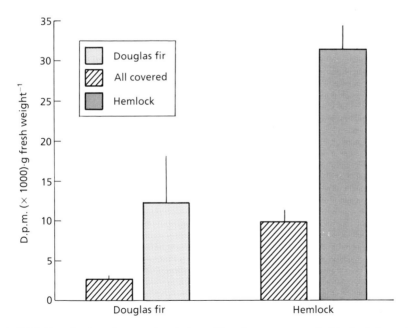

FIG. 6. $^{14}$C in Douglas fir and western hemlock seedlings that were covered with plastic bags when $^{14}$C was introduced into chambers. All data are from mixed species pots. For each species, the left bar refers to pots with all seedlings covered (controls) and the right bar to pots with the other species uncovered.

from Douglas fir to the smaller hemlock seedlings supports studies cited earlier showing that C moves from sunlit to shaded plants (Read, Francis & Finlay 1985; Finlay & Read 1986a).

A crude estimate of the ecological significance of transfers can be made by comparing the amount of labelled C apparently transferred to covered seedlings with that contained in uncovered seedlings: this comparison suggests that about 10% as much C was transferred as was fixed in photosynthesis by uncovered hemlock seedlings—not a great deal, but this could represent significant transfers of N P, or S attached to C skeletons. There is little evidence that the apparent transfers of C (or C plus nutrient) benefited hemlock growth in mixture. Average hemlock biomass in mixed pots was more similar to that in pure pots when seedlings were inoculated with forest soils than when they were not inoculated (Fig. 5b); however, in neither case were differences between mixed and pure pots statistically significant.

### Model of effect of mycorrhizal fungi on nutrient uptake

From the previous studies, it is clear that mycorrhizal fungi have the potential to decrease competition between plant species and to increase overall growth in mixed plantings. But how is this accomplished? The evidence suggests that resources are more evenly distributed when trees growing in mixture are mycorrhizal than when they are not—perhaps because nutrients and C are transferred through hyphal links, although this does not appear to be the full explanation. There is also evidence, especially from Puga (1985), that mycorrhizae increase the total amount of P available to plants in mixture.

A simple model shows that, under certain conditions, smoothing the distribution of resources between plant species can explain the patterns seen in these studies (Fig. 7). Consider two hypothetical plant species—$X$ and $Y$. Assume for simplicity that both are limited by phosphorus and that the response of both to changes in the supply of phosphorus is described by the same Mitscherlich curve of diminishing returns. Further assume that, when both are non-mycorrhizal, $Y$ outcompetes $X$ for available phosphorus. This means that when non-mycorrhizal plants are growing together, $Y$ seedlings gain phosphorus and $X$ seedlings lose it, relative to their status in monoculture. However, because of the shape of the Mitscherlich curve, $X$ loses more growth in this exchange than $Y$ gains (Fig. 7a), as happened in non-inoculated pots in the Perry *et al.* (1989b) replacement experiment, where Douglas fir lost more growth than ponderosa pine gained when the two species were growing together. Transfer of phosphorus from $Y$ to $X$ by mycorrhizal fungi would have the reverse effect—$X$ would gain more than $Y$ loses (Fig. 7b). The same argument can be applied to any resource whose effect on growth follows a law of diminishing returns (which includes about everything). This model critically depends on one species (the superior competitor) being in the region of diminishing returns on the resource–response curve and might be tested by altering resource availability.

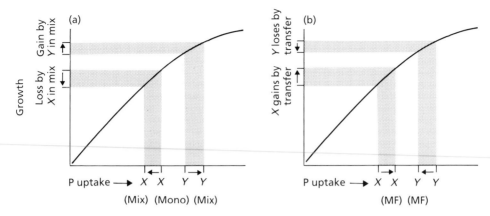

FIG. 7. Model of how phosphorus (P) transfers between species might affect growth in mixture. Assumptions: species $Y$ outcompetes $X$ for phosphorus; both species follow the same curve of declining yields with increasing P. (a) No transfer of P in mixture. $Y$ gains phosphorus at the expense of $X$, but $Y$ gains less growth than $X$ loses. In result, RYTs drop below 1.0 in mixture. (b) P transfered from $Y$ to $X$. $Y$ loses less growth than $X$ gains, resulting in a RYT greater than 1.0.

The degree to which the potential for mycorrhizal fungi to reduce competition between species is realized in nature is an open question. Even in controlled experiments the situation is complex and one should be cautious about generalizations. In both studies conducted in our laboratory, simple hypotheses about the benefits of mycorrhizal diversity and interactions with soil heterogeneity failed to capture the complexities of the systems under study. Rather than diversity *per se*, or heterogeneity *per se*, there are complex interactions among tree species, the particular types of mycorrhizae they form, and the soil in which they are growing. In nature these complexities are undoubtedly multiplied many-fold. Different soils are likely to produce different results, especially when nutrient limitations differ. Mycorrhizae may be especially beneficial to seedling interactions in the P-limited systems that commonly occur in the tropics. The complexity of the situation is illustrated by Ae *et al.* (1990) who found that VA mycorrhizae benefit agronomic plants to a much greater extent on Alfisols—in which P is bound primarily with Fe, than on Vertisols—whose P is bonded primarily with Ca.

## MYCORRHIZAL FUNGI LINK PLANT SPECIES IN TIME

In both forests and shrublands a so-called 'island' effect has been recognized in early succession, in which seedlings establish near existing plants. In south-west Oregon, for example, Douglas fir seedlings preferentially associate with Pacific madrone trees (*Arbutus menziesii* Pursh) (Amaranthus, Molina & Perry 1989). Various people have suggested that mycorrhizal links between plants may facilitate seedling establishment by providing a ready source of inocula (Wilcox 1983; Amaranthus & Perry 1987, 1989; Newman 1988; Read 1988). In his review,

Newman (1988) concluded that the evidence to support this hypothesis was scant and equivocal.

Studies in south-west Oregon and north-west California, conducted in mixed conifer/hardwood communities that comprise numerous species of ectomycorrhizal trees and shrubs, have a bearing on this issue. When stands are destroyed by wildfire or clear-cutting, hardwoods are usually the first to recover, either from sprouting rootstocks or buried seeds. Experiments demonstrate that soils beneath several, though not all ectomycorrhizal hardwood species in these communities enhance the survival and growth of Douglas fir seedlings. Mycorrhizal fungi are probably involved, but the emerging story is more complex than can be accounted for by invoking only mycorrhizal fungi.

The first experiment suggests that soil organisms mediate synergistic inter-actions among different species within these communities (Amaranthus & Perry 1989). Douglas fir seedlings were planted in cleared patches within three adjacent plant communities: a dense stand of the ectomycorrhizal shrub whiteleaf manzanita (*Arctostaphylos viscida* Parry) that had been established by wildfire; an annual grass meadow established during the 1920s when forest was cleared for agri-culture (the grasses would likely be either facultatively VA mycorrhizal or non-mycorrhizal); and an open stand of ectomycorrhizal Oregon white oak (*Quercus garryana* Dougl. ex Hook.). Within each community Douglas fir seedlings received one of three treatments: (i) unpasteurized soil from a nearby ectomycorrhizal Pacific madrone stand (150 ml of soil added to the planting hole at the time of planting); (ii) pasteurized soil from the same madrone stand; or (iii) no added soil.

By the end of the second growing season >90% of seedlings planted in the manzanita stand had survived, while in the meadow <50% survived and in the oak stand <20% (Fig. 8a). By the end of the third growing season all seedlings planted beneath the oaks were dead and only one seedling remained alive in the meadow, whereas survival in the manzanita stand was still >90%. Madrone soil transfers did not influence survival, but, when unpasteurized, more than doubled basal area growth of seedlings plnted amidst the manzanita (Fig. 8b). Pasteurized madrone soil had no effect; and neither pasteurized nor unpasteurized soil altered growth of seedlings in the pasture or beneath the oaks. Analyses of soil macro-nutrients and seasonal water content revealed no abiotic differences among sites that could explain these results. Seedlings formed quite different types of mycor-rhizae when planted amidst the manzanita than in the meadow or oak stand, and soon after planting had dense hyphal development around their roots that we believe came from mycorrhizae on the surrounding manzanita. Work on another site shows that Douglas fir seedlings planted within 0.3 m of a manzanita form root tips more rapidly than do seedlings planted 1.2 m or further from a manzanita (Amaranthus, Molina & Perry 1989).

Biotic effects associated with non-pasteurized madrone soil varied with where the seedlings were planted. In the manzanita stand, seedlings with unpasteurized madrone soil formed twice as many mycorrhizal (and total) root tips as those that

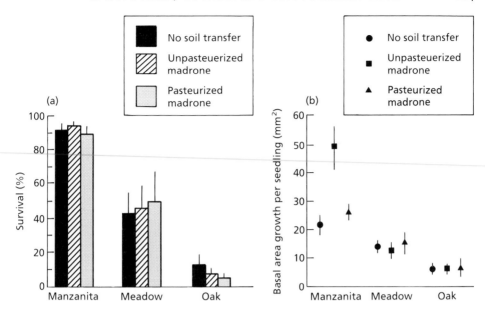

FIG. 8. Per cent survival (a) and basal area growth (b) for Douglas fir seedlings 2 years after outplanting to one of three vegetation areas. Seedlings received no soil transfer or were inoculated with unpasteurized or with pasteurized madrone soil (Amaranthus & Perry 1989).

received pasteurized madrone soil or none. One of those mycorrhizae was unique to the non-pasteurized soil treatment. Non-pasteurized madrone soil induced this same mycorrhiza type in the meadow, but not under the oaks. Total mycorrhizae formation was not influenced by soil treatment in either the meadow or the oaks.

Unpasteurized madrone soil produced the most strikingly different effects among vegetation communities with regard to $N_2$-fixation within seedling rhizospheres (estimated by acetylene reduction). When added to seedlings growing amidst the manzanita, it increased the amounts of acetylene reduced in rhizospheres by over 500% relative to seedlings receiving pasteurized madrone soil or none (Amaranthus, Li & Perry 1990). But when added to seedlings growing in the meadow, unpasteurized madrone soil *lowered* the amount of acetylene reduced by 80%. (Seedlings planted in oak were not tested.) Apparently there is positive synergism between the biota of madrone and manzanita soils, and negative synergism between the biota of madrone and meadow soils, but at this point we do not understand the biology of these interactions.

In another study conducted in the mixed communities of south-west Oregon, greenhouse-grown Douglas fir seedlings were inoculated with soils collected at various distances from ectomycorrhizal hardwoods in two 5-year-old clear-cuts (Borchers & Perry 1990). The hardwoods, averaging about 1.5 m in height, had sprouted from roots after clear-cutting. Of the several species present, the study focused on three: Pacific madrone, tanoak (*Lithocarpus densiflora* (Hook & Arn.)

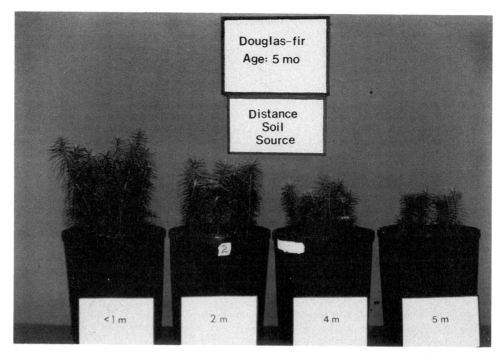

F IG. 9. Randomly selected 5-month-old Douglas fir seedlings grown in soil collected at different distances (left to right: <1 m, 2 m, 4 m and 5 m) from hardwood trees in an Oregon clear-cut (from Borchers & Perry 1990).

Rehd.), and canyon live oak (*Quercus chrysolepis* Liebm.). Conifer seedlings were sparse and soils were collected well away from those that were present.

Five-month-old Douglas fir grown in soil from beneath hardwood crowns (hardwood soils) were two times larger in total dry weight ($P < 0.01$) and had 60% more mycorrhizal root tips ($P < 0.05$) than those grown in open soil >4 m from a hardwood tree. The differences among seedlings grown in the various soils is apparent in Fig. 9. Those grown in hardwood soils also formed different mycorrhizal associations than those grown in open soils (Fig. 10). The three hardwood species were quite consistent in their effects.

Soils from beneath hardwoods had lower bulk densities, greater litter depths, greater moisture content, and were less acid than were open soils. Although there were no consistent differences in total soil nutrients between hardwood and open soils, available N (measured by anaerobic incubation) was two to seven times greater beneath hardwoods than in the open. Seedlings grown in hardwood soils averaged higher foliar N concentration than did those grown in open soil, but differences were not significant at $P = 0.10$. The only difference in foliar nutrients that was consistently significant was in the ratio of Fe to Mn, which ranged from two to four times greater in seedlings grown in hardwood soils than in seedlings in

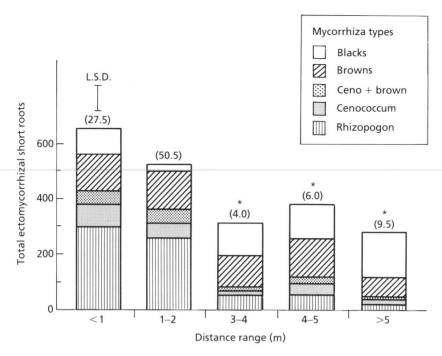

FIG. 10. Mycorrhiza formation by Douglas fir seedlings grown in soil collected at different distances from hardwood trees in an Oregon clear-cut. (From Borchers & Perry 1990.) (L.S.D., *$P \leq 0.05$.)

open soils. This did not, however, reflect differences in soil Fe and Mn—in fact, quite the opposite. Fe averaged lower and Mn higher in hardwood than in open soils.

Four lines of evidence suggest that the positive effect of hardwood soils on Douglas fir that was found in the Borchers and Perry (1990) study was primarily biological.

1 Examination of hardwood roots revealed mycorrhiza types similar to those formed by Douglas fir grown in hardwood soils, supporting the hypothesis that Douglas fir mycorrhizae were formed from inocula derived from hardwood mycorrhizae.

2 The higher foliar Fe:Mn ratio in seedlings grown in hardwood soils cannot be explained by soil contents of these nutrients, nor by pH differences (Fe should be more soluble in the relatively low pH open soils). In fact, Fe uptake in most plants depends on Fe-chelators that are produced by mycorrhizal fungi and rhizosphere bacteria (Bossier, Hofte & Verstraete 1988). Perry *et al.* (1984) found that one important class of these chelators, the hydroxymate siderophores, occurred in significantly lower concentrations in clear-cut than in forest soils, and that seedlings grown in clear-cut soil deficient in hydroxymate siderophores were Fe limited (Perry *et al.* 1984). Siderophores were not measured in the Borchers and Perry (1990) study; however, the possibility exists that hardwoods stabilized

the microflora that produce siderophores, thereby enhancing Fe uptake in seedlings.

**3** High levels of anaerobically determined mineralizable N in hardwood soils reflect greater microbial activity (Myrold 1987); this can be explained in part by the relatively low acidity and high water content in hardwood soils, but the primary factor is likely to be the greater availability of labile carbon in the form of litter and root exudates. Most microbial activity in soils occurs within rhizospheres (Katznelson, Lochhead & Timonin 1948; Starkey 1958; Atlas & Bartha 1987).

**4** Although hardwood soils had greater microbial biomass, preliminary isolations by C.Y. Li (unpublished) show that open soils have ten times more colonies of the common soil actinomycete *Streptomyces* spp. These organisms, which are among the few that can degrade lignin, are relatively tolerant of environmental extremes. Their abundance in open soils that are low in labile carbon and relatively harsh microclimatically is, therefore, not surprising. *Streptomyces* spp. are known to produce chemicals (e.g. the drug streptomycin) that inhibit plants and other micro-organisms, including both mycorrhizal and pathogenic fungi. *Streptomyces* spp. may play a role in reforestation failures in south-west Oregon (Perry & Rose 1983; Friedman *et al.* 1989), but some types that occur in rhizospheres benefit plants by inhibiting pathogens (Secilia & Bagyaraj 1987) and stimulating symbiotic N fixation (Rojas Melo 1989). Interestingly, the adverse effects of streptomycin on plants can be reversed by manganese (Rosen 1954; Gray 1955), which was present in greater concentrations in hardwood than in open soils.

## Summary of links through time

Clearly, soils beneath some ectomycorrhizal hardwood species do benefit Douglas fir seedlings. This phenomenon has been associated in one or more studies with accelerated root tip formation, greater numbers of total and mycorrhizal root tips, shifts in mycorrhizal type, increased nitrogen-fixation in seedling rhizospheres, or increased ratios of Fe to Mn in seedling foliage. Direct transfer of mycorrhizal inocula probably plays a role, but the story is more complex than this and apparently involves a range of biological, chemical and physical factors. At least some of the effects that we have seen on Douglas fir seedlings are probably the result of interactions among the various soil factors (Bowen & Theodorou 1979). For example, certain types of microbes that occur commonly in rhizospheres and within the mycorrhizal 'mantle' stimulate mycorrhizal formation (Garbaye & Bowen 1989) and mycorrhizal roots, through their influence on exudates, shape the composition of the surrounding rhizosphere community (Katznelson, Rouatt & Peterson 1962; Rambelli 1973; Meyer & Linderman 1986; Secilia & Bagyaraj 1987). Compared to non-mycorrhizal roots, mycorrhizal roots and rhizospheres have fewer pathogens, more nitrogen fixers, more bacteria that require complex nutritional factors (especially amino acids) from their environment and more actinomycete isolates that inhibit pathogens. Early successional hardwoods are probably stabilizing entire rhizosphere communities.

## DOES RANDOMNESS CATALYSE COOPERATION IN NATURE?

The picture that emerges is of plant species linked through mycorrhizal fungi and rhizosphere microbes in a network of cooperation and mutual benefit: carbon and nutrients pass from one species to another through hyphae, reducing conflict and promoting diversity. Entire sets of below-ground mutualists that are needed by one plant species are protected and nurtured by another. Can this be reconciled with a neo-Darwinian world view? If fungi are in the business of acquiring carbon, then why do they pass it from one plant to another? Why should one plant species support a mycorrhizal fungus that has the potential to benefit a competitor? Viewed in the short term, this contradicts the idea of communities (at least in the plant kingdom) being structured primarily by competition. When viewed within an evolutionary time-scale, however, and within the context of variable and unpredictable environments, cooperation among potentially competing plants can be explained as reciprocal altruism (Axelrod & Hamilton 1981); for any given species, the disadvantages of competition are outweighed by the insurance and buffering capacity that result from having diverse neighbours.

Two general requirements, that can be applied not only to plants and mycorrhizal fungi, but also to a wide variety of interactions within ecosystems, underly this premise: namely, positive feedback and hiatus. Positive feedback occurs when plants alter their biotic and abiotic environments in ways that benefit the plant's growth. As in any system maintained in disequilibrium by energy inputs, the stability of mycorrhizal fungi and all other components of the ecosystem that are linked to plants (or particular groups of plants) through positive feedback depends on the presence of the plants. During hiatus, a given plant species is periodically unable to maintain links to those components of the ecosystem with which it participates in positive feedback and must depend on others to maintain those links.

These conditions are widely met in forests (discussions on positive feedback in ecosystems include DeAngelis, Post and Travis (1986), Perry *et al.* (1989a, 1990)). Mutualisms are a common form of positive feedback in ecosystems. Virtually all trees participate in one or more mutualisms that always include mycorrhizal fungi, and might also include various other relationships with pollinators, seed dispersers, ants, obligate rhizosphere organisms and fungal endophytes in leaves and stems (Boucher 1985; Carroll 1988).

Hiatus is common in forests. Large and small disturbances periodically alter the species composition on a given piece of ground, creating strong selection pressures for generality in mycorrhizal fungi and other obligate rhizosphere organisms. It is clearly advantageous to the fungus to have as many potential partners as possible, but what does the plant gain by sharing its mutualists with others? Plants are not prisoners of the fungi and readily shed them when they become disadvantageous, suggesting that plants can withdraw support from partners that bestow favours on rivals. However, plants may from time to time rely on these same rivals to support their common mutualists or to ship resources through hyphal links. In such a case the cost–benefit trade-offs associated with cooperation may be quite different when evaluated in the long run rather than the short run. This can be illustrated from our studies in south-west Oregon, where wildfire has been the common disturbance. The intensity of any given fire varies widely across the landscape depending on numerous factors, such as regional and local climate, time of day, topography, stand structure and the direction of the fire in relation to slope. The resulting burn pattern can be very patchy with some areas completely destroyed and others only underburned. Yet others may not be touched. Because of the essentially random nature of the forcing factors, especially climate, the burn pattern has a large random component.

In this milieu, considerable uncertainty exists as to which plant species will have the adaptation necessary for survival or recovery. Thick-barked conifers survive ground fires quite well, but after a crown fire the hardwoods recover quickly by sprouting from rootstocks or by germinating from buried seeds. Where hardwoods survive, conifers that share mycorrhizal fungi with them gain an advantage because their mutualists will be maintained in their absence. Roles are reversed after an underburn: here the conifers have the upper hand, while the hardwoods—usually in a subdominant position in mature forests—are burnt back to the ground and must resprout. A sprout could have a tough time succeeding if the overstorey conifers were not only shading it, but also actively competing for nutrients and water. By sharing mycorrhizal fungi with the conifers, the hard-woods gain a ready source of inocula and perhaps some carbon and nutrients by way of hyphal links.

Randomness, or perhaps more accurately, stochasticity—expressed as unpredictability in the nature and timing of disturbance—is a key element of this scenario. Where disturbances are predictable, a single adaptation has a high probability of succeeding and the selective pressure for cooperation is lessened. For example, Oregon white oak (which is ectomycorrhizal, but in contrast to other oaks in south-west Oregon is antagonistic to Douglas fir) grows in open stands with little understorey to carry fires into the crowns. Under these conditions white oak primarily experiences ground fires, which it can survive due to its thick bark. In ponderosa pine forests where frequent light underburns are the norm, understorey plants are VA mycorrhizal, whereas in high elevation forests of western North America, where fires are severe but patchy, understories frequently contain ericaceous plants that share at least some ectomycorrhizae with conifers (G. Hunt, pers. comm.). Both the change in predominant mycorrhizal types as trees age, and the changing mycorrhizal dependency of plant species through succession, probably play some role in the efficacy with which one plant species replaces another during succession (Allen & Allen 1990); however, little is known about this. Plant composition alone may tell us little about the degree to which mycorrhizal fungi are shared, or not shared, among different species: ectomycorrhizal fungi have been recently reported to colonize herbs and grasses formerly thought to be strictly VA mycorrhizal or non-mycorrhizal (Theodorou &

Bowen 1971; Kope & Warcup 1986; Trappe 1988; Warcup 1988), while Cazares (pers. comm.) recently found VA mycorrhizae on Douglas fir, a species thought to be strictly ectomycorrhizal.

Another kind of unpredictability occurs in moist tropical rain forest, where the most common disturbance is tree-fall. Which of the many candidate species is positioned to colonize a tree-fall gap depends on such factors as the time of year when the gap is created and how large it is. The stochasticity associated with when and how much, along with the large number of potential colonizers in these species-rich forests, produces considerable uncertainty in the composition of the early successional community. Perhaps it is no coincidence that most trees in these systems form mycorrhizae with the highly general VA mycorrhizal fungi.

The validity of our argument hinges on, among other things, how much one plant species really depends on another to support its below-ground mutualists in its absence. This is related to how successfully the mycorrhizal fungi or other soil mutualist can maintain itself in the absence of a plant host—which it might do by entering a resting stage, by becoming saprophytic, or by maintaining an effective input of propagules onto the site. Much remains to be learned about all of these possibilities. At least some ectomycorrhizal fungi have limited saprophytic capability, probably as a way of extracting nutrients from soil organic matter (e.g. Read 1988), but whether or not a free-living mycorrhizal fungus can compete successfully with saprophytes for C is unclear. Many ectomycorrhizal fungi produce abundant airborne spores, but an important subset, along with all VA mycorrhizal fungi, fruit below ground and are thought to be dispersed primarily by animals. Even those fungal species that fruit above ground are unlikely to form abundant mycorrhizae through airborne spores because a critical number of spores (as yet ill defined, but certainly more than a few) is required to initiate an infection. All available evidence indicates that the mycorrhizal inoculation potential of soils declines in the absence of hosts, sometimes quite rapidly (Perry, Molina & Amaranthus 1987; Jasper, Abbott & Robson 1989a–c). Also, establishing seedlings form more mycorrhizae and form them more rapidly when their roots are in contact with the mycorrhizal roots of an established plant than when they are not.

Decline in mycorrhizal inoculation potential is only one of a number of changes that might occur in soils where indigenous plants are weakened or removed. These changes include nutrient loss, significant changes in soil biology, and degradation of soil physical structure that includes loss of surface clays and decline in the numbers of large soil aggregates (Low 1955; Adejuwon & Ekanade 1987; Perry, Molina & Amaranthus 1987; Meyer et al. 1988; Oren et al. 1988; Jasper, Abbott & Robson 1989a,b; Perry et al. 1989a; Schlesinger et al. 1990). Soils altered in this way are less hospitable for establishing plants, creating a positive feedback that can lead to rapid degradation (Perry et al. 1989a). The implication here is that plants have much more in common than shared mycorrhizal fungi—the integrity of the entire soil is at stake.

More work is needed on how quickly such changes occur and how readily they are reversed. At a time when ecosystems are subjected to increasing anthropogenic stresses of one kind or another, this becomes urgent. The potential for such pervasive and difficult-to-reverse changes in soils has significant implications for ecosystem stability during climate change (Perry *et al.* 1990)

## THOUGHTS ON DISEQUILIBRIUM ECOSYSTEMS AND IMPLICATIONS FOR SYSTEM STABILITY IN A RAPIDLY CHANGING WORLD

The tree–soil interactions discussed here are part of a larger system structure characterized by strong positive and negative feedbacks and ecosystems that exist far from any kind of equilibrium, whether it be with local environments, soils or among the species that compose the system (DeAngelis & Waterhouse 1987). Positive feedbacks include not only the mutualisms that pervade ecosystems, but various interactions between organisms and their physical environment of which plant–soil interactions are only one example. For instance, tree crowns rake water and nutrients that contribute to the growth of more crowns. Recent models suggest that evapotranspiration by the forests of Amazonia creates the rainfall that allows forests to persist in the region (Lean & Warrilow 1989; Shukla, Nobre & Sellers 1990).

In ecosystems, strong positive feedback is both a blessing and a curse because it enhances productivity and health of the participants while creating the potential for threshold degradation (DeAngelis, Post & Travis 1986; Perry *et al.* 1989a, 1990). Systems are buffered against such threshold changes by redundancies—of which diffuse mutualisms are a prime example.

Such a system structure sheds new light on the old debate about whether diversity confers stability: it does, but the relation is non-linear—species can be lost from a system with no discernible effect on processes but with a progressive weakening of resiliency until even a minor disturbance can produce an abrupt change in system behaviour (Holling 1988; Perry *et al.* 1990). There is no question that such threshold changes have occurred in terrestrial ecosystems (Perry *et al.* 1989a; Schlesinger *et al.* 1990). If ecologists are to provide policy makers and the general public with the information needed to avoid 'disagreeable surprises' (Street-Perrot & Perrot 1990) in today's world of unprecedented stresses, much more must be learned about how ecosystems are structured, particularly with regard to stabilizing redundancies and the potential for threshold changes.

## ACKNOWLEDGMENTS

Research was supported by National Science Foundation Grants BSR-8400403 and 851.4325 (Long-term Ecological Research). This is Paper 2709, Forest Research Laboratory, Oregon State University, Corvallis.

# REFERENCES

**Adejuwon, J.O. & Ekanade, O. (1987).** Edaphic component of the environmental degradation resulting from the replacement of tropical rain forest by field and tree crops in SW Nigeria. *International Tree Crops Journal*, **4**, 269–82.

**Ae, N., Arihara, J., Okada, K., Yoshihara, T. & Johansen, C. (1990).** Phosphorus uptake by pigeon pea and its role in cropping systems of the Indian subcontinent. *Science*, **248**, 477–80.

**Allen, E.B. & Allen, M.F. (1990).** The mediation of competition by mycorrhizae in successional and patchy environments. *Perspectives on Plant Competition* (Ed. by J.B. Grace & D. Tilman), pp. 367–89. Academic Press, New York.

**Amaranthus, M.P. & Perry, D.A. (1987).** The effect of soil transfers on ectomycorrhizal formation and the survival and growth of conifer seedlings on old, nonreforested clearcuts. *Canadian Journal of Forest Research*, **17**, 944–50.

**Amaranthus, M.P. & Perry, D.A. (1989).** Interaction effects of vegetation type and Pacific madrone soil inocula on survival, growth, and mycorrhiza formation of Douglas-fir. *Canadian Journal of Forest Research*, **19**, 550–6.

**Amaranthus, M.P., Molina, R. & Perry, D.A. (1989).** Soil organisms, root growth and forest regeneration. *Forestry on the Frontier*, pp. 1–5. Proceedings, Society of American Foresters National Convention, 1989, Spokane, Washington. Society of American Foresters, Bethesda, Maryland.

**Amaranthus, M.P., Li, C.Y. & Perry, D.A. (1990).** Influence of vegetation type and madrone soil inoculum on associative nitrogen fixation in Douglas-fir rhizospheres. *Canadian Journal of Forest Research*, **20**, 368–71.

**Ashton, P.S. (1988).** Dipterocarp biology as a window to the understanding of tropical forest structure. *Annual Review of Ecology and Systematics*, **19**, 347–70.

**Atlas, R.M. & Bartha, R. (1987).** *Microbial Ecology: Fundamentals and Applications*. Benjamin/Cummings Publishing, Menlo Park, California.

**Axelrod, R. & Hamilton, W.D. (1981).** The evolution of cooperation. *Science*, **211**, 1390–6.

**Bjorkman, E. (1960).** *Monotropa hypopitys* L.—an epiparasite on tree roots. *Physiologia Plantarum*, **13**, 308.

**Borchers, S.L. & Perry, D.A. (1990).** Growth and ectomycorrhiza formation of Douglas-fir seedlings grown in soils collected at different distances from pioneering hardwoods in southwest Oregon clear-cuts. *Canadian Journal of Forest Research*, **20**, 712–21.

**Bossier, P., Hofte, M. & Verstraete, W. (1988).** Ecological significance of siderophores in soil. *Advances in Microbial Ecology*, **10**, 385–414.

**Boucher, D.H.** (Ed.) **(1985).** *The Biology of Mutualism*. Oxford University Press, New York.

**Bowen, G.D. (1980).** Misconceptions, concepts, and approaches in rhizosphere biology. *Contemporary Microbial Ecology* (Ed. by D.C. Ellwood, J.N. Hedger, M.J. Latham, J.M. Lynch & J.M. Slater), pp. 283–304. Academic Press, London.

**Bowen, G.D. & Theodorou, C. (1979).** Interactions between bacteria and ectomycorrhizal fungi. *Soil Biology & Biochemistry*, **11**, 119–26.

**Carroll G. (1988).** Fungal endophytes in stem and leaves: from latent pathogen to mutualistic symbiont. *Ecology*, **69**, 2–9.

**Chiariello, N., Hickman, J.C. & Mooney, H.A. (1982).** Endomycorrhizal role for interspecific transfer of phosphorus in a community of annual plants. *Science*, **217**, 941–3.

**Crafts, A.S. & Yamaguchi, S. (1958).** Comparative tests and the uptake and distribution of labeled herbicides by *Zebrina pendula* and *Tradescantia fluminensis*. *Hilgardia*, **27**, 421–54.

**Cromack, K., Jr., Sollins, P., Graustein, W.C., Speidel, K., Todd, A.W., Sphycher, G., Li, C.Y. & Todd, R.L. (1979).** Calcium oxalate accumulation and soil weathering in mats of the hypogeous fungus *Hysterangium crassum*. *Soil Biology & Biochemistry*, **11**, 463–8.

**DeAngelis, D.L. & Waterhouse, J.C. (1987).** Equilibrium and nonequilibrium concepts in ecological models. *Ecological Monographs*, **57**, 1–21.

**DeAngelis, D.L., Post, W.M. & Travis, C.C. (1986).** *Positive Feedback in Natural Systems*. Springer, Berlin.

Dueck, T.A., Visser, P., Ernst, W.H.O. & Schat, H. (1986). Vesicular-arbuscular mycorrhizae decrease zinc-toxicity to grasses growing in zinc-polluted soil. *Soil Biology & Biochemistry*, **18**, 331–3.

Finlay, R.D. & Read, D.J. (1986a). The structure and function of the vegetative mycelium of ectomycorrhizal plants. I. Translocation of $^{14}C$-labeled carbon between plants interconnected by a common mycelium. *New Phytologist*, **103**, 143–56.

Finlay, R.D. & Read, D.J. (1986b). The structure and function of the vegetative mycelium of ectomycorrhizal plants. II. The uptake and distribution of phosphorus by mycelial strands interconnecting host plants. *New Phytologist*, **103**, 157–65.

Fitter, A.H. (1977). Influence of mycorrhizal infection on competition for phosphorus and potassium by two grasses. *New Phytologist*, **79**, 119–25.

Fitter, A.H. (1989). The role and ecological significance of vesicular-arbuscular mycorrhizas in temperate ecosystems. *Agriculture, Ecosystems, and Environment*, **29**, 137–51.

Fleming, L.V. (1984). Effects of soil trenching and coring on the formation of ectomycorrhizas on birch seedlings grown around mature trees. *New Phytologist*, **98**, 143–53.

Francis, R. & Read, D.J. (1984). Direct transfer of carbon between plants connected by vesicular-arbuscular mycorrhizal mycelium. *Nature*, **307**, 53–6.

Friedman, J., Hutchins, A., Li, C.Y. & Perry, D.A. (1989). Actinomycetes inducing phytotoxic or fungistatic activity in a Douglas-fir forest and in an adjacent area of repeated regeneration failure in southwestern Oregon. *Biologia Plantarum*, **31**, 487–95.

Garbaye, J. & Bowen, G.D. (1989). Stimulation of ectomycorrhizal infection of *Pinus radiata* by some microorganisms associated with the mantle of ectomycorrhizas. *New Phytologist*, **112**, 383–8.

Gartlan, J.S., Newberry, D. McC., Thomas, D.W. & Waterman, P.G. (1986). The influence of topography and soil phosphorus on the vegetation of Korup Forest reserve, Cameroun. *Vegetatio*, **65**, 131–48.

Gray, R.A. (1955). Inhibition of root growth by streptomycin and reversal of the inhibition by manganese. *American Journal of Botany*, **42**, 327–31.

Hall, I.R. (1978). Effects of endomycorrhizas on the competitive ability of white clover. *New Zealand Journal of Agricultural Research*, **21**, 509–15.

Harley, J.L. & Smith, S.E. (1983). *Mycorrhizal Symbiosis*. Academic Press, London.

Harper, J. (1977). *Population Biology of Plants*. Academic Press, London.

Hayman, D.S. (1978). Endomycorrhizae. *Interactions Between Non-pathogenic Soil Microorganisms and Plants* (Ed. by Y.R. Dommergues & S.V. Krupa), pp. 401–42. Elsevier, Amsterdam.

Hogberg, P. (1986). Soil nutrient availability, root symbioses and tree species composition in tropical Africa: a review. *Journal of Tropical Ecology*, **2**, 359–72.

Holling, C.S. (1988). Temperate forest insect outbreaks, tropical deforestation, and migratory birds. *Memoirs of the Entomological Society of Canada*, **146**, 21–32.

Holm, L.R., Plucknett, D.L., Pancho, J.V. & Herberger, J.P. (1977). *The World's Worst Weeds*. University of Hawaii Press, Honolulu.

Janos, D.P. (1980). Mycorrhizae influence tropical succession. *Biotropica*, **12**, 56–64.

Janos, D.P. (1983). Tropical mycorrhizas, nutrient cycles and plant growth. *Tropical Forest: Ecology and Management* (Ed. by S.L. Sutton, T.C. Whitmore & A.C. Chadwick), pp. 327–45. Blackwell Scientific Publications, Oxford.

Janos, D.P. (1987). VA mycorrhizas in humid tropical ecosystems. *Ecophysiology of VA Mycorrhizal Plants* (Ed. by G.R. Safir), pp. 107–34. CRC Press, Boca Raton, Florida.

Janos, D.P. (1988). Mycorrhiza applications in tropical forestry: are temperate-zone approaches appropriate? *Trees and Mycorrhiza* (Ed. by F.S.P. Ng), pp. 133–88. Forest Research Institute, Kuala Lumpur, Malaysia.

Jasper, D.A., Abbott, L.K. & Robson, A.D. (1989a). The loss of mycorrhizal infectivity during bauxite mining may limit the growth of *Acacia pulchella*. *Australian Journal of Botany*, **37**, 33–42.

Jasper, D.A., Abbott, L.K. & Robson, A.D. (1989b). Soil disturbance reduces the infectivity of external hyphae of vesicular-arbuscular mycorrhizal fungi. *New Phytologist*, **112**, 93–9.

Jasper, D.A., Abbott, L.K. & Robson, A.D. (1989c). Hyphae of a vesicular-arbuscular mycorrhizal fungus maintain infectivity in dry soil, except when the soil is disturbed. *New Phytologist*, **112**, 101–7.

Jones, M.D., Dainty, J. & Hutchinson, T.C. (1988). The effect of infection by *Lactarius rufus* or *Scleroderma flavidum* on the uptake of [63]Ni by paper birch. *Canadian Journal of Botany*, **66**, 934–40.

Katznelson, H., Lochhead, A.G. & Timonin, M.I. (1948). Soil microorganisms and the rhizosphere. *Botanical Review*, **14**, 543–87.

Katznelson, H., Rouatt, J.W. & Peterson, E.A. (1962). The rhizosphere effect of mycorrhizal and non-mycorrhizal roots of yellow birch seedlings. *Canadian Journal of Botany*, **40**, 257–76.

Kope, H.H. & Warcup, J.H. (1986). Synthesized ectomycorrhizal associations of some Australian herbs and shrubs. *New Phytologist*, **104**, 591–9.

Lean, J. & Warrilow, D.A. (1989). Simulation of the regional climatic impact of Amazon deforestation. *Nature*, **342**, 411–13.

Low A.J. (1955). Improvements in the structural state of soils under leys. *Soil Science*, **6**, 179–99.

Lynch, J.M. & Bragg, E. (1985). Microorganisms and soil aggregate stability. *Advances in Soil Science*, **2**, 133–71.

Malloch, D.W., Pirozynski, K.A. & Raven, P.H. (1980). Ecological and evolutionary significance of mycorrhizal symbioses in vascular plants. *Proceedings of the National Academy of Science*, **77**, 2112–18.

Marx, D.H. & Krupa, S.V. (1978). Ectomycorrhizae. *Interactions Between Non-pathogenic Soil Microorganisms and Plants* (Ed. by Y.R. Dommergues & S.V. Krupa), pp. 373–400. Elsevier, Amsterdam.

Mason, P.A., Wilson, J., Last, F.T. & Walker, C. (1983). The concept of succession in relation to the spread of sheathing mycorrhizal fungi in inoculated tree seedlings growing in unsterile soil. *Plant and Soil*, **71**, 247–56.

Meyer, J.R. & Linderman, R.G. (1986). Selective influence on populations of rhizosphere or rhizoplane bacteria and actinomycetes by mycorrhizas formed by *Glomus fasciculatum*. *Soil Biology & Biochemistry*, **18**, 191–6.

Meyer, J., Schneider, B.U., Werk, K.S., Oren, R. & Schulze, E.D. (1988). Performance of two *Picea abies* stands at different stages of decline. V. Root tip and ectomycorrhiza development and their relation to above-ground and soil nutrients. *Oecologia (Berlin)*, **77**, 7–13.

Myrold, D.D. (1987). Relationship between microbial biomass nitrogen and a nitrogen availability index. *Soil Science Society of America Journal*, **51**, 1047–9.

Newbery, D.M., Alexander, I.J., Thomas, D.W. & Gartlan, J.S. (1988). Ectomycorrhizal rain-forest legumes and soil phosphorus in Korup National Park, Cameroon. *New Phytologist*, **109**, 433–50.

Newman, E.I. (1988). Mycorrhizal links between plants: their functioning and ecological significance. *Advances in Ecological Research*, **18**, 242–71.

Newman, E.I. & Ritz, K. (1986). Evidence on the pathways of phosphorus transfer between vesicular-arbuscular mycorrhizal plants. *New Phytologist*, **104**, 77–87.

Oren, R., Schulze, E.D., Werk, K.S. & Meyer, J. (1988). Performance of two *Picea abies* stands at different stages of decline. VII. Nutrient relations and growth. *Oecologia (Berlin)*, **77**, 163–73.

Perry, D.A. & Rose, S.L. (1983). Soil biology and forest productivity: opportunities and constraints. *IUFRO Symposium on Forest Site and Continuous Productivity* (Ed. by R. Ballard & S.P. Gessel), pp. 229–38. United States Department of Agriculture Forest Service General Technical Report PNW-163.

Perry, D.A., Rose, S.L., Pilz, D. & Schoenberger, M.M. (1984). Reduction of natural ferric iron chelators in disturbed forest soils. *Soil Science Society of America Journal*, **48**, 379–82.

Perry, D.A., Molina, R. & Amaranthus, M.P. (1987). Mycorrhizae, mycorrhizospheres, and reforestation: current knowledge and research needs. *Canadian Journal of Forest Research*, **17**, 929–40.

Perry, D.A., Amaranthus, M.P., Borchers, J.G., Borchers, S.L. & Brainerd, R.E. (1989a). Bootstrapping in ecosystems. *BioScience*, **39**, 230–7.

Perry, D.A., Margolis, H., Choquette, C., Molina, R. & Trappe, J.M. (1989b). Ectomycorrhizal mediation of competition between coniferous tree species. *New Phytologist*, **112**, 501–11.

Perry, D.A., Borchers, J.G., Borchers, S.L. & Amaranthus, M.P. (1990). Species migrations and ecosystem stability during climate change: the below-ground connection. *Conservation Biology*, **14**, 266–74.

Pilz, D.P. & Perry, D.A. (1984). Impact of clearcutting and slash burning on ectomycorrhizal associations of Douglas-fir. *Canadian Journal of Forest Research*, 14, 94–100.

Pirozynski, K.A. (1981). Interactions between fungi and plants through the ages. *Canadian Journal of Botany*, 59, 1824–7.

Puga, C. (1985). *Influence of vesicular-arbuscular mycorrhizas on competition between corn and weeds in Panama*. M.S. thesis, University of Miami.

Rahteenko, I.N. (1958). On the transfer of mineral nutrients from one plant to another owing to interaction between the root systems. [transl. title]. *Botanicheskii Zhurnal*, 43, 695–701 [in Russian]. (In Forestry Abstracts 20, 1433, 1959.)

Rambelli, A. (1973). The rhizosphere of mycorrhizae. *Ectomycorrhizae: their Ecology and Physiology* (Ed. by A.C. Marks & T. Kozlowski), pp. 229–49. Academic Press, London.

Read, D.J. (1988). Development and function of mycorrhizal hyphae in soil. *Mycorrhizae in the Next Decade* (Ed. by D.M. Sylvia, L.L. Hung & J.H. Graham), pp. 178–80. Proceedings, 7th North American Conference on Mycorrhizae. University of Florida Press, Gainesville, Florida.

Read, D.J., Francis, R. & Finlay, R.D. (1985). Mycorrhizal mycelia and nutrient cycling in plant communities. *Ecological Interactions in Soil* (Ed. by A.H. Fitter, D.J. Read & M.B. Lusher), pp. 193–217. Blackwell Scientific Publications, Oxford.

Reid, C.P.P. & Woods, F.W. (1969). Translocation of $^{14}$C labeled compounds in mycorrhiza and its implications in interpreting nutrient cycling. *Ecology*, 50, 179–81.

Rojas Melo, N. (1989). *The influence of actinomycetes isolated from soil, root, and nodule surfaces on nodulation and nitrogen fixation of red alder seedlings*. MS thesis, Oregon State University.

Rose, S.L., Perry, D.A., Pilz, D. & Schoenberger, M.M. (1983). Allelopathic effects of litter on the growth and colonization of mycorrhizal fungi. *Journal of Chemical Ecology*, 9, 1153–62.

Rosen, W.G. (1954). Plant growth inhibition by streptomycin and its prevention by manganese. *Society for Experimental Biology*, 85, 385–8.

Rosendahl, S., Rosendahl, C.N. & Sochting, U. (1989). Distribution of VA mycorrhizal endophytes amongst plants from a Danish grassland community. *Agriculture, Ecosystems, and Environment*, 29, 329–35.

Schlesinger, W.H., Reynolds, J.F., Cunningham, G.L., Huenneke, L.F., Jarrell, W.M., Virginia, R.A. & Whitford, W.G. (1990). Biological feedbacks in global desertification. *Science*, 247, 1043–8.

Schoenberger, M.M. & Perry, D.A. (1982). The effect of soil disturbance on growth and ectomycorrhizae of Douglas-fir and western hemlock seedlings: a greenhouse bioassay. *Canadian Journal of Forest Research*, 12, 343–53.

Secilia, J. & Bagyaraj, D.J. (1987). Bacteria and actinomycetes associated with pot cultures of vesicular-arbuscular mycorrhizas, pp. 1069–73. University of Agricultural Sciences, Bangalore, India.

Shukla, J., Nobre, C. & Sellers, P. (1990). Amazon deforestation and climate change. *Science*, 247, 1322–5.

Sieverding, E. (1989). Ecology of VAM fungi in tropical agrosystems. *Agriculture, Ecosystems, and Environment*, 29, 369–90.

Slankis, V. (1974). Soil factors influencing formation of mycorrhizae. *Annual Review of Phytopathology*, 12, 437–57.

Starkey, R.L. (1958). Interrelations between microorganisms and plant roots in the rhizosphere. *Bacteriological Review*, 22, 154–72.

St John, T.V. & Coleman, D.C. (1982). The role of mycorrhizae in plant ecology. *Canadian Journal of Botany*, 61, 1005–14.

Street-Perrott, F.A. & Perrott, R.A. (1990). Abrupt climate fluctuations in the tropics: the influence of Atlantic Ocean circulation. *Nature*, 343, 607–11.

Sutton, J.C. & Sheppard, B.R. (1976). Aggregation of sand-dune soil by endomycorrhizal fungi. *Canadian Journal of Botany*, 54, 326–33.

Theodorou, C. & Bowen, G.D. (1971). Effects of non-host plants on growth of mycorrhizal fungi of radiata pine. *Australian Forestry*, 35, 17–22.

Trappe, J.M. (1987). Phylogenetic and ecological aspects of mycotrophy in the angiosperms from an evolutionary standpoint. *Ecophysiology of VA Mycorrhizal plants* (Ed. by G.R. Safir), pp. 5–25. CRC Press, Boca Raton, Florida.

Trappe, J.M. (1988). Lessons from alpine fungi. *Mycologia*, **80**, 1–10.

van Kessel, C., Singleton, P.W. & Hoben, H.J. (1985). Enhanced N-transfer from a soybean to maize by vesicular arbuscular mycorrhizal (VAM) fungi. *Plant Physiology*, **79**, 562–3.

Warcup, J.H. (1988). Mycorrhizal associations and seedling development in Australian Lobelioideae (Campanulaceae). *Australian Journal of Botany*, **36**, 461–72.

Whittingham, J. & Read, D.J. (1982). Vesicular-arbuscular mycorrhizas in natural vegetation systems, III. Nutrient transfer between plants with mycorrhizal interconnections. *New Phytologist*, **90**, 277–84.

Wilcox, H.E. (1983). Fungal parasitism of woody plant roots from mycorrhizal relationships to plant disease. *Annual Review of Phytopathology*, **21**, 221–42.

Woods, F.W. & Brock, K. (1964). Interspecific transfer of Ca-45 and P-32 by root systems. *Ecology*, **45**, 886.

# Canopies and microclimate of tree species mixtures

J.J. BARKMAN*

*Kampsweg 29, 9418 PD Wijster, The Netherlands*

## SUMMARY

The types and distribution of mixed and monospecific forests in Europe, and their relation to seminatural forests and afforestation, are discussed. Differences among monospecific forests of different species in light climate (intensity, quality, pattern), temperature (minima, maxima), precipitation (amount, distribution, pattern) and litter microclimate are described as a function of crown density, foliage periodicity, leaf size, branch angles and bark relief. Effects on the herb and moss layer and macrofungi are briefly discussed. Some prognoses are made of the effect of mixing trees on microclimate and undergrowth. As a rule, homogeneous mixtures of tree species are expected to produce intermediate conditions, except when the species differ in either leaf size or time of leaf shedding. Heterogeneous mixtures are likely to create a patchwork of microclimates typical of the constituent species, with one exception: clusters (small stands) of deciduous trees within an evergreen forest are predicted to have a special gap climate, called here the 'blue–green climate'.

## INTRODUCTION

Although measurements of microclimate in woods have been made for over 100 years, longer than in any other vegetation type, measurements in mixed forests are rare. In Mitscherlich's well-known book *Waldklima und Wasserhaushalt* (1971) one looks in vain for the word 'Mischwald' in the subject index. Indeed, there appears to be no published treatise on the possible special features of the microclimate of mixed woods as contrasted to monospecific stands. If differences exist, they should be studied by comparing mixed stands with 'pure' stands of all of its component species. These stands should have the same age, stem density, management, soil type and general climate as the mixed stands. They should be situated close to the mixed stands and be studied in the same years, both summer and winter. Little of this kind of research has been done, neither is it known whether a clustered distribution of different tree species creates a microclimate different from that of a homogeneous mixture.

There may be, however, enough direct or indirect data on monospecific stands to make prognoses for mixed ones. These data may come from the following

---

*It was with great sadness that we learned that Professor Barkman died in 1990 while on holiday in Scotland following this symposium. (Editors.)

sources: (i) direct observations on the microclimate of single-species stands of different trees, (ii) data on foliage density, leaf periodicity, leaf size and leaf angle, branch angle and bark relief of different species, and (iii) empirical and theoretical knowledge of the effects of these parameters on microclimate. I shall first discuss these variables and their effects, and then make prognoses of the microclimates of some mixtures and their possible bearing on undergrowth and soil life.

Data concerning (i) and (ii) are taken from Geiger (1961), Mitscherlich (1971) and Barkman and Stoutjesdijk (1987) and from references cited in these works. Data on (iii) are derived from Barkman (1979, 1988, 1989).

## MIXED FORESTS IN EUROPE

First, it is worth noting the natural situation; which kinds of tree mixtures occur in natural and seminatural forests in Europe? For more detailed data see Barkman (1990).

Although the temperate zone is much poorer in tree species than are the tropics, and within the temperate zone Europe is particularly poor, monospecific forests are probably rare. Mediterranean climax forests on limestone apparently consist of *Quercus ilex* only, which by its heavy shade throughout the year excludes all other trees. Similarly, in *Larix decidua/Pinus cembra* forests in the Alps, in the rare cases where they are untouched by man, the evergreen *Pinus cembra* with its dense crown may completely eliminate *Larix*. *Picea abies* may behave in the same way. But in old primeval forests one sees that natural gaps caused by fallen spruces and pines are temporarily filled by rowan and larch (Alps) or birch (Scandinavia). And since these gaps will be formed again and again, a mixed forest is likely to be the normal situation, although one species (the evergreen) is by far the dominant one. The same applies to virginal high-boreal pine forests (*Pinus sylvestris* var. *lapponica*) of northern Lapland. Here *Betula pubescens* is always present in natural gaps, germinating on fallen pine boles.

The monospecific oak woods (*Quercus robur*) in The Netherlands are not natural, but were mixed with beech in the early Middle Ages. The beech was systematically eliminated and oak favoured by frequent coppicing for iron-foundry charcoal.

Natural monospecific forests are probably found only near the timber line (e.g. *Sorbus aucuparia* in the Massif Central of France, *Fagus sylvatica* in the Vosges, *Pinus mugho* in the Carpathians, *Betula tortuosa* in Scandinavia) and in extreme habitats such as water-logged fen peat (*Alnus glutinosa*) and bog peat (*Betula pubescens* in western Europe, *Pinus sylvestris* in eastern Europe). Even here, one may occasionally encounter mixtures of alder and birch, or birch and pine, and frequently *Frangula alnus* is admixed.

The richest mixed forests in Europe are the alluvial hardwood forests of the upper Rhine valley (near Strasbourg) with twenty to thirty woody plants and eight or more tree species in the upper canopy alone. The non-flooded deciduous forests on rich, mineral soil from eastern Poland to Estonia (*Tilia/Carpinetum*) are also fairly rich in tree species (about six). Generally, however, the mixture consists of two or three species. They may form one or several storeys. If in one storey, it may be (i) mixed deciduous/evergreen, as in the *Fagus/Abies* woods of Jura and Prealps and in the *Quercus/Pinus* woods of Poland and Russia, (ii) mixed evergreen (e.g. *Pinus/Picea* forests in Scandinavia) or (iii) mixed deciduous (*Quercus/Fagus/Carpinus* in central Europe, *Quercus/Betula* in western Europe). If two storied, the overstorey is usually deciduous (*Quercus, Fagus*) and the lower is evergreen (*Ilex, Arbutus, Taxus*). In the Mediterranean there may be an evergreen layer of *Quercus ilex* under a deciduous canopy of *Quercus pubescens*.

Whether there are one or two storeys, they are usually irregular and often not sharply delimited from the shrub layers, because mixed woods usually contain trees of different age-classes. In this respect they resemble natural forests with one tree species. Therefore, in practice the comparison between mixed and monospecific forests cannot be separated readily from the comparison between natural forest and plantations.

## STRUCTURE OF NATURAL AND PLANTED FORESTS

Natural forests have several, not well-separated tree and shrub layers with uneven canopies and numerous crown gaps. They have many lianas and—before air pollution in large parts of Europe—many cryptogamic epiphytes. These woods are rich in dead wood, including standing dead trees. Rejuvenation is usually excellent. The undergrowth is varied and heterogeneous. Plantations usually have one, even-tree layer without gaps, consisting of one tree species of a single age-class. There are no lianas, few epiphytes and only one—sharply separated—lower layer, mostly a herb layer, sometimes a shrub or moss layer, which forms a homogeneous, not to say monotonous, undergrowth. They have little dead wood and no dead standing trees, while regeneration is poor.

Since natural woods have a variable canopy and complex structure, horizontal differences in microclimate are probably larger and probably stable, because horizontal exchange of air masses is hampered by the dense undergrowth. Unfortunately, adequate data are lacking. Planted woods, on the other hand, have a homogeneous canopy, and local variation in microclimate may be eliminated by horizontal air currents.

Mixed woods may be homogeneous or heterogeneous. The latter structure is found when one species is a shade-intolerant pioneer, filling gaps in the stand of other species. If these trees are large, the gaps will be wide and the pattern of the pioneers clustered. In this case, microclimate can be expected to be heterogeneous too, if the pioneer and the climax trees differ in structure or periodicity.

## THE LIGHT CLIMATE OF VARIOUS STAND TYPES

Tree species differ strongly in crown density. Compare the thin crowns of *Betula*, *Robinia* and *Larix* with the intermediate crowns of *Quercus robur*, *Tilia* and *Alnus glutinosa*, and the dense crowns of *Fagus, Carpinus* and *Castanea*. Even within one genus in Europe one may encounter large differences, witness the series *Pinus halepensis*, *P. sylvestris*, *P. cembra* and *P. pinea*.

Trees differ also in leaf size which influences the size of sun and shade flecks. Many small sun flecks favour photosynthesis of leaves in the lower crown and in the undergrowth. Flecks tend to move with the sun and wind and, when oscillations in light intensity become more frequent than $36\,h^{-1}$, the rate of photosynthesis increases with constant average flux density. This means that small-leaved trees, especially conifers, can both develop denser crowns (higher leaf area indices) and enable the undergrowth to photosynthesize at a low average light intensity. This also applies to a lesser degree to deciduous trees with small leaves or leaflets, such as *Betula*, *Sorbus aucuparia*, *Fraxinus*, *Robinia* and *Nothofagus*, compared to large-leaved trees, the extremes of which in Europe are *Acer pseudoplatanus*, *Aesculus hippocastanum*, *Castanea sativa*, *Tilia platyphyllos* and the introduced *Quercus rubra*.

Apart from the overall light intensity and the light pattern on the forest floor, the red/far red ratio is extremely important for the undergrowth. Tree species may differ widely in this. For example, in a spruce forest this ratio is twice that of a beech forest in summer.

Finally, the light relations of forests differ in their periodicity. Even within deciduous species there may be considerable differences. Beech and hornbeam flush in early spring and many forest plants of these woods are specialized geophytes which unfold their leaves long before the trees, and die off in early summer. Their leaves are linear and erect (the monocots) or broad and horizontal (the dicots). Pedunculate oak flushes a month later and casts a less dense shade. Here, ground vegetation consists mainly of summer-green grasses, ferns and some forbs, most of them with arcuate leaves. Birch not only flushes a month earlier than oak, but also sheds its leaves about 2 months earlier. Whereas the light period in spring is important for geophytic herbs, a light period in autumn is important for bryophytes and lichens, for which spring and summer are often too dry for them to benefit from warmth and light. For these plants, birch woods would seem ideal but when light on the forest floor increases with the fall of the leaves, the low cryptogams are covered by fallen leaves. Terrestrial bryophytes in woods therefore thrive best on rocky outcrops and elevated tree bases, where leaves do not lie. Here again, birch is the better host, for its leaves are lighter than oak leaves and more readily carried away. This is important because the moss layer is highly favourable for mycorrhizal fungi: both number of fruiting species and number of fruit bodies of these fungi are strongly positively correlated with the total cover of the moss layer (Jansen 1981).

Homogeneous mixtures may be expected therefore (and the few measurements confirm it) to have a light climate intermediate between component species. Mixed oak/beech forests will have an intermediate light climate during one month in spring, when beech has flushed and oak has not. A similar situation can be expected during two months in autumn in mixed oak/birch woods. The quantitative effect of these situations on the undergrowth has not yet been studied.

In heterogeneous mixtures, the largest difference is between deciduous and evergreen species. In evergreen forests gaps filled with deciduous trees (e.g. *Picea* with *Betula*) have, in winter, the special microclimate typical of real forest gaps, but in summer they have a forest microclimate. Thus, the microclimate is cool and shady in summer, with green light and a low red/far red ratio but very cold in winter with mainly blue light and with high humidity throughout the year. One might call it a 'blue–green climate'. In these forests in The Netherlands, *Usnea* species were abundant in the last century, and at present epiphytic mosses occur high on the tree boles.

## THE TEMPERATURE CLIMATE OF VARIOUS STAND TYPES

Day temperatures in woods are largely controlled by the solar radiation flux densities reaching the forest floor, i.e. by the crown densities. Night temperatures are governed by the nocturnal net radiation, which is equally a function of crown density, albeit in another direction (mainly towards the zenith). Dense forests have low temperature maxima and high minima while deciduous forests are much colder in winter than evergreen ones, which is often reflected in a greater share of boreal and/or montane species in their undergrowth.

Next to crown density the physical condition of the forest soil is very important. Wet and clayey soils are much less heated in daytime and much less cooled during the night than dry, sandy and peaty soil surfaces. If the soil is covered with a thick layer of litter and raw humus, as is the case in old oak, beech, pine and spruce forests on very poor, podzolized soils in western and northern Europe, the forest floor is heated strongly by the sun. This phenomenon is particularly marked in oak woods, because they remain leafless until mid-May. In combination with shelter from wind, high temperatures may be reached. The same applies to open pine woods in summer. However, the low heat conductivity of litter and humus also insulates the subsoil, which therefore remains cool. Alluvial forests of ash and elm on base-rich clay, loam and calcareous sand have high rates of litter decomposition and hence hardly any humus. Thus, these soil surfaces remain fairly cool in spring, but the subsoil, where the bulbs and rhizomes of the vernal geophytes are situated, gets some warmth, because of a high heat conductivity. Therefore the situation in various tree mixtures may be predicted approximately.

## PRECIPITATION IN VARIOUS STAND TYPES

Precipitation in forests is divided into interception, throughfall and stemflow. Interception varies considerably, from 32% in *Quercus robur* to 74% in *Pinus cembra* and even 85% in the densest crowns of *Juniperus communis*. This means a scanty water supply for certain forests that are above groundwater influences. For the same reason, some forests remain free from snow in winter, with a marked effect on soil temperature.

The non-intercepted rain is also quite differently distributed over the forest floor. Centrifugal crowns (drooping branches) have the highest throughfall at the crown periphery. In spruce it may be twice as much as that falling through the crowns and is often more than the free precipitation in forest gaps. Centripetal crowns (erect branches) have the highest precipitation at the stem base, especially if the trees have smooth bark, such as in beech. In oak, larch and Scots pine, where the branches are mainly horizontal, precipitation is more evenly distributed but for gaps in the crowns. Stemflow varies from about 1% in oak and spruce to 17% in beech. In beech the stemflow, enriched with dust and minerals from the bark and sometimes with pollutants collected from deposits on leaves, twigs and stems, removes litter from the soil near the stem base creating special micro-communities of bryophytes and algae (Wittig & Neite 1985). In Lapland, open sprucewoods have a more xerophytic vegetation under the crowns and a more hygrophytic one, in spite of higher insulation, in the open spaces. A similar pattern might be expected in mixed pine/spruce-woods, although generalizations should be avoided as it may be that, in warmer climates, the shade of the trees is more important in reducing evapotranspiration than crown interception.

## MICROCLIMATE IN LITTER AND HUMUS

Litter has an enormous influence on the ground flora and the underlying soil. Bryophytes and lichens may be suppressed, especially by large, heavy leaves. On the other hand, large leaves create cavities filled with saturated air, which provide an ideal habitat for drought-sensitive organisms such as sow-bugs (Isopoda) and beetle larvae (Carabidae). Litter of small leaves, and in particular of conifer needles, dries out much more rapidly and contains fewer representatives of these animal groups.

Litter layers are thick where leaf production per surface area is high, rate of decomposition low, and removal by wind negligible. Deeper litter layers are therefore found in small depressions in many-layered woods of trees with large leaves that are slowly decomposed, either because they have high tannin (*Quercus*) or alkaloid (e.g. *Prunus serotina*) content or are thick, dry and leathery (*Ilex, Quercus ilex, Rhododendron ponticum,* etc.) or where decomposition is retarded by dry or acid soil conditions.

High litter production may create a thick humus layer which is an effective heat insulator, allowing large temperature fluctuations at the surface, and small fluctuations in and under the humus. This affects the entire soil flora and fauna, including the mycorrhizal symbionts of the trees. The latter may be negatively correlated with the thickness of the humus layer (De Vries, Jansen & Barkman 1985).

When considering homogeneous mixtures of tree species, it was assumed that their microclimate would be intermediate and equally (at the same scale) homogeneous. With litter, however, a new phenomenon is observed—segregation in space of the litter of different tree species. This segregation may be vertical, horizontal or both. In mixed oak/birch woods the birch litter, falling in September, is entirely covered by the oak litter in November. This means that in winter and spring the birch litter is subjected to high humidity, high $CO_2$ and low oxygen concentrations, whereas the oak litter is more exposed to frost and desiccation. Decomposition processes in these mixed forests might therefore differ fundamentally from those in monospecific stands of oak and birch. This point has not been fully investigated.

Another point is that birch litter, being lighter, is carried by wind beyond the forest farther than oak litter. One may readily observe this in adjacent grassland or heath, but litter may also be blown, for instance, into a pine forest, which then has an admixture of oak and birch litter near the mixed stand, and of only birch litter somewhat deeper in the pinewood. Such admixtures could be very important for the ecosystem. In juniper woods, for instance, an admixture of oak litter drastically changes its chemical composition and the composition of the moss layer. Plagiothecium species are limited to this microhabitat within the juniper ecosystem.

## GENERAL CONCLUSIONS

Mixing trees leads to a more diversified structure, which will change the microclimate and consequently the ground flora and fauna as well as soil inhabiting organisms. The extent of these changes are, however, hardly known. Homogeneous mixtures probably lead to intermediate, homogeneous ecosystems, with the exception of the litter and humus subsystems: if the trees differ in leaf size, there is a spatial separation of their litter; if they differ in time of defoliation, a vertical stratification results.

Heterogeneous mixtures may increase biological diversity and are therefore attractive for nature conservation, provided, of course, that tree species are used on appropriate sites. These mixtures create patterns in microclimate, the elements of which correspond with 'pure' stands of the constituent species. In addition they may create special gap situations, not encountered in pure stands, the so-called 'blue–green climate gaps'.

## REFERENCES

Barkman, J.J. (1979). The investigation of vegetation texture and structure. *The Study of Vegetation* (Ed. by M.J.A. Werger), pp. 125–60. W. Junk, The Hague.

Barkman, J.J. (1988). Some reflections on plant architecture and its ecological implications. A personal view demonstrated on two species of *Quercus*. *Plant Form and Vegetation Structure* (Ed. by M.J.A. Werger, P.J.M. van der Aart, H.J. During & J.T.A. Verhoeven), pp. 1–7. SPB Academic Publishing, The Hague.

Barkman, J.J. (1989). Some remarks on the texture and structure of forests and their implications for the functioning of forest ecosystems. *Proceedings of the Workshop on Unification of European Forest Pattern Research, Strasbourg*, pp. 37–44.

**Barkman, J.J. (1990).** A tentative typology of European scrub and forest communities, based on vegetation texture and structure. *Vegetatio*, **86**, 131–41.

**Barkman, J.J. & Stoutjesdijk, Ph. (1987).** *Microklimaat, Vegetatie en Fauna.* Pudoc, Wageningen.

**Geiger, R. (1961).** *Das Klima der bodennahen Luftschicht.* Vieweg, Braunschweig.

**Jansen, A.E. (1981).** *The vegetation of macrofungi of acid oakwoods in the north-east Netherlands.* Ph.D. thesis, University of Wageningen.

**Mitscherlich, G. (1971).** *Wald, Wachstum und Umwelt.* II. Band: Waldklima und Wasserhaushalt. J.D. Sauerländer's Verlag, Frankfurt am Main.

**Vries, B.W.L. de, Jansen, A.E. & Barkman, J.J. (1985).** Changes in the species composition of fungi in coniferous woods of Drenthe 1958–1983 (in Dutch). *Changes in the Mycoflora* (in Dutch) (Ed. by E. Arnolds), pp. 74–83.

**Wittig, R. & Neite, H. (1985).** Acid indicators around the trunk base of *Fagus sylvatica* in limestone and loess beechwoods: distribution pattern and phytosociological problems. *Vegetatio*, **64**, 113–19.

# Ecology and management of semi-natural tree species mixtures

K.J. KIRBY* AND G. PATTERSON†

*Nature Conservancy Council, Northminster House, Peterborough, PE1 1UA, UK; and †Forestry Commission, Northern Research Station, Roslin, Midlothian EH25 9SY, UK

## SUMMARY

**1** Only 25% of the woodland in Great Britain is semi-natural (two-thirds in England). Its composition reflects climatic and soil differences across the country. Management has had more effect on structure; as well as high forest, distinct structural types occur in woods derived from former coppice or wood pasture treatments.

**2** Timber can be produced from semi-natural mixtures, but in Great Britain the main reasons for retaining them are for non-timber benefits, particularly nature conservation, landscape and game. This is recognized in national forestry policies. Implementing these policies to integrate timber production and nature conservation may be more or less easy according to the type of owner. Most semi-natural woods are privately owned.

**3** The potential for minimum intervention, former management practices (such as coppice) and high forest treatments to retain the nature conservation value of pine, oak/birch and lime/hornbeam woods is discussed. The main issues raised, and needs for further research are outlined.

## INTRODUCTION

We discuss the management of mixed semi-natural woods in Great Britain, concentrating on the integration of nature conservation and timber production. The first section describes some of the variation in the composition and structure of semi-natural stands. The second part deals with why they are managed to retain such diversity. The third section describes management options that are used in practice to achieve this aim using contrasting examples.

### What is semi-natural woodland?

Semi-natural woods are those where the trees and shrubs are mainly native to the site and have grown up naturally, i.e. they have not been planted (Peterken 1977). Many of the associated plants and animals might also have been found in a natural woodland on that site. There is not an absolute division between planted and semi-natural stands. Some oakwoods, for example, meet the above definition of semi-natural with respect to the tree layer, but have an understorey composed of *Rhododendron ponticum*, an introduced shrub. Conversely a mature plantation of trees native to a site, e.g. old oak (*Quercus* spp.) areas in the Forest of Dean, may contain many species and features associated with semi-natural stands. However, most stands can be assigned reasonably and unambiguously to either semi-natural or plantation categories.

Semi-natural woodland occurs on both ancient woodland sites (those believed to have been continuously wooded since A.D. 1600) and as stands that have grown up more recently on open ground (Peterken 1981). In Great Britain only about 25% of woodland is semi-natural, two-thirds of which is in England (Fig. 1).

### Differences between natural, semi-natural and plantation stands

In Scandinavia and North America, species whose abundance declines when natural forests are brought into management include those requiring old logs, over-mature (veteran) trees and particularly species that need such conditions in abundance (Gustafsson & Hallingback 1988; Heliovaara & Vaisanen 1984; Niemela *et al.* 1988; Soderstrom 1988; Doak 1990; Virkkala & Liehu 1990).

In Britain direct comparison between natural and semi-natural stands is not possible—no natural woodland survives. Most semi-natural woods are small and isolated (Kirby *et al.* 1984; Walker & Kirby 1989); they probably have fewer large veteran trees, less dead wood and its associated invertebrates (Girling 1982), but more of the species associated with young trees and recently cut stands than natural woodland (Warren & Key 1991). The understorey in lowland woods may be more dense because shrubs such as hazel (*Corylus avellana*) were favoured by coppicing. Mammals such as grey squirrel (*Sciurus carolinensis* Gmelin) have been introduced to semi-natural woodland, while others, such as wild boar (*Sus scrofa* L.), have become extinct. Large areas of woodland have been used for extensive grazing by domestic stock (Putman 1986).

Compared to most plantations, however, and particularly to those established in the twentieth century, semi-natural stands contain few introduced trees; the distribution of trees, both as species and individuals, tends to reflect local environmental variation; and there may be considerable genetic differences between populations (Forrest 1980.) The structure and composition of both stands and the woodland as a whole are generally more varied than those of plantations, and the soils may retain profile patterns lost from adjacent land (Collins 1978).

### Variations in mixed semi-natural woodland

Semi-natural woodland is normally mixed, although the scale and arrangement of the mixture varies. There may be groups of 0.1–0.5 ha where one species predominates or one species may be abundant throughout the canopy, for example sessile oak (*Quercus petraea*) in North Wales or Scots pine (*Pinus sylvestris*) in Deeside (Scotland), but with a more varied composition in the understorey

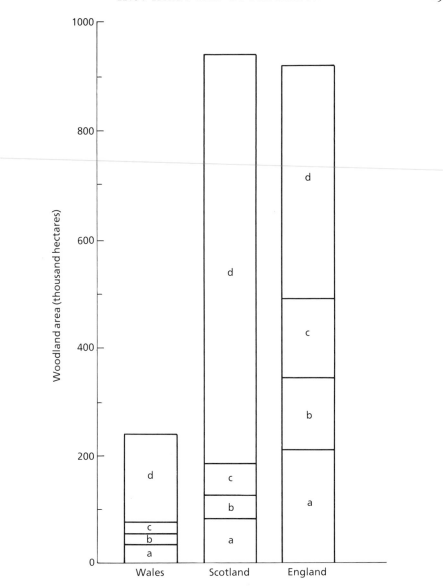

FIG. 1. Extent of semi-natural woodland in England, Scotland and Wales (based on data in Locke (1987) and unpublished NCC material): (a) ancient semi-natural woodland; (b) recent semi-natural woodland; (c) plantations on ancient woodland sites; and (d) plantations on recent woodland sites.

(Peterken 1981; Rodwell 1991). The shift to a simpler composition in the canopy may be partly owing to intervention by the forest manager and partly to natural processes. The distribution of woodland types in semi-natural stands still reflect regional and local trends in climate and soil (Rodwell 1991) (Fig. 2).

On ancient woodland sites managed as coppice, the composition may have

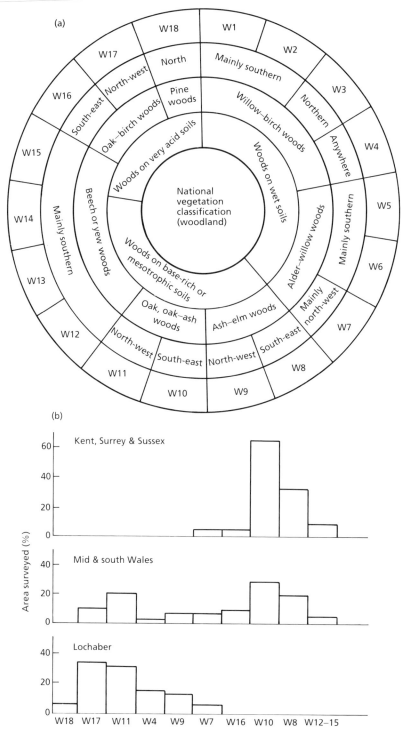

FIG. 2. Variations in the composition of semi-natural mixtures in Great Britain. (a) Summary of the main woodland types in relation to soils and regions based on descriptions in Rodwell (1991). (b) Per cent area found in different regions in three surveys (NCC unpubl. material).

been 'frozen' in terms of the abundance and distribution of species at the point when coppicing began, because subsequent restocking was largely from existing stools. This stability was used by Peterken (1981) and Rackham (1980) in their classifications of woodland types according to the coppice layer species. Once coppicing stops, natural disturbance and regeneration may lead to fresh variation in the woodland composition even over quite short periods (Peterken & Jones 1987, 1989). The location of parent trees, seed dispersal and seedling survival are further factors which may determine woodland composition. Oak (*Quercus robur*) and hawthorn (*Crataegus monogyna*) for example have spread more in recent semi-natural woods than small-leaved lime (*Tilia cordata*) and hazel (*Corylus avellana*) (Peterken 1974).

Semi-natural stands exist in three broad structural types: coppice (with or without standards and now frequently neglected), wood pasture with pollards, and high forest. Coppice was widespread and remains most distinct in the south-east where the majority of worked coppice survives (Crowther & Evans 1986; Fuller & Warren 1990). Wood pasture with pollards is similarly most abundant in the south where sites such as the New Forest contain internationally important populations of lichens and saproxylic invertebrates (Harding & Rose 1986). High forest structures are found in many recent woods, in the native pine woods (Steven & Carlisle 1959) and in large areas of former coppice promoted by neglect and deliberate thinning.

## OBJECTIVES OF MIXTURE MANAGEMENT

### Why manage semi-natural mixtures as mixtures?

### Wood production

In North America and northern Europe, much timber comes from semi-natural forests which are large compared to those in Britain, where natural regeneration of a wide range of native commercial timber species is more reliable.

In Britain, most woods were mixed when first brought into management (Rackham this volume) and under a coppice system would be likely to remain mixed. Different species could supply different markets, for instance oak for building timbers, hornbeam (*Carpinus betulus*) for firewood, and it was often convenient to grow all the species on one site. Mixed species may still provide flexibility in harvesting and marketing for a few owners, but the main arguments for maintaining semi-natural mixtures are connected with nature conservation, landscape, game and other non-timber benefits. Only the first of these is considered further.

### Nature conservation

Three strands in the practice of woodland nature conservation in Britain are (i) the maintenance and encouragement of the more natural elements of woodland ecosystems, including local genetic variation, (ii) the maintenance and en-hancement of populations of rare native woodland species, and (iii) the promotion of diversity of native woodland species and communities throughout their natural range. Semi-natural mixtures contribute more to all three strands than mono-specific plantations through their generally more varied structure as well as the presence of several tree and shrub species.

Semi-natural mixtures on ancient sites may reflect the original natural wood-land type for that area (Peterken 1977; Birks 1982). These tree and shrub popu-lations are no less a part of the nature conservation value of the wood than the ground flora, invertebrates or birds. Semi-natural mixtures may be of historic, cultural and research interest (Rackham 1975).

Ancient semi-natural woods contain a high proportion of rare and endangered woodland species (Peterken 1983). In some cases, a rare species may prefer a mixture of tree species: the purple emperor butterfly (*Apatura iris* L.) for example, is associated, as adults, with tall trees such as oak, but as a larva feeds on sallow (*Salix cinerea*, *S. caprea*) (Thomas 1986). More often, rare species occur because other factors such as a long history as woodland or the irregularity of structure and species are associated with semi-natural mixtures.

A mixture of trees is not always richer in associated plants and animals than a pure stand, but this is generally the case with long-established semi-natural mixtures (Mitchell & Kirby 1989). Little research has been done specifically on the differ-ence between semi-natural mixtures and semi-natural pure stands in Britain, but variations in litter type and turnover, shade and throughfall, affect the distribution of individual ground flora species. Each new tree species may support an addi-tional set of lichens or invertebrates. Trees and shrubs fruit at different seasons so that a wide range of species may provide a more regular supply of food for berry-eating birds throughout the year (Snow & Snow 1988). However, the effects of different proportions of individual species and different arrangements of the same species are difficult to assess.

### Management objectives in mixed semi-natural stands as influenced by type of ownership

National forestry policy is to increase the area of woodland in Britain and the production of timber. Other objectives are recognized (Table 1), but the priority given to them is influenced by the type of owner.

### Management primarily for nature conservation

Financial returns from non-timber benefits are limited so that few organizations or individuals can afford to manage primarily for nature conservation. These organ-izations include government bodies such as the Nature Conservancy Council (NCC) which manages *c.* 17 000 ha of woodland as National Nature Reserves (mainly semi-natural mixtures). Timber may be harvested (e.g. as coppice) to

TABLE I. National forest policy in Britain, relevant to semi-natural woodland. These aims are applied to the management of both state- and private-sector woodlands

*Broad-leaved woodland extract from policy aims* (Forestry Commission 1985a)

To encourage greater use of all types of broad-leaved woodland for conservation, recreation, sport
 and landscape as well as for wood production
To ensure that the special interest of the ancient semi-natural woodlands is recognized and
 maintained
Essential to these aims is the need to increase the quality and value of timber produced by the broad-
 leaved woodlands as a whole

*Native pinewoods extracts from policy aims* (Forestry Commission 1989a)

To maintain and enhance the pinewood ecosystems and their aesthetic value
To maintain the genetic integrity of the native pine population and so far as practicable, maintain
 identifiable sub-populations
To produce utilizable timber

create conditions favoured by particular species or, where this is not incompatible with maintaining the nature conservation value of the site, to demonstrate management systems that might be followed elsewhere. This is also true for reserves managed by bodies such as the Forestry Commission who have identified several hundred areas on their estate where conservation takes priority. There are also at least sixty non-governmental organizations in Britain involved in woodland conservation, responsible for more than 20 000 ha (although not all of this is semi-natural mixed species) (Stokes & Goldsmith 1989). Private owners are encouraged to manage primarily for nature conservation on other sites identified as of national importance (Sites of Special Scientific Interest). Compensation may be paid where such management departs from the owner's wishes (Vincent 1989). However, only about 20% of ancient semi-natural woodland (and a small proportion of recent semi-natural woodland) falls within any of the above categories (NCC unpublished material).

## Management mainly for timber production

In semi-natural mixed stands that are privately owned, timber production was the highest priority for most applicants for grant in the first 3 years of the Broad-leaved Woodland Policy introduced in 1985 (Table 2a). Owners and managers are still constrained however by national policy objectives and guidelines (Forestry Commission 1985a,b, 1989a, Tables 1 and 3).

## 'Unmanaged woods'

Most privately-owned broad-leaved woodland, and a high proportion of semi-natural mixtures, are not within grant schemes administered either by the Forestry Commission (Forestry Commission 1984) or by others, such as the Countryside

TABLE 2. Attitudes of British private woodland owners. (a) Objectives of management of applicants for grants under the Broad-leaved Woodland Grant Scheme (Forestry Commission 1989b). (b) Farmers (Blyth *et al.* 1987)

(a) Objectives of management (ancient semi-natural woodland only)

| | Stated priority as a percentage of all objectives | | |
| | 1st priority | 2nd priority | 3rd priority |
| --- | --- | --- | --- |
| Timber production | 68 | 14 | 20 |
| Landscape/amenity | 17 | 45 | 24 |
| Sporting/game management | 11 | 26 | 34 |
| Nature conservation | 2 | 10 | 14 |
| Recreation (other than sporting) | 1 | 3 | 5 |
| Shelter | 1 | 2 | 3 |
| Total | 100 | 100 | 100 |

(b) Response to questionnaire on benefits and disadvantages of farm woodlands: percentage of farmers replying that particular factors were relevant

| Factor | All farms | No trees | Farms with woods |
| --- | --- | --- | --- |
| *Benefits* | | | |
| Private shooting | 38 | 20 | 55 |
| Firewood production | 65 | 60 | 76 |
| Control of snow drift | 48 | 34 | 53 |
| Crop shelter | 47 | 33 | 52 |
| Steading shelter | 64 | 44 | 70 |
| Stock shelter | 73 | 32 | 79 |
| Pleasure for farmer | 68 | 45 | 75 |
| Landscape value | 77 | 50 | 85 |
| Wildlife habitats | 77 | 56 | 86 |
| *Disadvantages* | | | |
| Worse snow drifting | 32 | 25 | 36 |
| Financial costs | 36 | 20 | 42 |
| Crop lodging | 41 | 29 | 45 |
| Crop shading | 54 | 35 | 59 |
| Leaves, debris, etc. | 54 | 36 | 58 |
| Foxes, rabbits, etc. | 65 | 45 | 72 |
| Flies | 69 | 51 | 75 |

Commissions. Often these woods are on farms and are small, with poor access and have little timber value, although they have other benefits (Table 2b).

In Wales, northern England and western Scotland woods are used for grazing and shelter by stock. Current grazing levels are often too high to allow regeneration, so that this usage is incompatible in the long-term with either timber production or nature conservation (Kirby 1990; Mitchell & Kirby 1990; Mitchell 1990). However, controlled grazing might provide an alternative form of production (hence income) from the woods and encourage the maintenance of mixed semi-natural stands.

Semi-natural woods have very high nature conservation values, but most

TABLE 3. Extracts from the operational guidelines produced for the Native Pinewoods Policy (Forestry Commission 1989a)

1   The aim should be a diversity of stand structure

2   A proportion of trees should be allowed to reach natural maturity, senescence and death

3   There will be a strong preference for regeneration by natural means

4   A pinewood ecosystem is incomplete without other native tree and tall wood shrub species . . . and their regeneration should be encouraged

5   The adequate reduction or complete removal of grazing animals, whether deer or domestic stock, is a pre-requisite of successful regeneration

6   The aim is to obtain adequate natural regeneration with minimum and preferably no cultivation . . . where cultivation is proven to be necessary, scarifying and preferably patch scarifying will be the strongly preferred method

7   Wet and boggy areas are integral components of the pinewood ecosystem . . . draining will only be approved in exceptional circumstances

owners are unwilling or unable to make conservation their highest priority. What systems might allow some timber production without the stand losing many of the features that are associated with it being semi-natural or mixed?

## MANAGING TO KEEP SEMI-NATURAL MIXTURES

Management for timber production often leads to changes in the tree and shrub composition and in the stand and woodland structures; drainage, cultivation and fertilizers change soil conditions; pesticides and herbicides alter species abundance in the short or long term (Mitchell & Kirby 1989). Small-scale and short-term changes can often be absorbed by a semi-natural wood without significantly changing its character, because the changes are similar in type and scale to disturbances found in natural woods. Single events may, however, have long-term consequences.

There is insufficient research in Britain to predict precisely how different treatments will affect the species and communities in any particular semi-natural wood. Recommendations are made, based on experience, as to preferred methods of management to retain the semi-natural character and composition of a wood, but there may be practical problems in their implementation; natural regeneration, for example, is often suggested but is not always easy to achieve (Newbold & Goldsmith 1981; Pryor & Savill 1986; Evans 1988). Views on nature conservation also change in the light of new knowledge; the full importance of ancient woodland became widely appreciated only in the 1970s.

Three contrasting types of mixture (pine/birch, oak/birch and woods with lime or hornbeam) have been selected to explore the possibilities and problems of integrating nature conservation and timber production. For each mixture the treatments considered are minimum intervention, reviving the former management, and modifications which may be more suited to current conditions and owners.

Minimum intervention treatments allow the woodland to develop towards a natural structure and composition with as little interference by people as possible. A currently mixed stand could become less so with time; individual species might become extinct. Such areas are of value in nature conservation terms because they become more natural. They can also be sites for the study of natural woodland processes such as disturbance and succession; areas where old growth stands and deadwood in all its forms may accumulate; and the soils remain undisturbed by machinery or timber extraction. These areas provide 'controls' against which the effects of management can be measured.

In Britain, human influence is too pervasive to allow complete non-intervention. Commonly, decisions have to be made as to whether grazing is to be controlled (and at what level), whether non-native species are to be removed (i) at the outset, and (ii) subsequent to the start of the minimum intervention regime, whether the initial structure or composition of the site should be altered to speed the natural development process, and how to react to catastrophic events that affect large parts of a minimum intervention block.

The former management systems permitted the development of the present semi-natural mixture on a site, so are likely to allow its maintenance. Changes which permit greater production from the site, however, may be acceptable, even desirable, in nature conservation terms, because the former management was to meet the local market needs (Rackham 1990), not for wildlife. Some bryophytes, for example, may have declined under former coppice treatments (Edwards 1986).

### Pine/birch woods

The native pinewoods of Scotland (Steven & Carlisle 1959; Bunce & Jeffers 1977) are western outliers of the European boreal forests. Birches (*Betula* spp.) and rowan (*Sorbus aucuparia*) are present with pine to varying degrees, and juniper (*Juniperus communis*) is a characteristic understorey species in the eastern stands. Other trees and shrubs tend to be scarce. Areas may previously have fluctuated between pine and birch dominance with consequent changes in soil fertility and the field layer (Miles 1985). The pinewoods (National Vegetation Classification (NVC) type W18 (Rodwell 1991)) usually have a species-poor ground flora, but with a number of characteristic higher plants, e.g. *Pyrola* spp., *Linnaea borealis* as well as distinctive bird, invertebrate and lichen communities. They often abut other semi-natural vegetation, providing an opportunity to study natural forest dynamics at the landscape scale.

The pinewoods retain a high forest structure and in sites such as Abernethy veteran trees are not uncommon. Some of the sites are very large by British standards (more than 1000 ha) and contain naturally open glades and boggy hollows. Like other boreal forests, regeneration has been in large fairly even-aged patches following fire or large scale wind-throw (Anderson 1967; Sykes & Horrill 1981; Sykes 1987; Miller & Ross in press) often on the edge of the wood (Watson & Hinge 1989). Red deer (*Cervus elaphus* L.) are part of the ecosystem, but the numbers of deer using the woods are often too high to permit regeneration except within enclosures (Sykes & Horrill 1985). Broad-leaved (e.g. birch) regeneration may have been particularly affected by heavy browsing with implications for soil fertility and likelihood of pine regeneration (Malcolm 1957), especially in west Scotland (Booth 1984). However, the lack of regeneration may be over-emphasized on some sites, because 'moribund' trees survive for considerable periods and slow recruitment of seedlings may be sufficient to provide potential replacement trees (Fenton 1985; Peterken & Stace 1987) although not necessarily an acceptable timber yield.

## Minimum intervention

There are minimum intervention areas (Forster & Morris 1977; Bain 1987) in several of the major pinewoods. Some control of grazing and browsing by red deer is usual (either by fences or culling); fires may be controlled even if they start naturally, to avoid destroying adjacent property; and non-native trees, even non-local provenances of pine, have been removed from sites. Native species known to have been present on the site previously have also been reintroduced to some areas as seed sources.

This management option is difficult to apply to small sites and, while desirable in nature conservation terms, does not provide for any timber production. Alternatives with varying degrees of harvesting and active promotion of re-generation have been investigated (Baylis & Matthews 1981; Edwards 1981; Low 1988) and are encouraged under the new pinewood policy (Tables 1 and 3).

## Harvesting of timber/restocking by natural regeneration

Within some pinewoods, harvesting of timber takes place already, or is planned to occur in future. Restocking is usually by natural regeneration. Small-scale clear-felling with groups or strips, 1.5 tree widths across, have been recommended (Booth 1984) or shelterwood felling by 3 ha blocks (Forestry Commission 1989a) leaving fifty mother trees per hectare. (These would normally be felled once regeneration is established, but for nature conservation purposes some might be left unharvested.) Larger coupes, however, where successive waves of patchy regeneration are allowed to develop without infill planting (Dunlop 1983; Peterken & Stace 1987), may result in more irregularity of stocking and structure, perhaps more closely mimicking the results of natural fire and storm (Miller & Ross in press).

When trees are felled the root plate remains in the soil. Disruption of the vegetation may occur during extraction, but possibly less exposure of mineral soil than occurs when trees blow over to form pit-and-mound microtopography. Nor are there abundant rotting logs to act as a seed-bed (McKee, Le Roi & Franklin

1982). Natural fires may also expose mineral soil and reduce competition from the ground vegetation. Can and should these processes be mimicked?

Fires to encourage regeneration are difficult to control at the right intensity and cannot be timed to coincide with good seed years, because seed-bed conditions are optimal some years after the fire (Booth 1984). Cultivations can stimulate regeneration, for example, at Glen Tanar (Edwards 1981; Low 1988). However, there is always a risk that a particular rare species might be damaged by such processes. There is not the long-term research to judge what other consequences there might be for the flora and fauna. In some natural broad-leaved woods 10% of ground may be covered by logs and uprooted root plates (Falinski 1986). A higher level of disturbance might be tolerable in pinewoods on some dry mineral soils, but if wet peaty areas were cultivated the vegetation patterns would take much longer to be restored and often such ground is unlikely to support a high density of trees.

Some control of grazing/browsing animals (sheep/deer) is needed for natural regeneration to occur as in minimum intervention treatments. This is rarely feasible over the whole forest. Fencing individual coupes or existing open areas is more expensive per unit area the smaller the block, but it is easier to ensure that a small block (2–3 ha) is free and secure from grazing animals. If there is a break-in on one enclosure the loss of trees is less significant to the wood as a whole.

A varied age structure through the forest can provide a more even supply of timber and greater continuity of habitat conditions for species associated with one particular age-class. However, the value of veteran trees for nature conservation is such that, if there is only a low density of them, it would be better to expand the age structure by regeneration in existing gaps on mineral soils and adjacent open ground, rather than by felling areas of old trees.

## Pinewood management involving planting

Establishment of a crop by ploughing and/or closed-spaced planting are discouraged within the pinewoods if they are to maintain their semi-natural composition and structure (Forestry Commission 1989a).

Small-scale planting to supplement natural regeneration using individuals of local stock might be no more of a disruption to the natural system than extensive scarification, although a more uniform crop is likely to result. There is a need for more research on ways of improving and speeding up the natural regeneration process. Planting may be the only way of re-establishing a seed source for tree or shrub species other than pine that may have been eliminated from the site by past treatments, and some 15% of other species may be included where completely new areas of pinewood are planted (Forestry Commission 1989a).

## Hornbeam and lime coppices

Hornbeam is found as a native tree in Britain only in south-east England, where it is abundant as neglected coppice. In the woods north of London it may have been favoured in the past because of its value for firewood (Rackham 1980). Lime was formerly much more widespread, probably being the dominant tree in much of southern and central England and parts of Wales (Rackham 1990). Sites with both lime and hornbeam are rare, although it is a common mixture in central Europe, for example in the Białowieża Forest (Poland) (Falinski 1986). More often, they occur with oak, birch or ash, but with the lime or hornbeam dominating over much of the stand (NVC types W8, W10 (Rodwell 1991)).

The boundaries of lime and hornbeam woods (unlike those of the pinewoods), and perhaps even the limits of the different stands of trees, have sometimes been fixed for centuries (Rackham 1980, 1990). Regeneration has largely been by stump regrowth following coppicing. The small scale of past working plus the presence of standard trees (frequently oak) lead to considerable structural diversity in even small woods. Species which live in open woodland, notably butterflies, were favoured by coppicing but dead wood, over-mature trees and the species associated with them were discouraged (Warren & Key, 1991). In neglected stands, there is some accumulation of dead wood, but the structure of the wood becomes more uniform and opportunities for open phase species are reduced (Kirby, in press; Mitchell & Kirby 1989).

Few lime and hornbeam woods are grazed by domestic stock, but deer (fallow (*Dama dama* L.), muntjac (*Muntiacus reevesi* Ogilvy) and roe (*Capreolus capreolus* L.)) numbers are increasing to greater levels than ever before (Rackham 1989). Deer fencing or heavy culls may become needed in lowland as well as upland woods.

*Minimum intervention*

The allocation of some areas of lime (e.g. in the Lincolnshire lime woods (Table 4)) and hornbeam (Wormley Wood, Hertfordshire) to minimum intervention treatments benefits nature conservation as with the pinewoods. However the lime and hornbeam woods are smaller and more isolated, their structure more modified by past treatment. Naturally-formed gaps are likely to be mainly small and caused by the death of one or a few trees, rather than occurring as extensive swathes. This may not provide the regular supply of large gaps within small sites needed for some open phase species to survive. Five species of fritillary butterfly have been lost from lime woods in Lincolnshire for example following the cessation of coppicing and consequent reduction in young stands (Nature Conservancy Council unpubl. material). Over-mature stands and the species associated with them, which might benefit most from minimum intervention treatments, are scarce. The balance of conservation advantage is therefore often towards maintaining the rich open-stage species through woodland management.

Sites where minimum intervention may be considered for nature conservation purposes are usually long neglected (>50 years) coppice which has already developed towards a high forest structure. Specialist open stage species are likely to have been lost already and the soil seed-bank depleted (Brown & Warr, in press).

TABLE 4. Key points in the management of (a) coppice, (b) high forest, and (c) minimum intervention areas in the Lincolnshire lime woods

   (a) *Coppice/coppice with standards*
      Coupes no more than 2 ha
      Rotation length 20–40 years
      20–30 standards ha$^{-1}$ to be maintained
      Some trees left to die on site
      Felling restricted to September–March
      Restrictions on extraction routes and timing
      Establishment new lime stools by layering

   (b) *High forest*
      Rotation length 120–200 years
      Stands to be thinned on a 10–15 year cycle
      Not more than 25% of high forest in a wood to be worked on a 5 year period
      Thinning to favour (for retention) particular tree species
      Felling restricted to September–March
      Restrictions on extraction routes and timing
      Rides to be widened

   (c) *Minimum intervention*
      Access to be maintained along existing rides and way-marked paths
      Possibility of setting up deer-fenced plots

No timber production is possible. Conflict may also develop where there is public access because the over-mature trees may become a safety hazard.

## *Restore coppice or coppice-with-standards*

Coppice and coppice-with-standards can provide roundwood suitable for pulp and firewood and in small quantities for specialized markets such as turnery or carving on rotations of 10–40 years (Crowther & Evans 1986). Small quantities of larger dimension wood are available from the standards but their quality may be poor if there is extensive epicormic growth on oak after the coppice is cut (Watkins 1990). Access is often difficult and with the small-scale working that is generally preferred by nature conservation organizations (Fuller & Warren 1990) there may be insufficient produce for a full load of timber. The system is appropriate for some sites and owners, e.g. at Chalkney Wood, Essex, because it is simple to operate and can use volunteer labour; the value of the non-timber benefits including historical interest is high; and often the material cut can be used on the same estate.

    Success depends on the rate of regrowth from the stools which is affected by the tree species, the vigour (age) of the stool and the time since it was last cut. Lime grows well from stumps almost whatever their size or time since last cut, but is sensitive to browsing; hornbeam is less sensitive to browsing, but regrowth, particularly from old or long-neglected coppice is initially slow (Rackham 1980). On some sites this may be offset by hornbeam regeneration from seed. Birch (less often oak) regenerates well from seed in these woods so that the composition of the recently cut stand changes. Dense growth of bramble (*Rubus fruticosus*) may follow coppicing (Mason & Long 1987) and be a problem either because it competes with stump regrowth or interferes with other management in the wood, e.g. for shooting.

## Conversion to high forest

Lime and hornbeam stands may be converted to high forest by singling and thinning, although conversion is more successful if young rather than neglected coppice is treated. Once a high forest structure is achieved group felling may be used as both species are reasonably shade-tolerant, but the lime component may need to be maintained by stump regrowth or layering as regeneration from seed is poor (Pigott 1969, 1985; Pigott & Huntley 1981). However, high forest may not be an attractive commercial proposition because neither species is of particularly high value. The value of the high forest crop can be increased by encouraging some oak in the mixture although this requires larger gaps for regeneration than for lime or hornbeam alone.

In small woods it is difficult to transform even-aged coppice to a high forest structure where all ages of trees are present (Mitchell & Kirby 1989). In nature conservation terms the age structure through much of the wood may be compressed if recently cut and over-mature woodland conditions are maintained separately by, for example, ride side coppice strips (Warren & Fuller 1990) and groups of trees set aside to grow on to over-maturity and death on site (Kirby, in press).

Retaining local genetic variety is desirable for lime, hornbeam and other tree species that have not been widely planted in the past (Soutar & Spencer, 1991) and, if planting is proposed, then use of local provenance is desirable. Planting is, however, usually on a smaller scale than in the pinewoods and this may make it more acceptable in nature conservation terms.

Control of weeds (both herbaceous and woody regrowth) and disturbance to the soil surface during extraction (either of coppice or high forest products) are other factors that cause concern in lime and hornbeam stands. The ground flora is often rich and not necessarily adapted to large-scale disturbance (cf. pinewoods). The woods are often on heavy clays so that damage done to the soil during extraction may be long-lasting. The sites are small so any disruption of the ground flora affects a larger percentage of the whole wood.

## Upland oak/birch woods

The commonest semi-natural woodland type in western Britain consists of mixtures of oak and birch with a range of ground flora. NVC types W11, 16, 17 are common (Rodwell 1991) according to the degree of oceanicity and the acidity of the soil. Most woods are heavily grazed by stock so that an understorey is lacking (which favours a characteristic bird community of pied flycatcher (*Ficedula hypoleuca* Pallas), redstart (*Phoenicurus phoenicurus* L.) and woodwarbler (*Phylloscopus sibilatrix* Bechstein)). The woods include some of the richest for bryophytes and lichens with species which are rare and restricted in European terms (Ratcliffe 1977; British Lichen Society 1982).

Many woods were coppiced in the past although in parts of Wales and Scotland only for short periods. The woods are frequently small and their boundaries are less fixed than eastern mixed coppices. Many have contracted in the past century but locally there may be expansions, often apparent as areas of birch around a predominantly oak core as at Letterewe (Ross) (Peterken 1981).

## Minimum intervention

Some Atlantic bryophytes are sensitive to desiccation such that their survival may be threatened by large-scale disturbance to the canopy (Ratcliffe 1968; Edwards 1986) in the less favourable climatic areas. Minimum intervention treatment may be proposed, but as in the North Wales oak wood reserves, with limits, so that there is (i) control of introduced species, particularly rhododendron, which is also a threat to agricultural and forestry interests, and (ii) control of grazing and browsing animals (sheep and goats in North Wales, elsewhere usually combinations of sheep and deer) (Mitchell & Kirby, 1990) (Table 5). Some grazing may be desirable although on a seasonal or rotational basis, and was reintroduced at Coed-y-Rhygen National Nature Reserve, to maintain particular bryophyte communities (Mitchell & Kirby, 1990). In the long term oak might become less abundant, and birch, rowan and holly more so.

## Former management: coppice or coppice-with-standards

Charcoal is still made from oak coppice in the Lake District and a tannery in Cornwall uses oak bark, but what were the major markets for oak coppice have largely disappeared. Woods might be managed as coppice for historical reasons, perhaps in conjunction with sites where old furnace buildings exist (e.g. Bonawe (Argyll), Duddon Valley (Cumbria)), but the nature conservation reasons for restoring this system are limited. There is seldom a distinctive rich post-coppice flora such as exists in more mixed coppices on richer soils and sufficient open glades are usually present without the need to create more to benefit species which thrive in recently cut stands. Structural diversity within the wood can be increased by excluding stock for periods to allow regeneration and understorey development.

## High forest treatments

Singling and thinning of the coppice to high forest may be successful, particularly at lower altitudes and on the better soils. If grazing is controlled the wood can be regenerated by shelterwood or small-scale clear-fell systems (0.25–2 ha) as at

TABLE 5. Reasons for intervention (*) in some North Wales oak wood reserves

Removal/control of

| Site | Rhododendron | Conifers | Other introduced plants | Feral goats | Management for demonstration purposes | Management to benefit particular species/groups |
|---|---|---|---|---|---|---|
| Coedydd Maentwrog | * | * | * | * | * | |
| Coed Cymerau | * | | | * | | |
| Coed Camlyn | * | * | * | * | | *Cirsium dissectum |
| Cuenant Cynfal | * | | | | | *If necessary for Atlantic bryophytes |
| Cuenant Llennyrch | * | * | | * | | |
| Coed y Rhygen | | | | | | *Atlantic bryophytes |
| Coed Lletywalter | * | | * | | | *Osmunda regalis |
| Coed Ganllwyd | * | | | | * | *Atlantic bryophytes |
| Coedydd Aber | * | * | | | * | *Lichens, red squirrel, pine marten |
| Coed Gorswen | * | | * | | | |

Grizedale, Cumbria. The natural regeneration in such circumstances usually contains a much higher proportion of birch where the ground surface is disturbed than is present in the original canopy and of rowan and holly where the vegetation is undisturbed. These species which may have once been more common were cut out as undesirable by the woodsmen. Enclosing natural oak seedlings in tree shelters (Evans 1988) increases the likelihood that there are enough for a reasonable timber crop. In future, birch may become more acceptable as a commercial tree and reduce the 'need' for dense oak regeneration which is often scarce in upland woods.

If oak regeneration is very poor owners may wish to plant oak in gaps. In woods such as the Forest of Dean (Gloucestershire), where previous planting has occurred, this may be an acceptable compromise in terms of nature conservation, depending on the number of trees involved. It would be less desirable in remote sites (e.g. Wistman's Wood, Devon), where there is no such history of planting.

## DISCUSSION

Recent changes in attitudes and policies to land use in Great Britain mean that it may be easier to encourage the management of semi-natural mixtures in future. Certain key issues emerge which should be the subject of further research and debate.

**1** Should non-local provenance or species be introduced into existing semi-natural mixtures by planting? The nature conservation ideal may be to stick to natural regeneration but other objectives and pragmatic considerations often lead to some planting. Guidelines could be developed based on, amongst other factors: (i) the scale of the planting proposed, (ii) the degree of past planting or selection for a particular species, and (iii) the degree of existing variation.

**2** In many of these mixtures, achieving adequate natural regeneration for wood production purposes is not easy. How much ground disturbance or other preparation techniques are necessary to achieve an acceptable density; will this be acceptable in terms of the other objectives on the site; what are the long-term consequences for ground flora of scarification? Are there circumstances where small-scale planting would be less disruptive than measures to enhance natural regeneration?

**3** Most mixtures have a species composition influenced by past management. Where they are to be managed in future, should limits be set on the degree of variation permitted from the present mixture and if so, how? Is there a minimum level of lime, oak or pine that must be maintained in the examples discussed earlier to retain their character and how should this be distributed; is there a minimum level of other species that should be encouraged?

In pine/birch woods on Deeside, for maximum songbird richness, the minor component needed to be at least 20% of a mixture and distributed as clumps of 0.2–0.8 ha (French, Jenkins & Conroy 1986), but different scales may apply for other woodland types and taxa.

**4** How and when should the abundance of a species be manipulated—at the regeneration stage, by for example varying coupe size—or during the later stages of the rotation in thinning? Coupe size influences the structure of the wood and the likely extent of disturbance during harvesting. To what extent may varying woodland structures compensate for some simplification of species compositon for commercial reasons?

**5** How are grazing and browsing levels best controlled, given that these influence species composition as well as abundance of regeneration and may provide an alternative use for the wood?

The technical problems of managing mixtures for specific objectives can be solved, but there is a need to bring together the practical experience of those who are currently managing these woods; to undertake long-term studies of some of the issues raised above and persuade society that semi-natural mixtures are worth the cost of maintaining them.

## REFERENCES

Anderson, M.L. (1967). *A History of Scottish Forestry*. Nelson, London.

Bain, C. (1987). *Native Pinewoods in Scotland*. Royal Society for the Protection of Birds, Sandy.

Baylis, N.T. & Mathews, J.D. (1981). *Options for forest management in the native pinewoods*. Unpublished, Nature Conservancy Council (CSD Research Report 546), Peterborough.

Birks, H.J.B. (1982). Mid-Flandrian forest history of Roudsea Wood, National Nature Reserve, Cumbria. *New Phytologist*, **90**, 339–54.

Blyth, J., Evans, J., Mutch, W. & Sidwell, C. (1987). *Farm Woodland Management*. Farming Press, Ipswich.

Booth, T.C. (1984). Natural regeneration in the native pinewoods of Scotland: a review of principles and practice. *Scottish Forestry*, **38**, 33–42.

British Lichen Society (1982). *Survey and assessment of epiphytic lichen habitats*. Unpublished, Nature Conservancy Council (CSD Research Report 384).

Brown, A.H.F. & Warr, S. (in press). Buried seed in coppice woods. BES Symposium proceedings. *The ecology and management of coppicewoods* (Ed. by G.P. Buckley).

Bunce, R.G.H. & Jeffers, J.N.R. (1977). *Native Pinewoods of Scotland*. Institute of Terrestrial Ecology, Grange-over-Sands.

Collins, M.A. (1978). *History and soils on the South Downs*. Kings College (Rogate Papers 1), London.

Crowther, R.E. & Evans, J. (1986). *Coppice*. Forestry Commission Leaflet No. 83. HMSO, London.

Doak, D. (1990). Spotted owls and old growth logging in the Pacific Northwest. *Conservation Biology*, **3**, 389–96.

Dunlop, B.M.S. (1983). The natural regeneration of Scot's pine. *Scottish Forestry*, **37**, 259–63.

Edwards, I.D. (1981). The conservation of Glen Tanar native pinewood, near Aboyne, Aberdeenshire. *Scottish Forestry*, **35**, 173–8.

Edwards, M.E. (1986). Disturbance histories of four Snowdonian woodlands and their relation to Atlantic bryophyte distributions. *Biological Conservation*, **37**, 301–20.

Evans, J. (1988). *Natural Regeneration of Broadleaves*, Forestry Commission Bulletin 78. HMSO, London.

Falinski, J.B. (1986). *Vegetation Dynamics in Temperate Lowland Primeval Forests*. Junk, The Hague.

Fenton, J. (1985). Regeneration of native pine in Glen Affric. *Scottish Forestry*, **39**, 104–16.

Forestry Commission (1984). *Broadleaves in Britain: a Consultative Paper*. Forestry Commission, Edinburgh.

Forestry Commission (1985a). *The Policy for Broadleaved Woodland*. Forestry Commission, Edinburgh.

Forestry Commission (1985b). *Management Guidelines for Broadleaved Woodland*. Forestry Commission, Edinburgh.

Forestry Commission (1989a). *Native Pinewood Grants and Guidelines*. Forestry Commission, Edinburgh.

Forestry Commission (1989b). *Broadleaves Policy—Progress 1985–1988*. Forestry Commission, Edinburgh.

Forrest, G.I. (1980). Genotypic variation among native Scots pine populations in Scotland based on monoterpene analysis. *Forestry*, **53**, 101–28.

Forster, J.A. & Morris, D. (1977). The conservation of native pinewoods. *The Native Pinewoods of Scotland* (Ed. by R.G.H. Bunce & J.N.R. Jeffers), pp. 116–20. Institute of Terrestrial Ecology, Grange-over-Sands.

French, D.D., Jenkins, D. & Conroy, J.W.H. (1986). Guidelines for managing woods in Aberdeenshire for song birds. *Trees and wildlife in the Scottish uplands* (Ed. by D. Jenkins), pp. 129–43. Institute of Terrestrial Ecology Symposium 17, Banchory.

Fuller, R.J. & Warren, M.S. (1990). *Coppiced woodlands: their management for wildlife*. Nature Conservancy Council, Peterborough.

Girling, M.A. (1982). Fossil insect faunas from forest sites. *Archaeological Aspects of Woodland Ecology* (Ed. by M. Bell & S. Limbrey), pp. 129–46, British Archaeological Report 146, Oxford.

Gustafsson, L. & Hallingback, T. (1988). Bryophyte flora and vegetation of managed and virgin coniferous forests in south west Sweden. *Biological Conservation*, **44**, 283–300.

Harding, P.T. & Rose, F. (1986). *Pasture woodlands in lowland Britain, a review of their importance for wildlife conservation*. Institute of Terrestrial Ecology, Abbots Ripton.

Heliovaara, K. & Vaisanen, R. (1984). Effects of modern forestry on north western European forest invertebrates: a synthesis. *Acta Forestalia Fennici*, **189**, 1–32.

Kirby, K.J. (1990). Conservation management of upland woods: some objectives. *Grazing Research and Nature Conservation in the Uplands* (Ed. by D.B.A. Thompson & K.J. Kirby), pp. 57–63. Nature Conservancy Council (Research and survey in nature conservation 31), Peterborough.

Kirby, K.J. (in press). Accumulation of dead wood: a missing ingredient in coppice? BES Symposium proceedings. The ecology and management of coppicewoods (Ed. by G.P. Buckley).

Kirby, K.J., Peterken, G.F., Spencer, J.W. & Walker G. J. (1984). *Inventories of Ancient and Semi-natural Woodland*. Nature Conservancy Coucil (Focus on nature conservation 6), Peterborough.

Locke, G.M.L. (1987). *Census of woodlands and trees 1979–82*, Forestry Commission Bulletin 63. HMSO, London.

Low, A.J. (1988). Scarification as an aid to natural regeneration in Glen Tanar native pinewood. *Scottish Forestry*, 42, 15–20.

Malcolm, D.C. (1957). Soil degradation in stands of native pine in Scotland. *Bulletin of the Forestry Department (University of Edinburgh)*, 4, 1–38.

Mason, C. & Long, S. (1987). Management of lowland broadleaved woodland, Bovingdon Hall, Essex. *Conservation Monitoring and Management* (Ed. by R. Matthews), pp. 37–42. Countryside Commission, Cheltenham.

McKee, A., Le Roi, G. & Franklin, J.F. (1982). Structure, composition and reproductive behaviour of terrace forests, South Fork Hoh River, Olympic National Park. *Ecological Research in National Parks of the Pacific North West* (Ed. by E.E. Starkey, J.F. Franklin & J.W. Matthews), pp. 22–9. National Park Service Co-operative Studies Unit, Corvallis.

Miles, J. (1985). The pedogenic effects of different species and vegetation types and the complications of succession. *Journal of Soil Science*, 36 571–84.

Miller, H.G. & Ross, I. (in press). Management and silviculture of the forests of Deeside. *Silvicultural systems proceedings of the ICF Easter Symposium 1990*.

Mitchell, F. (1990). Effects of grazing on upland woods. *Grazing Research and Nature Conservation* (Ed. by D.B.A. Thompson & K.J. Kirby), pp. 50–6. Nature Conservancy Council (Research & Survey in Nature Conservation 31), Peterborough.

Mitchell, F.J.G. & Kirby, K.J. (1990). The impact of large herbivores and the conservation of semi-natural woods in the British uplands. *Forestry*, 63, 333–53.

Mitchell, P.L. & Kirby, K.J. (1989). *Ecological effects of forestry practices in long-established woodland and their implications for nature conservation*. Oxford Forestry Institute Occasional Paper 39, Oxford.

Newbold, A.J. & Goldsmith, F.B. (1981). *The regeneration of oak and beech: a literature review*. Discussion Papers in Conservation No. 33. University College, London.

Niemela, J., Haila, Y., Halme, E., Lahti, T., Pajunen, T. & Punttila, P. (1988). The distribution of carabid beetles in fragments of old coniferous taiga and adjacent managed forest. *Annales Zoologici Fennici*, 25, 107–19.

Peterken, G.F. (1974). A method for assessing woodland flora for conservation using indicator species. *Biological Conservation*, 6, 239–45.

Peterken, G.F. (1977). Habitat conservation priorities in British and European woodlands. *Biological Conservation*, 11, 223–36.

Peterken, G.F. (1981). *Woodland Conservation and Management*. Chapman & Hall, London.

Peterken, G.F. (1983). Woodland conservation in Britain. *Conservation in Perspective* (Ed. by A. Warren & F.B. Goldsmith), pp. 83–100. Wiley, London.

Peterken, G.F. & Jones, E.W. (1987). Forty years of change in Lady Park Wood: the old growth stands. *Journal of Ecology*, 75, 477–512.

Perterken, G.F. & Jones, E.W. (1989). Forty years of change in Lady Park Wood: the young-growth stands. *Journal of Ecology*, 77, 401–29.

Peterken, G.F. & Stace, H. (1987). Stand development in the Black Wood of Rannoch. *Scottish Forestry*, 41, 29–44.

Pigott, C.D. (1969). The status of *Tilia cordata* and *T. platyphyllos* on the Derbyshire limestone. *Journal of Ecology*, 57, 491–504.

Pigott, C.D. (1985). Selective damage to tree seedlings by bank voles (*Clethrionomys glareolus*). *Oecologia*, 67, 367–71.

Pigott, C.D. & Huntley, J.P. (1981). Factors controlling the distribution of *Tilia cordata* at the northern limits of its geographical range. III. Nature and causes of seed sterility. *New Phytologist*, 87, 817–39.

Pryor, S.N. & Savill, P.S. (1986). *Silvicultural systems for broadleaved woodland in Britain*. Oxford Forestry Institute Occasional Paper 32, Oxford.

Putman, R.J. (1986). *Grazing in Temperate Ecosystems: Large Herbivores and the Ecology of the New Forest*. Croom Helm, London.

Rackham, O. (1975). *Hayley Wood*. Cambridgeshire & Isle of Ely Naturalist's Trust, Cambridge.

Rackham, O. (1980). *Ancient Woodland*. Edward Arnold, London.

Rackham, O. (1990). *Trees and Woodland in the British landscape*, 2nd edn. Dent, London.

Ratcliffe, D.A. (1968). An ecological account of the Atlantic bryophytes in the British Isles. *New Phytologist*, **67**, 365–439.

Ratcliffe, D.A. (ed.) (1977). *A Nature Conservation Review*. Cambridge University Press, Cambridge.

Rodwell, J. (1991). *British Plant Communities*. Cambridge University Press, Cambridge.

Snow, B. & Snow, D. (1988). *Birds and Berries*. Poyser, Calton.

Soderstrom, L. (1988). Sequence of bryophytes and lichens in relation to substrate variables of decaying coniferous wood in northern Sweden. *Nordic Journal of Botany*, **8**, 89–97.

Soutar, R.G. & Spencer, J.W. (1991). The conservation of genetic variation in Britain's native trees. *Forestry*, **64**, 1–12.

Steven, H.M. & Carlisle, A. (1959). *The Native Pinewoods of Scotland*. Oliver & Boyd, Edinburgh.

Stokes, J. & Goldsmith, F.B. (1989). *Woodland management by non-governmental organizations*. (Unpublished.) Forestry Commission, Alice Holt Research Station.

Sykes, J.M. (1987). Further observations on the recovery of vegetation in the Caledonian pinewood of Coille Creag-loch after fire. *Botanical Society of Edinburgh, Transactions*, **45**, 161–2.

Sykes, J.M. & Horrill, A.D. (1981). Recovery of vegetation in a Caledonian pinewood after fire. *Botanical Society of Edinburgh, Transactions*, **43**, 317–26.

Sykes, J.M. & Horrill, A.D. (1985). Natural regeneration in a Caledonian pinewood: progress after eight years of enclosure at Coille Coire Chuilc, Perthshire. *Arboricultural Journal*, **9**, 13–24.

Thomas, J.A. (1986). *Butterflies of the British Isles*. Country Life Books, London.

Vincent, M. (1989). A landowner's guide to SSSIs. *Country Landowner*, **December**, 14–15.

Virkkala, R. & Liehu, H. (1990). Habitat selection by the Siberian tit *Parus cinctus* in virgin and managed forests in northern Finland. *Ornis Fennica*, **67**, 1–12.

Walker, G.J. & Kirby, K.J. (1989). *Inventories of ancient, long-established and semi-natural woodland in Scotland*. Nature Conservancy Council (Research & Survey in nature conservation 22), Peterborough.

Warren, M.S. & Fuller, R.J. (1990). *Woodland rides and glades: their management for conservation*. Nature Conservancy Council, Peterborough.

Warren, M.S. & Key, R. (1991). Woodlands: past present and potential for insects. *The conservation of insects and their habitats* (Ed. by M. Collins & J. Thomas), pp. 155–212. Royal Entomological Symposium, 15.

Watkins, C. (1990). *Woodland Management and Conservation*. David & Charles, London.

Watson, A. & Hinge, M. (1989). *Natural tree regeneration on open upland in Deeside and Donside*. (Unpublished.) Nature Conservancy Council, Aberdeen.

# Ground vegetation under planted mixtures of trees

E.A. SIMMONS AND G.P. BUCKLEY
*Wye College, University of London, Wye, Ashford, Kent TN25 5AH, UK*

## SUMMARY

Differences in the ground vegetation communities of mixed and pure even-aged plantations were investigated at twenty sites in lowland England. Sites were selected which contained stands of common conifer/broad-leaved mixtures such as spruce/oak and pine/beech, which had good 'controls' of pure stands nearby, and were mostly on former ancient woodland sites.

In spruce/oak mixtures, correlation and multiple regression analyses indicated that high spruce densities depressed vascular plant cover and species number, as did beech in pine/beech mixtures. Tree litterfall and specific site factors such as pH also played a significant role. Crop influences on the ground flora were most marked in young stands at the thicket stage, but these tended to diminish after thinning, which rapidly removed the marked vegetation gradients across some crop interfaces. Seed-banks of vascular species were relatively unaffected by mixtures in comparison with pure stands.

There was some evidence to suggest that ground flora composition was the genuine result of interaction between crop species, as for example in the relative increase of bryophytes in oak stands mixed with spruce. However, in the context of the limited number of crop combinations examined, the flora was primarily determined by a single dominant crop species. Taking some of these factors into account, improvements in mixture design likely to benefit ground vegetation are briefly discussed.

## INTRODUCTION

Even-aged forest plantations are usually considered poor subjects for nature conservation compared with semi-natural woodland stands. The dictates of forest production and especially high forest silviculture, in which closed forest canopies are maintained for long periods of the rotation, have many ecological implications for the development of forest-floor vegetation.

The most obvious of these is the drastic reduction in light penetration below the canopy. During the thicket stage which forms as the canopy closes, levels of photosynthetically active radiation reaching the forest floor are frequently reduced to 1–5% of ambient during the growing season, with concomitant effects on light quality (notably the red/far red ratio) (Mitchell, in press). This is particularly marked under some conifers, leading to rapid reductions in the cover of vascular plants (Hill 1979), especially populations of vernal herbs which depend on high levels of radiation transmitted through deciduous canopies in the spring (Anderson 1979). Moreover, these conditions are reinforced by conservative thinning regimes which maintain closed canopy conditions for much of the remainder of the rotation, except in tree crops grown for extended rotations beyond the point of maximum yield production (e.g. pedunculate or sessile oak (*Quercus robur, Q. petraea**) and beech (*Fagus sylvatica*)), where continuous thinning gradually creates gaps in the canopy.

Forest crops have relatively simple canopy structures and the planted species are not necessarily native to the site. This species impoverishment extends to understorey trees and shrub layers, which are selectively removed by forest management practices such as weeding, cleaning and thinning. The crop is there-fore the chief source of litter, especially during the period of maximum litterfall from canopy closure to the middle stages of the rotation. This leaf litter affects not only the abundance of the ground flora through physical and perhaps chemical suppression due to litter breakdown, but also the character of forest-floor communities, depending on the effects of litter persistence, quantity and structure on individual plant species (Anderson 1979; Sydes & Grime 1981a,b).

The forest crop may also influence soil properties, such as pH and the type and abundance of organic matter (see Mitchell & Kirby 1989, for a review). Finally, soil type and the ground flora species already present are major determinants of the ground flora communities developing under plantations of different species. This is particularly true of the sensitive, species-rich floras of lowland ancient woodlands which contain many plants of limited colonizing ability (Peterken 1974) which have been overplanted with commercial tree crops, including chestnut coppice (*Castanea sativa*). Altogether, it has been estimated that more than 275 000 ha, nearly 12% of the total woodland area in Britain, consist of plantations of trees on ancient woodland or old parkland sites (Kirby, Peterken & Walker 1989).

## The effect of mixed-species stands

Most work on the ground flora of plantations has concentrated on pure, rather than mixed stands. Long-term studies by Ovington (1955) and Anderson (1979) on plots established in 1951 in the Forest of Dean, Bedgebury and Thetford Forests, and by Hill and Jones (1978) in Wales, have demonstrated how several crop species, especially Sitka and Norway spruce (*Picea sitchensis* and *P. abies*) may impoverish the ground vegetation after canopy closure. In one of the few accounts of mixture plantations in the UK, Brown (1982) reported effects of tree mixtures on the ground flora of an afforested upland site at Gisburn, in which Scots pine (*Pinus sylvestris*), Norway spruce, sessile oak and alder (*Alnus*

*Scientific names follow Clapham, Tutin & Moore (1987) for vascular plants, and Smith (1980) for mosses.

*glutinosa*) were planted in pure stands or as two-species mixtures 25 years previously. Scots pine interacted with the other tree crops; pine/oak canopies creating conditions favourable to the grasses *Agrostis capillaris* and *Deschampsia cespitosa* which were not present under pure oak, and increasing their frequency under alder. In Salcey forest, Northamptonshire, Kirby (1988) found much small-scale variation in ground flora species number and cover, corresponding to the alternating bands of young spruce and oak in mixture. Work in North America on semi-natural stands suggests that mixtures influence the spring and summer ground flora by increasing the patchiness of shade on the forest floor (Brewer 1980; Beatty 1984) and the variation in base-status of stem-flow water from different tree species (Crozier & Boerner 1984). In the specific case of mixed plantations, the following differences compared with pure stands might be expected:

1  more opportunities for the maintenance of vernal plant species under the deciduous component, or under lighter-canopied conifers such as pine and larch;

2  increases in the cover of shrub and field layers resulting either from the short-term nature of one or more crop species components (e.g. conifer 'nurse' crops) or differential thinning;

3  improved prospects for the establishment and spread of small plants owing to reduced covering by litterfall under some tree species, or physical interactions between different litter components, or increased breakdown of mixed litter layers; and

4  wider gradients in ecological conditions across each species' component interface, increasing the range of environmental conditions and allowing greater coexistence of different species in the ground flora.

In order to test these hypotheses, our investigation focused on commercial high forest mixtures which had been planted on former ancient woodland sites, where components of the original ground flora were still present. Another objective was to cover a range of species combinations, but in particular conifer/broad-leaved mixtures which might show contrasting ecological gradients, especially with respect to old woodland sites, compared with stands composed entirely of conifers or broad-leaves.

## METHODS

### The study area

Conifer/broad-leaved mixtures are common in lowland Britain, where soils are suitable for broad-leaved crops grown with conifer nurses, and amenity and conservation considerations have a high priority. In these relatively sheltered situations line thinnings of row mixtures are also less likely to suffer wind-throw than in the uplands. Data on planted mixtures in the UK and their silviculture are summarized by Kerr, Nixon and Matthews (this volume). Although this investigation was carried out primarily in Forestry Commission woods, it is important to emphasize here the role of the private sector in planting mixtures. Traditionally, private owners have planted a wider range of species and have experimented to a greater extent with mixed species planting patterns than the Forestry Commission. Post-war census data indicates that private estates accounted for 85% of the total high forest mixture acreage (Forestry Commission 1952), and, although methods of presenting census data have changed in recent years, there is much circumstantial evidence to suggest that the percentage owned by private estates is similar today (Locke 1970, 1987).

Forestry Commission subcompartment data from 1981 were obtained from Alice Holt Research Station and used to obtain a picture of the distribution of mixtures in lowland Britain. The figures are crude estimates, because the proportion of each species in the mixed stand canopy is assessed visually, and minor species components (e.g. occupying 5–10%) may be discounted. Areas of mixed crops were obtained for former Forestry Commission Conservancies (south-east,

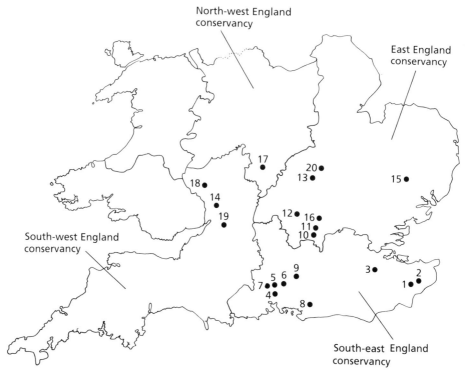

FIG. 1. Forestry Commission Conservancies from which data on plantation mixtures was drawn.
1 = Kings Wood, 2 = Denge Wood, 3 = Shipbourne, 4 = Ampfield Wood, 5 = West Wood, 6 = Crab Wood, 7 = Parnholt Wood, 8 = Stoughton Down, 9 = Chawton Park and Bushy Leaze, 10 = Homefield Wood, 11 = Wendover Woods, 12 = Bernwood Forest, 13 = Salcey Forest, 14 = Flaxley (Forest of Dean), 15 = Rougham Estate, 16 = Ashridge, 17 = Ragley, 18 = Whitfield Estate, 19 = Ebworth Estate, 20 = Boughton Park.

TABLE I. Components of the Forestry Commission stocked area in lowland England (data 1981). A limit of four species per mixture was used to calculate area totals

|  | % Total stocked area in English regions | | | | |
|  | South-east | East | South-west | North-west | Total |
|---|---|---|---|---|---|
| Pure conifer | 39.9 | 57.6 | 50.7 | 60.0 | 52.0 |
| Conifer/conifer | 9.0 | 10.9 | 10.1 | 17.8 | 12.0 |
| Conifer/broad-leaf | 19.8 | 15.3 | 17.6 | 11.9 | 16.2 |
| Broad-leaf/broad-leaf | 6.1 | 2.0 | 2.8 | 2.2 | 3.3 |
| Pure broad-leaf (high forest) | 16.2 | 6.4 | 14.0 | 2.2 | 9.7 |
| Two-storey crops | 2.7 | 2.0 | 1.1 | 2.3 | 2.0 |
| Low-grade broad-leaf | 3.3 | 5.3 | 3.2 | 3.0 | 3.7 |
| Worked coppice | 2.5 | 0.0 | 0.3 | 0.0 | 0.7 |
| Research plantation | 0.4 | 0.4 | 0.2 | 0.5 | 0.4 |

east and south-west), with a subtotal calculated for lowland regions of the north-west Conservancy (the Marches, Midlands and Sherwood Forest districts) (Fig. 1).

Although in lowland England overall pure conifer stands account for more than half the crop acreage (Table 1), mixtures are relatively common, occupying nearly one-third (31.5%) of the total area. The commonest mixture category was conifer/broad-leaf (16%), followed by all-conifer mixtures (12%): broad-leaved only mixtures and two-storied high forest (usually mixed stands) were relatively rare. All of the commonly planted broad-leaved and coniferous species appeared in mixture to varying extents. Most stands consisted of simple two-species mixtures planted in alternate rows or bands, but complex planting arrangements of five or six species were occasionally found, usually as intimate plantings.

The mixtures most frequently planted by the Forestry Commission in lowland Britain are shown in Fig. 2. Norway spruce/oak mixtures predominate, especially on heavy soils which may have previously supported coppice woodland, while stands of Corsican or Scots pine (*Pinus nigra maritima* or *P. sylvestris*) mixed with beech are almost as common, occurring on lighter, often more calcareous sites, particularly afforested downland. Besides spruce and pine, Douglas fir (*Pseudotsuga menziesii*) and larch (*Larix decidua*) are also common conifer components, while broad-leaves are mainly represented by oak and beech. Many stands contain 'low-grade broad-leaf' species, particularly where the crop was planted on a former coppice site, where coppice shoots and broad-leaved re-generation (especially birch (*Betula pendula*, *B. pubescens*)) and ash (*Fraxinus excelsior*) contribute to the canopy.

Although the subcompartment data included mixed crops at different stages in their rotation, it was clearly biased towards stands planted during the period 1950–1960 (Fig. 3). This trend, which cannot be explained by crop demography alone, is the same in all the common conifer/broad-leaved mixtures and is a legacy of postwar silvicultural thinking which favoured row mixtures on lowland sites. Recently, however, there has been a loss of interest in conifer/broad-leaved

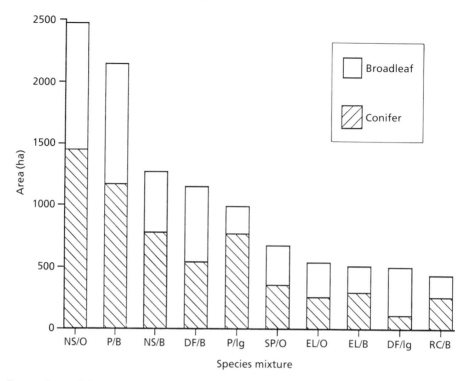

FIG. 2. Areas of the ten most common species mixtures in Forestry Commission lowland plantations. NS = Norway spruce, O = oak, B = beech, P = pine, lg = low-grade broad-leaf, EL = European larch, DF = Douglas fir, RC = Red cedar.

mixtures as policies have moved away from production towards pure broad-leaved crops on landscape and conservation grounds.

### Stratification of investigation sites

Having established the main spread of species and age-classes, the subcompartment database was searched for appropriate study sites. Initial stratification of the data was based on the following factors.

1   The most common species combinations were selected (e.g. spruce/oak and pine/beech), including other common conifer 'nurse' species such as Douglas fir and larch mixtures. Two-species combinations were considered intrinsically more likely to show clear ecological gradients than more complex planting arrangements involving several species.

2   Suitable 'control' stands were identified. If the site in question had nearby pure stands, similarly aged, of each crop species present in the mixture, the chance of selection was high. The presence on the site of fragments of old, original woodland adjacent to the mixture under investigation provided further

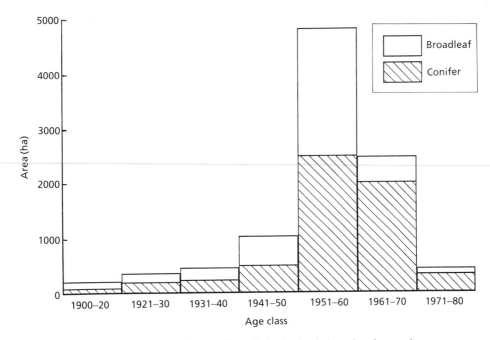

FIG. 3. Areas of species mixtures in Forestry Commission lowland plantations by age class.

'control' stands, but such instances were unfortunately rare: most of these fragments had been removed during replanting, or were confined to edges too narrow to sample.

3   A wide range of ground flora communities was sought for each crop mixture investigated. The majority of sites were old woodland sites listed on the Nature Conservancy Council's Ancient Woodland Register (Kirby, Peterken & Walker 1989) rather than sites recently afforested, where few true woodland species would have been established.

4   A variety of stands at different growth stages was selected in order to contrast young stands in the early stages of thinning with older, more open stands. Although some suitable older stands were found, row mixtures at a stage beyond the second thinning were relatively scarce in comparison with younger crops. Verification of older crops also proved difficult, because of their tendency to be recorded as pure crops after the nurse species had been fully or partially removed in thinning. For the same reason the integrity of 'pure' control stands adjacent to older mixtures was occasionally in doubt.

Uneven-aged mixtures were not included in the investigation although many were listed in the database. Such stands often originated during the practice of under-planting broad-leaves with shade-tolerant conifers or using conifers retrospectively to replace failures in a pure broad-leaved planting.

    Other potential investigation sites were the many different silvicultural exper-

TABLE 2. Mixture sites selected in lowland England showing crop types sampled in 1988. NS = Norway spruce; EL = European larch; DF = Douglas fir; SP = Scots pine; CP = Corsican pine; BE = beech; OK = oak; ok = supressed or thinned oak; HZ = hazel. *Numbers refer to Fig. 1

| Location* | Species | Planting year | 'Controls' | Soil series | Rows of each species |
|---|---|---|---|---|---|
| Kings Wood (1) | DF/BE | 1956 | BE 1951 | Paleo-argillic | 3:3 |
| | SP/BE | 1942 | BE 1941 | brown earth | 2:4 |
| | | | SP 1942 | (Batcombe) | |
| Denge Wood (2) | NS/BE | 1964 | NS 1964 | Brown rendzina (Andover) | 3:3 |
| Shipbourne (3) | NS/OK | 1953 | NS 1951 | Typical stagnogley | 3:3 |
| | | | OK 1951 | (Wickham) | |
| Ampfield Wood (4) | NS/OK | 1959 | NS 1960 | Pelo-stagnogley | 3:2 |
| | NS/OK | 1959 | | (Windsor) | 3:2 |
| West & Crab | CP/BE | 1956 | OK/HZ | Brown rendzina | 3:3 |
| Woods (5 & 6) | DF/BE | 1955 | Coppice | (Andover) | 3:3 |
| Parnholt (7) | EL/BE | 1916 | — | Typical brown earth (Carstens) | — |
| Stoughton Down (8) | CP/BE | 1954 | BE 1953 | Stagnogley-argillic | 3:3 |
| | CP/BE | 1954 | CP 1953 | brown earth | 1:1 |
| Chawton | NS/OK | 1953 | — | Typical brown | 3:3 |
| Park & Bushy | CP/BE | 1953 | CP 1950 | calcareous earth | 3:3 |
| Leaze (9) | DF/BE | 1951 | DF 1950 | | 3:3 |
| | EL/BE | 1954 | BE 1953 | | 3:3 |
| Homefield Wood (10) | DF/BE | 1955 | OK/BE | Stagnogley | 3:3 |
| | NS/BE | 1960 | pre 1900 | paleo-argillic | 3:3 |
| | CP/BE | 1958 | CP 1964 | brown earth (Batcombe) | 4:2 |
| Wendover (11) | NS/BE | 1934/36 | NS 1933 | Grey | 1:1 |
| | SP/BE | 1934 | — | rendzina | 1:1 |
| | CP/BE | 1951 | — | (Upton) | 2:2 |
| Bernwood | NS/OK | 1956 | Coppice | Pelo-stagnogley | 3:3 |
| Forest (12) | NS/ok | 1956 | plots | (Denchworth) | 3:3 |
| Salcey Forest (13) | NS/OK | 1908 | OK 1904 | Stagnogley | — |
| | NS/OK | 1945 | OK 1946 | | 3:6 |
| | SP/OK | 1915 | | | — |

iments carried out by the Forestry Commission since the 1920s, some involving mixed-species stands. Details of fifty relevant experiments, short-listed from Forestry Commission records, showed that most had been felled (often following wind-throw), or had been replanted, sold or modified. However, one series of investigations on the Rehabilitation of Devastated Woodlands dating from 1954 to 1958 were extant, including three at Bernwood (Oxfordshire), Ampfield Wood

(Hampshire) and Flaxley (Forest of Dean). Two included semi-natural 'control' areas adjacent to Norway spruce/oak mixtures, while at Flaxley the experiment consisted of larch/oak with 'controls' of mixed birch, lime (*Tilia* sp.), sweet chestnut and alder.

Table 2 gives details of thirteen sites (138 plots) selected for survey in 1988. Suitable sites in the south-east of England were relatively rare, owing to windthrow during the severe storm of October 1987. In 1989, a further seven sites were surveyed to include more old stands, larch mixtures and three-species mixtures.

## Survey methods

At each selected site stands of the mixed crops, together with any neighbouring pure species plantations or original coppice woodland, were surveyed in June and July of 1988 and 1989. Three replicate sample plots of 300 m², measuring 15 × 20 m in the pure stands, were randomly located within each chosen subcompartment. The plots were located at least 20 m into the stand to avoid edge effects. In

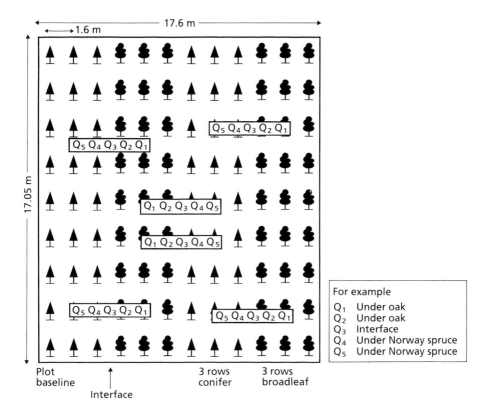

FIG. 4. A typical sample plot within a conifer/broad-leaf mixture, showing six minitransects across conifer/broad-leaf interfaces.

mixed stands, the plot dimensions were dictated by the plot width, which was drawn to include at least two bands of each species in a conventional row mixture (Fig. 4). In each plot the diameters of all living tree and shrub diameters were measured at breast height (1.3 m), and the heights of the six most dominant trees (i.e. three of each species) were recorded with a hypsometer. Slope, aspect and orientation of the species lines was noted.

Ground vegetation was recorded in thirty 1 × 1 m quadrats, arranged in six minitransects across the canopy interface between species in row mixtures (Fig. 4). In each transect, the spacing was invariably such that two quadrats in each transect were located under the broad-leaved species and two under the conifer. The same configuration was used regardless of whether the stands were pure or mixed. Within each quadrat, the percentage cover of all vascular plant species and mosses was estimated visually.

In November and early December of 1988 and 1989 all sites were revisited to obtain soil and litter samples. Twenty random locations were sampled in each plot, but in row mixtures these were stratified so that ten were taken from the row interfaces, with five each under the conifer and broad-leaf rows. Litter and humus was collected from a 15 × 15 cm square at each point. A soil core was then taken from the centre of each square to a depth of 15 cm where possible. Soil and litter samples were usually bulked for each plot, but in four stands those from under the different tree species were kept separate.

The bulked soil and litter samples were used to germinate buried seed and to grow plant fragments in seed trays in an unheated polythene tunnel. Layers of soil or litter were placed over a 2 cm layer of peat compost in seed trays, the surplus being retained for soil analysis. Emerging plants were identified and removed in four harvests between April and July.

## RESULTS

### *Correlations between stand components and ground flora abundance*

Results from the thirteen sites listed in Table 2 are given here. Most of the stands sampled were at the thicket stage, and therefore had much reduced vegetation cover, e.g. in Kings Wood the mean percentage cover was 33.6% for vascular plants and 3.2% for mosses. The number of species observed per m² was similarly low: 2.4 for vascular plants and 0.6 for mosses.

Relationships between the number and percentage cover of ground vegetation species were sought with all measured stand variables for the two main mixture combinations, Norway spruce/oak and pine/beech (Table 3). Data from all plots of each mixture type, irrespective of site and including pure species 'controls', were entered into a correlation matrix. Clear differences were observed between the behaviour of mosses and vascular plants. Moss cover was significantly increased by greater spruce dominance (number of stems, yield-class and basal area), in

TABLE 3. Significant correlations between (i) ground flora mean species number and cover m$^{-2}$ and (ii) tree stocking and site factors in mixed stands of (a) Norway spruce/oak and (b) pine/beech. Basch area volues exclude trees less that 7 cm in diameter at treast height. * $P > 0.05$; ** $P > 0.01$; *** $P > 0.001$

| | Mosses | | Vascular plants | |
| --- | --- | --- | --- | --- |
| (a) Norway spruce/oak | Mean % cover m$^{-2}$ (ln) | Mean No. species m$^{-2}$ | Mean % cover m$^{-2}$ (ln) | Mean No. species m$^{-2}$ |
| *Norway spruce* | | | | |
| Basal area (m$^2$ ha$^{-1}$) | +* | | −** | |
| Stem number ha$^{-1}$ | +** | | −*** | −*** |
| Yield-class (m$^3$ ha$^{-1}$ year$^{-1}$) | | | −** | |
| % Spruce basal area | +** | +* | −*** | |
| *Oak* | | | | |
| Basal area (m$^2$ ha$^{-1}$) | −* | | +** | |
| Stem number ha$^{-1}$ | −* | | | |
| Top height (m) | −* | | +*** | +*** |
| Yield-class (m$^3$ ha$^{-1}$ year$^{-1}$) | −* | | +* | |
| *Combined spruce/oak* | | | | |
| Stem number ha$^{-1}$ | +* | | −*** | −*** |
| *Plot data* | | | | |
| Crop age | | | +** | +*** |
| Soil pH | | | | +* |
| (b) Pine/beech | | | | |
| *Pine* | | | | |
| Basal area (m$^2$ ha$^{-1}$) | | | | +** |
| Stem number ha$^{-1}$ | | | | +* |
| % Pine basal area | | +* | | +** |
| *Beech* | | | | |
| Basal area (m$^2$ ha$^{-1}$) | | −* | −* | −*** |
| Stem number ha$^{-1}$ | −** | −*** | −*** | −*** |
| Top height (m) | −* | −** | −** | −*** |
| Yield-class (m$^3$ ha$^{-1}$ year$^{-1}$) | −* | −** | −*** | −*** |
| *Combined pine/beech* | | | | |
| Stem number ha$^{-1}$ | −*** | −*** | −*** | −*** |
| *Plot data* | | | | |
| Crop age | | | +* | |
| Litter depth | | −* | −* | −*** |
| Humus depth | +* | +** | | |
| Organic horizon depth | | | | −* |
| Soil pH | −*** | −** | | |
| % Carbon (loss on ignition) | −** | | −** | −** |

contrast to vascular plant cover which was strongly suppressed (Table 3a). Some of this increase may be explained by the lower light values under the spruce canopy compared with oak, which would tend to benefit shade-tolerant bryophytes at the expense of vascular species (Hill & Hays 1978), or by improved substrate conditions under spruce. Moss species numbers were also increased by greater dominance of spruce, while numbers of vascular species declined markedly.

Increasing the amount of oak in the mixture had the opposite effect. In this case, moss cover fell as the density, yield and basal area of oak increased, which in turn increased the cover of vascular plants. Oak top height was also positively correlated with vascular species number, possibly because increasing height brings forward the prospect of thinning and the opening up of the canopy. Combined crop data for both species gave similar results to the spruce component alone, and was in turn heavily influenced by the amount of spruce in the mixture. Older crops, which generally had lower stocking densities as a result of thinning, had more vascular plant species and cover. In pine/beech mixtures the effects of conifer and broad-leaved species on the ground flora were largely reversed, the pine component being correlated with more vascular plant species, while beech was associated with less vegetation cover and fewer species of mosses and vascular plants (Table 3b). The pine component was not correlated with the number of moss species. The balance of the mixed crop was in this case largely determined by the beech component. Crop age was again positively correlated with vascular plant cover, but several soil variables were also significant. The most important of these were litter depth and soil organic carbon, both of which were associated with fewer ground flora species and reduced cover.

## *Multiple regression analysis*

The objective of the multiple regression analysis was to produce a predictive equation with all variables showing a significant '$t$' value at $P < 0.05$. Factors were included on the basis of both correlation and simple linear regression results. Stepwise multiple regression procedures and manual factor selection were used to maximize the variance accounted for (adjusted $R^2$) in either the percentage cover or species number of mosses and vascular plants, respectively. Factors were dropped from the equations when they did not significantly increase the amount of variation explained or if multicollinearity occurred. The results were broadly similar to the correlation analysis.

In Norway spruce, high stocking levels reduced vascular plant cover (Fig. 5) and species number. The effect of soil pH was also important, more vascular plants being associated with higher base status. Together these factors explained 74% of the variation in vascular plant cover and 61% of the variation in vascular plant species number. Moss species number and cover increased with greater spruce stocking, but crop and soil factors explained relatively little of the overall variation in species number and cover.

In pine/beech mixtures both mosses and vascular plants were strongly inhibited by beech stocking, but were also influenced by a number of site factors. Higher base status again resulted in more vascular plant species and cover, which in turn probably reduced moss cover. Humus depth was a significant factor

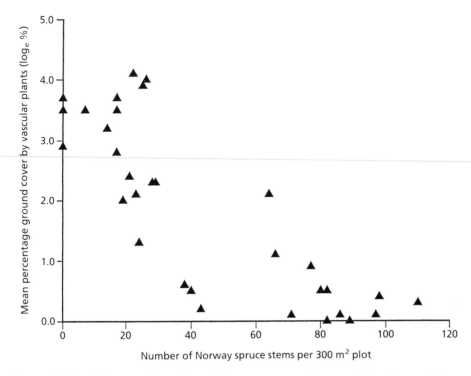

FIG. 5. Relationship between the number of Norway spruce stems in spruce/oak mixtures and the percentage of the ground covered with vascular plants.

increasing the range of moss species, but increasing soil organic carbon, possibly caused by increasing litterfall, reduced the diversity and cover of all vegetation. The four factors—number of beech stems, soil pH, soil carbon and humus depth—accounted for between 53 and 66% of the overall variation in both moss and vascular plant cover and species number.

### Vegetation differences between mixed and pure stands

Comparisons between the ground vegetation of mixed and pure stands at the same site showed considerable variation. However, out of eight possible comparisons, pure pine crops consistently had more vegetation cover than pine/beech mixtures or pure beech. The pure Norway spruce plots tended to have more mosses, but fewer vascular plant species than spruce/oak mixtures. There were also more vascular species in larch/beech stands at one site compared with pure beech, and under Scots pine/oak mixtures compared with pure oak at another. The specific effect of mixed-tree crops on the ground flora community varied according to site.

## *Soil seed-bank composition*

Typically, the composition and size of soil seed-banks within stands varied considerably over very short distances across the forest floor. Therefore, seed-bank data was highly variable at all sites, and comparisons of mixed stands with pure stands of either species component showed few significant differences. However, at two sites, Kings Wood and Chawton Park, higher densities of germinable seeds were found in pine plots than in adjacent beech plots of similar age. Similarly, higher seed-bank densities were found in pure oak than in pure spruce at Shipbourne, and also at Ampfield where the oak lines of thinned spruce/oak plots had more germinable seeds than those of spruce. These trends correspond broadly to the vascular vegetation, which was relatively more abundant in mixtures containing higher proportions of pine or oak.

Table 4 shows examples of seed-bank data for (i) Shipbourne Forest and (ii) Kings Wood. At Shipbourne, the pure oak stands contained more germinable seeds than those of spruce but not more than the corresponding mixture, in which germinable seed densities and species numbers were similar in oak and spruce rows. In beech/Scots pine plots at Kings Wood the pure beech stands had the lowest germinable seed densities, differing significantly from Scots pine stands of similar age. However, there was no significant difference between pure beech and mixed pine/beech stands, nor any between the younger Douglas fir/beech and pure beech stands nearby.

## *Vegetation gradients at mixture interfaces*

In many row mixture combinations, considerable variation in ground flora abundance was found corresponding to the alternating bands of tree species. For the mixed plots of each site, the percentage cover and species number of vascular plants and mosses was compared in transects across the interface of the alternating bands of each tree species (see Fig. 4), using analysis of variance. The results are given in Table 5. The greatest number of significant differences were found in Norway spruce/oak combinations, where quadrats placed directly under the conifer had consistently fewer plant species and less vegetation cover than under oak. The only exception to this was at Chawton Park, where there was greater moss cover under the central spruce row, most of which had been removed in earlier thinnings. The effect of thinning was also evident in the three Norway spruce:two oak planting arrangement at Ampfield. In the unthinned stands the ground flora was more abundant under oak, but the thinning of the two outer lines of the spruce, together with suppressed individuals of oak (in a silvicultural thinning), effectively removed this vegetation zonation.

The pine/beech interface showed less ground flora zonation than spruce/oak, but at two sites, Chawton and Homefield, the dominance of the beech rows over the ground flora was clearly seen. However, at Westwood there was a greater moss cover under beech than under pine, possibly due to increased vascular cover

TABLE 4. Numbers of species and numbers of germinable seeds in seed-banks at (a) Shipbourne and (b) Kings Wood, showing differences between mixed-species crops and their pure-species components. At each site the mixed and pure stands of beech and pine or oak and spruce were in close proximity

| Tree crop and planting year | Species number | Seed density (No. m$^{-2}$) |
|---|---|---|
| **(a) Shipbourne** | | |
| Oak (pure) (1951) | 12 | 9097 |
| Norway spruce (pure) (1951) | 9 | 2682 |
| Oak/spruce mixture (1953) | | |
|     (i) spruce rows | 14 | 12 978 |
|     (ii) oak rows | 18 | 9053 |
|     Grand mean | | 8452 |
|     L.S.D. ($P = 0.05$) | | 7297 |
| **(b) Kings Wood** | | |
| Beech (pure) (1941) | 10 | 1493 |
| Scots pine (pure) (1942) | 14 | 4236 |
| Beech/Scots pine mixture (1942) | 11 | 3693 |
| Douglas fir/beech mixture (1956) | 10 | 4429 |
| Beech (pure) (1952) | 14 | 2532 |
|   Grand mean | | 3277 |
|   L.S.D. ($P = 0.05$) | | 3354 |

in thinned pine rows. In combination with Douglas fir or Norway spruce, the effect of beech was less conclusive, the conifer lines tending to suppress the ground flora to a greater extent. However, at Chawton, thinning once again improved the abundance of vascular plants where Douglas fir lines were removed from beech in mixture.

## DISCUSSION

For reasons already noted, the majority of the mixtures stands sampled were young, having been planted in the 1950s. Many were therefore at the thicket stage, or had reached the pole stage following initial line-thinning or (rarely) silvicultural thinnings. At this point in the rotation, when canopy closure is virtually complete, the effects of tree crops on light interception, litterfall and soil chemistry tend to reach their maximum (Ovington 1955; Page 1968). The dominance of the crop at this stage restricts most ground flora species, as was evident in dense mixtures with large numbers of spruce and beech, as well as in the pure stands. Tree stem density was selected by multiple regression and correlation analysis as the dominant factor influencing the ground flora, suggesting that light transmission through thicket-stage and early pole-stage stands was the major limiting factor, rather than basal area, which tends to increase steadily beyond the thicket stage.

Conversely, stands with reasonable ground flora abundance had lower stem densities resulting from (i) crop failures (for example due to the suppression of a

TABLE 5. Ground flora composition compared in different quadrats within 5 m minitransects across crop interfaces. In any one stand there were eighteen estimates of per cent cover and species number for each quadrat position. Means were calculated and compared using the L.S.D. ($P < 0.05$). Data from 6 NS/OK and 6 pine/beech stands was analysed. This table indicates the number of stands out of six in which the quadrat in the position at the top of the table contained significantly *more* vegetation than the quadrat in the position indicated in the left-hand column. Comparisons should be read in the direction of the arrow in (a)

*Norway spruce/oak*

(a) Moss % cover

| Quadrat position | Oak | Oak | Interface | Spruce | Spruce |
|---|---|---|---|---|---|
| Oak | | | | | |
| Oak | | | | | I |
| Interface | I | 2 | | | I |
| Spruce | 2 | 3 | 3 | | |
| Spruce | 2 | 4 | 3 | I | |

(b) Vascular plants % cover

| Quadrat position | Oak | Oak | Interface | Spruce | Spruce |
|---|---|---|---|---|---|
| Oak | | | | | |
| Oak | | | | | |
| Interface | I | I | | | |
| Spruce | 2 | I | | | |
| Spruce | 2 | 2 | | | |

(c) Moss species number

| Quadrat position | Oak | Oak | Interface | Spruce | Spruce |
|---|---|---|---|---|---|
| Oak | | I | | | |
| Oak | | | | | |
| Interface | I | 2 | | | |
| Spruce | 2 | 3 | 2 | | |
| Spruce | 2 | 4 | 3 | | |

(d) Vascular plants species number

| Quadrat position | Oak | Oak | Interface | Spruce | Spruce |
|---|---|---|---|---|---|
| Oak | | | | | |
| Oak | | | | | |
| Interface | | | | | |
| Spruce | I | | I | | |
| Spruce | I | 2 | I | | |

*Pine/beech*

(e) Moss % cover

| Quadrat position | Beech | Beech | Interface | Pine | Pine |
|---|---|---|---|---|---|
| Beech | | | | I | I |
| Beech | | | | I | I |
| Interface | | | | | |
| Pine | I | | | | |
| Pine | I | | | I | |

(f) Vascular plants % cover

| Quadrat position | Beech | Beech | Interface | Pine | Pine |
|---|---|---|---|---|---|
| Beech |  |  | I | I | I |
| Beech |  |  | I | I | I |
| Interface |  |  |  |  |  |
| Pine |  |  |  |  |  |
| Pine |  |  |  |  |  |

(g) Moss species number

| Quadrat position | Beech | Beech | Interface | Pine | Pine |
|---|---|---|---|---|---|
| Beech |  |  |  |  |  |
| Beech |  |  |  |  |  |
| Interface | I |  |  |  | I |
| Pine | I |  |  |  |  |
| Pine | I |  |  | I |  |

(h) Vascular plants species number

| Quadrat position | Beech | Beech | Interface | Pine | Pine |
|---|---|---|---|---|---|
| Beech |  |  | I | I | I |
| Beech |  |  |  |  |  |
| Interface |  |  |  |  |  |
| Pine |  |  |  |  |  |
| Pine |  |  |  |  |  |

broad-leaved crop by its conifer companion, or the failure of conifer rows through lime-induced chlorosis on chalky soils) or (ii) the thinning of spruce in spruce/oak mixtures, or beech in pine/beech stands. Ground flora abundance was also positively related to crop age, top height and yield-class, all of which were related to the age at first thinning. Thinning appeared not only to promote the recovery of the ground flora after long periods in the thicket stage, but also to reduce the zonation of plant species abundance across the canopy interfaces of mixed tree crops, as observed at Chawton Park. This recovery is consistent with the findings of several investigations which have compared the ground vegetation of forest stands at different stages in their development, and was observed in most of Ovington's plots after periods of thinning (Anderson 1979).

As the gaps in the crop occur beyond the thicket stage, the influence of site factors such as climate and soil chemistry become more important. In the mixed stands of pine and beech, multiple regression showed a positive relationship between vascular plant species number and soil base status. In the same stands, there was a negative relationship between plant species number and soil organic carbon. These factors may have been influenced by litterfall and canopy structure, suggesting that the crop may continue to exert an indirect effect on the ground flora in later stages of the rotation.

The number of vascular species tended to be reduced by increasing amounts of spruce and beech. However, when spruce or beech were present in mixture with their respective companions, oak and pine, the number of ground flora species was not necessarily depressed. Rather, combinations of the two crop species tended to create a range of environmental conditions which benefited species diversity, both in space and time. Moderate proportions of spruce with oak for example, tended to increase the performance of shade-tolerant bryophytes in the darker conditions created by the conifer, by reducing the dominance of vascular plants and possibly by increasing the variety of suitable substrate in mixed stands. These bryophytes were mostly present under the oak lines, but at the same time most vascular plant species managed to persist in parts of the stand in small numbers.

The relative similarities of seed-bank abundance and composition in pure and mixed stands were not unexpected in these young plantations. Insufficient time will have elapsed for differences in seed-bank decay rate to manifest themselves if, as Brown and Warr (in press) suggest, many species can maintain persistent seed populations under these conditions for up to 50 years. Nevertheless, the occurrence of larger seed-banks under pine and oak compared with beech and spruce found at four different sites suggests that opportunities for recharging the seed-bank are greater under light canopies where the ground flora is more abundant.

Compared with mixtures, pure crops may have a detrimental effect on the ground flora. Under pure spruce or beech, for example, plant cover and species numbers were often severely restricted by poor illumination and heavy litterfall, whereas under the lighter canopies of pure oak or pine the dominance of understorey shrubs, or plants such as *Rubus fruticosus* and *Pteridium aquilinum*, could themselves restrict the abundance of other ground flora species. It follows that some optimum percentage of each crop species component and pattern of admixture will give the maximum species diversity of the ground flora, but this will change according to the management and the growth stage reached by the crop.

### Silvicultural practices in relation to ground vegetation diversity

A primary factor to be considered at the outset is the tree species composition of the mixture. Vigorous stands of beech or spruce are clearly detrimental to the ground flora in the thicket stage, even in row mixtures where, for example, the vegetation may be sharply restricted under lines of spruce. Similar effects might be expected with other species casting heavy shade, such as hornbeam (*Carpinus betulus*), Douglas fir and western red cedar (*Thuja plicata*). The danger here is that some species might be virtually eliminated from a compartment (for example *Hyacinthoides non-scripta* under Norway spruce at Shipbourne), so that they fail to increase in abundance when conditions improve after thinning. Localized populations of specialist woodland species might disappear altogether. On the other hand, pine, oak and, under some circumstances, larch, tend to allow greater persistence of the ground flora, especially of spring perennials, either through increased illumination levels or by virtue of the deciduous habit.

One factor which may also improve the situation is the 'low-grade broad-leaf' component, present as old coppice regrowth or as natural regeneration. In the mixed stands surveyed, the basal area of volunteer broad-leaves sometimes accounted for 10% of the stand. Although this level is too low to dictate the characteristics of the ground flora, neglecting or modifying weeding practices may increase the influence of this component. Similarly, the admixture of other broad-leaved species with beech, or broad-leaved mixtures *per se*, might improve ground flora diversity by influencing light transmission or the quality and quantity of litterfall.

Several planting patterns and silvicultural treatments are advocated for mixtures; however, all but a few are limited by commercial realities. The most viable are likely to remain even-aged patterns of two or three crop species, arranged as intimate plantings, as alternating bands of different species, or as small groups of one species within a matrix of another. Each of these planting patterns has its advantages and disadvantages regarding the persistence and maintenance of the ground flora.

Intimate mixtures of two or three species tend to have short gradients between the crowns of adjacent young trees, so that the 'averaging' of conditions created by the composite canopy might also tend to produce uniformity in the ground vegetation. Although at any one time conditions at the forest floor may benefit the vegetation, such arrangements are inherently unstable if one or more crop species is subordinate to another, leading to changes in environmental conditions and to difficult choices at thinning. Moreover line thinning, which opens up the canopy and which may allow recovery or persistence of a ground flora suppressed during the thicket stage, is not appropriate in these types of mixture.

Group mixtures, on the other hand, create longer gradients across the forest floor and a more patchy environment of light and litterfall which may encourage greater numbers of ground flora species. Later thinning may increase this patchiness, particularly at the interfaces of each group. However, if one species component tends to eliminate the vegetation underneath it, the denuded area may only partially be recolonized from adjacent vegetation sources when conditions improve.

Row mixtures, such as those investigated above, have a number of advantages not shared by either intimate or group mixtures. The rows are eminently suitable for line thinnings, which may allow some sensitive ground flora species sufficient periods of growth and flowering to carry them through the rest of the rotation. Barren rows under species such as spruce and beech also appear to be colonized rapidly, as the distances across species bands are relatively short. However, for the same reason the gap sizes created by row thinning may be insufficient to favour the ephemeral species which form large, persistent seed-banks.

Conventional mixtures may be manipulated by the timing and nature of operations for the benefit of the ground flora. Line thinnings could be earlier and more severe, for example by routinely removing both outer rows of the nurse crop in 3:3 row mixtures. Less reliance could be placed on high initial stockings of all planted species, with correspondingly more emphasis given to natural regeneration or coppice regrowth of broad-leaved species native to the site. Lower ratios of nurse crops could be tolerated, particularly where these are not considered 'benign', but if they are omitted altogether this may actually lower diversity in the ground vegetation. Policy guidelines (e.g. Forestry Commission 1985) which suggest reducing the conifer component of mixtures planted on ancient woodland sites need to be sufficiently detailed to take account of the consequences for the ground flora of growing pine with beech or oak with spruce. Removal of the nurse crops needs to be planned so as to create the maximum opportunities for recolonization, but the retention of some stems to maturity in order to promote variation in the canopy may also be an advantage.

## ACKNOWLEDGMENTS

We are grateful to the Wildlife & Conservation Branch of the Forestry Commission, Alice Holt Lodge, for supporting this study.

## REFERENCES

Anderson M. (1979). The development of plant habitats under exotic forestcrops. *Ecology and Design in Amenity Land Management* (Ed. by S.E. Wright & G.P. Buckley), pp. 87–108. Wye College, Wye.

Beatty S.W. (1984). Influence of microtopography and canopy species on spatial patterns of forest understorey plants. *Ecology*, **65**, 1406–19.

Brewer R. (1980). A half-century of changes in the herb layer of a climax deciduous forest in Michigan. *Journal of Ecology*, **68**, 823–32.

Brown A.F.H. (1982). The effects of tree species, planted pure and in mixtures, on vegetation and soils at Gisburn. *Institute of Terrestrial Ecology Annual Report 1981*, 74–5.

Brown A.F.H. & Warr S.J. (in press). The effects of changing management on seedbanks in ancient coppices. *The Ecology of Coppice Management* (Ed. by G.P. Buckley). British Ecological Society Symposium 1990.

Clapham A.R., Tutin T.G. & Moore D.M. (1987). *Flora of the British Isles*, 3rd edn. Cambridge University Press, Cambridge.

Crozier C.R. & Boerner R.E.J. (1984). Correlations of under-storey herb distribution patterns with micro habitats under different tree species in a mixed mesophyte forest. *Oecologia*, **62**, 337–43.

Forestry Commission (1952). Census of Woodlands 1947–49. Woodlands of five acres and over. *Census Report No. 1.* HMSO, London.

Forestry Commission (1985). *Guidelines for the Management of Broadleaved Woodland*. Forestry Commission, Edinburgh.

Hill M.O. (1979). The development of a flora in even-aged plantations. *The Ecology of Even-aged Plantations* (Ed. by E.D. Ford, D.C. Malcolm & J. Atterson), pp. 175–92. Institute of Terrestrial Ecology, Cambridge.

Hill M.O. & Hays J.A. (1978). Ground flora illumination under differing crop species in a Forestry Commission experiment in North Wales. *Chief Scientist Team Report No. 171.* Nature Conservancy Council, Peterborough.

Hill, M.O. & Jones E.W. (1978). Vegetation changes resulting from afforestation of rough grazings in Caeo Forest, South Wales. *Journal of Ecology*, **66**, 433–56.

Kirby K.J. (1988). Changes in the ground flora under plantations on ancient woodland sites. *Forestry*, **61**, 317–38.

Kirby K.J., Peterken G.F. & Walker G.L. (1989). Inventories of ancient semi-natural woodland. *Focus on Nature Conservation No. 6.* Nature Conservancy Council, Shrewsbury.

**Locke G.M.L. (1970).** *Census of Woodlands 1965–67.* A report on Britain's resources. Forestry Commission, HMSO, London.

**Locke G.M.L. (1987).** *Census of Woodland and Trees* 1979–82. Forestry Commission Bulletin, **63**, HMSO, London.

**Mitchell P.L.** (in press). Growth stages and microclimate in coppice and high forest. *The Ecology of Coppice Management* (Ed. by G.P. Buckley). British Ecological Society Symposium. Chapman & Hall, London.

**Mitchell P.L. & Kirby K.J. (1989).** Ecological effects of forestry practices in long-established woodland and their implications for nature conservation. *Oxford Forestry Institute Occasional Papers No. 39*, University of Oxford.

**Ovington J.D. (1955).** Studies of the development of woodland conditions under different trees. III. The ground flora. *Journal of Ecology*, **43**, 1–21.

**Page G. (1968).** Some effects of conifer crops on soil properties. *Commonwealth Forestry Review*, **47**, 52–62.

**Peterken G.F. (1974).** A method for assessing woodland flora for conservation using indicator species. *Biological Conservation*, **11**, 223–36.

**Smith A.J.E. (1980)** *The Moss Flora of Britain and Ireland.* Cambridge University Press, Cambridge.

**Sydes C. & Grime J.P. (1981a).** Effects of tree leaf litter on herbaceous vegetation in deciduous woodland. I. Field investigations. *Journal of Ecology*, **69**, 237–48.

**Sydes C. & Grime J.P. (1981b).** Effects of tree leaf litter on herbaceous vegetation in deciduous woodland. II. An experimental investigation. *Journal of Ecology*, **69**, 249–62.

# Bird populations: effects of tree species mixtures

K.W. SMITH

*Royal Society for the Protection of Birds, The Lodge, Sandy, Bedfordshire SG19 2DL, UK*

## SUMMARY

The breeding bird community of a mixed oak and beech high forest in the Forêt de Crécy, north-west France was studied by counting birds at 150 randomly selected points. Habitat measures were collected at each point and used to examine the effects of the ratio of oak to beech trees at each point on the breeding bird community. In univariate and multivariate analyses no significant correlation was found between the fraction of oak trees and the overall breeding bird density although, for some species, there were significant correlations with the individual counts. It was found to be impossible to divorce entirely the fraction of oak trees from other differences in the vegetation which were thought to contribute to the trends in counts for some species. For these particular woods no overall differences were found between the mixtures and pure stands. However, other studies have shown that admixtures of broad-leaves in predominantly coniferous forests are of considerable value to breeding songbird populations.

## INTRODUCTION

In Britain, tree species mixtures as habitats for birds have received little attention despite their silvicultural importance. It is well established that both woodland structure and tree species affect the breeding bird community (MacArthur & MacArthur 1961; Willson 1974; Moss 1978a,b; James & Wamer 1982; Erdelen 1984) and that individual bird species show distinct preferences for particular tree species both within and outwith the breeding season (Gibb 1954; Ulfstrand 1975; Peck 1989). Although a number of studies have looked at the breeding bird assemblages of mixed stands (Tomialojc, Wesolowski & Walankiewicz 1984; French, Jenkins & Conroy 1986; Muller 1985, 1986) in only one case has there been any attempt to generalize the conclusions beyond those of the particular study.

Compared with single-species stands, mixtures may be important for many reasons, including the availability of particular food resources, foraging niches, nest sites or shelter. They could also lead to resources being available at different times of year — for instance seeds in autumn and winter in addition to invertebrates in spring and summer (Watt this volume; Young this volume). Tree species mixtures may also have an indirect influence on breeding bird

233

communities because of their structural diversity. Different tree species have different growth rates and forms and the mixtures may, in themselves, produce a more complex structure than the equivalent monoculture.

The overall differences in the breeding bird communities of coniferous and broad-leaved woodlands have been well documented (Yapp 1952; Simms 1971; Adams & Edington 1973; Moss 1978b). However, only three studies have looked in any detail at the effects of admixtures of broad-leaves in a predominantly coniferous woodland (French, Jenkins & Conroy 1986; Bibby, Aston & Bellamy 1989; Peck 1989). Although they took different approaches, all three studies concluded that broad-leaves could have a major effect on the breeding bird community of a coniferous forest.

In this paper, the results of a study of the breeding birds of mixtures of mature oak *Quercus robur* and beech *Fagus sylvatica* are described. This study was conducted to determine if, even with two broad-leaf species, there were any differences between the breeding birds in mixtures compared with pure stands. These two tree species would be expected to carry distinct invertebrate communities (Southwood 1961) with a considerably higher biomass on oak than beech (K. Peck unpubl.; Southwood, Moran & Kennedy 1982; Peck 1989); both of these factors would be expected to influence the composition and density of the breeding bird community.

## METHODS

The data reported here were collected as part of a wider study of the breeding bird communities of managed oak and beech woodlands of the Forêt de Crécy, north-west France (Smith 1988). This site was selected for its extensive areas of broad-leaved woodland under relatively uniform management, and its nearness to southern England. No comparable study site could be found in lowland England. The forest covered a total area of 4300 ha, the majority of which was managed as oak or beech high forest. The study sites were twenty-nine compartments of the forest containing mixtures of mature oak and beech trees, covering a total area of 345 ha. The mean compartment area was 11.9 ha with a range of 6.5–20.0 ha. The oak and beech standards were between 150 and 200 years old with a sparse understorey of mainly hornbeam *Carpinus betulus* and beech.

Breeding birds were surveyed using the point count method (Bibby, Phillips & Seddon 1985) at 150 sites distributed randomly within the twenty-nine compartments. Counts were made at each of the 150 points for a period of 5 minutes on two visits during the breeding season (mid April until early June). Birds were recorded in two categories, those within 25 m, when first observed, and those beyond 25 m. A team of three trained observers carried out all the counts and, to avoid bias, points were swapped between visits. Relative densities were estimated using the maximum count of the two visits for each species and assuming a half-normal detection function (Bibby, Phillips & Seddon 1985) with standard errors estimated by jackknife (Millar 1974).

A series of habitat measurements, mainly relating to the structure of the woodland, were made for the area within 25 m of each point. Ten of these variables were used in the subsequent analyses. Stem counts of oak and beech trees within the 25 m were made and expressed as the fraction of the total trees which were oak. The height of the tallest tree in the circle was estimated, the observers having already been trained on trees of known height. Basal area was measured using a relascope (Hamilton 1975). To estimate the vegetation cover, two transects, running north–south and east–west, were made across the circle. At every metre on these transects the observers used a sighting rod to determine whether there was foliage cover in four height bands: 0.5–2 m, 2–4 m, 4–10 m and >10 m. These assessments of presence or absence were then converted to percentage cover for the whole circle. The ground cover was also assessed at each metre on the transects.

The effects of tree species mixtures were examined by sorting the points into fifteen groups of ten on the basis of the fraction of oak trees within the 25 m circle and examining the trends in bird density, species diversity (Shannon & Weaver 1949) and counts. In addition, the habitat variables were simplified using Principal Components Analysis with varimax rotation (Cooley & Lohnes 1971), and trends against the new variables were examined.

## RESULTS

A total of thirty-seven species were recorded at the points. The overall bird community, derived from the counts at all the points, was dominated by five species (Table 1)—chaffinch, marsh tit, robin, blue tit and great tit making up 66% of the overall density. Such simple bird communities have previously been reported for other mature deciduous woodlands (Smith, Averis & Martin 1987). The values of the habitat variables are summarized in Table 2. The study plots had very low foliage cover up to a height of 10 m, with the ground cover largely dominated by leaf litter. Canopy cover, stem count and basal area were very variable, with some plots falling in areas of high stem count and others in almost fully open glades. The fraction of oak trees ranged from 0 to 1, i.e. pure beech to pure oak.

The points were ranked according to the fraction of oak trees within the count circle, and the density and bird species diversity recalculated for fifteen groups of ten points (Fig. 1). There was no significant trend in overall density or diversity against the fraction of oak trees ($r = 0.04$, d.f. $= 13$; $r = 0.00$, d.f. $= 13$). Using the total counts for each of the groups of ten points, there were significant correlations for eight species (Table 3). Great tit and short-toed treecreeper were positively correlated with the fraction of oak, whilst counts of blackbird, blackcap, garden warbler, willow warbler, chiffchaff and wood warbler were negatively correlated with the fraction of oak.

There were significant correlations between the individual habitat variables and they were therefore simplified using Principal Components Analysis with

TABLE 1. Counts of breeding birds and density estimates at 150 randomly selected points in beech and oak high forest in the Forêt de Crécy, north-west France. Counts are given of bird numbers within 25 m of the selected points, and totals within and beyond 25 m

| Species | Counts 0–25 m | Counts 0–∞ | Density ± S.E. (no. ha$^{-1}$) |
|---|---|---|---|
| Chaffinch, *Fringilla coelebs* | 86 | 530 | 3.18 ± 0.41 |
| Marsh tit, *Parus palustris* | 36 | 148 | 1.39 ± 0.29 |
| Robin, *Erithacus rubecula* | 35 | 240 | 1.28 ± 0.20 |
| Blue tit, *Parus caeruleus* | 30 | 180 | 1.11 ± 0.22 |
| Great tit, *Parus major* | 29 | 170 | 1.08 ± 0.23 |
| Crested tit, *Parus cristatus* | 20 | 92 | 0.76 ± 0.20 |
| Short-toed treecreeper, *Certhia brachydactyla* | 16 | 128 | 0.58 ± 0.16 |
| Blackcap, *Sylvia atricapilla* | 11 | 106 | 0.39 ± 0.13 |
| Nuthatch, *Sitta europea* | 11 | 206 | 0.38 ± 0.13 |
| Chiffchaff, *Phylloscopus collybita* | 10 | 115 | 0.35 ± 0.11 |
| Willow warbler, *Phylloscopus trochilus* | 8 | 126 | 0.28 ± 0.10 |
| Great spotted woodpecker, *Dendrocopus major* | 7 | 90 | 0.25 ± 0.09 |
| Long-tailed tit, *Aegithalos caudatus* | 5 | 7 | 0.22 ± 0.23 |
| Hawfinch, *Coccothraustes coccothraustes* | 5 | 56 | 0.18 ± 0.10 |
| Wood warbler, *Phylloscopus sibilatrix* | 5 | 86 | 0.17 ± 0.08 |
| Blackbird, *Turdus merula* | 4 | 111 | 0.14 ± 0.09 |
| Willow tit, *Parus montanus* | 1 | 2 | 0.09 ± 0.07 |
| Goldcrest, *Regulus regulus* | 2 | 4 | 0.07 ± 0.08 |
| Jay, *Garrulus glandarius* | 2 | 39 | 0.07 ± 0.05 |
| Song thrush, *Turdus philomelos* | 2 | 64 | 0.07 ± 0.05 |
| Grasshopper warbler, *Locustella naevia* | 1 | 5 | 0.03 ± 0.04 |
| Golden oriole, *Oriolus oriolus* | 1 | 23 | 0.03 ± 0.03 |
| Yellowhammer, *Emberiza citrinella* | 1 | 24 | 0.03 ± 0.03 |
| Garden warbler, *Sylvia borin* | 1 | 26 | 0.03 ± 0.03 |
| Wren, *Troglodytes troglodytes* | 0 | 39 | |
| Nightingale, *Luscinia megarhynchos* | 0 | 13 | |
| Starling, *Sturnus vulgaris* | 0 | 7 | |
| Linnet, *Acanthis cannabina* | 0 | 5 | |
| Lesser spotted woodpecker, *Dendrocopus minor* | 0 | 4 | |
| Redstart, *Phoenicurus phoenicurus* | 0 | 4 | |
| Green woodpecker, *Picus viridis* | 0 | 3 | |
| Tree sparrow, *Passer montanus* | 0 | 2 | |
| Greenfinch, *Carduelis chloris* | 0 | 2 | |
| Dunnock, *Prunella modularis* | 0 | 1 | |
| Whitethroat, *Sylvia communis* | 0 | 1 | |
| Icterine warbler, *Hippolais icterina* | 0 | 1 | |
| Treecreeper, *Certhia familiaris* | 0 | 1 | |

varimax rotation (Cooley & Lohnes 1971). Table 4 shows the loadings of the ten original variables on axes 1 and 2 of the PCA. Axis 1 was identified as the gradient from open canopy with high coverage in the three shrub layers to tall, closed canopy woodland. Axis 2 was positively weighted by the percentage of litter on the ground and negatively by the percentage of grass and the fraction of oak trees. It therefore represented the gradient from beech to oak dominated woodland although, in addition, it also contained positive weighting for the

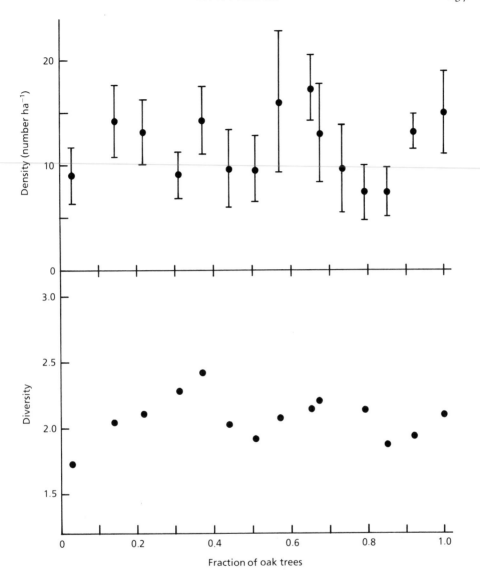

FIG. 1. Overall breeding bird density and species diversity calculated for groups of ten points ranked by the fraction of trees within 25 m of the point that were oak. Error bars indicate ±1 S.E.

coverage in the shrub layers and negative for basal area and stem count.

To examine the trends in the bird community along these two axes the points were ranked, and density, species diversity and total count calculated for fifteen groups of ten points (Table 5). There were no significant correlations with overall density along either axis although the species diversity was positively correlated with the score on axis 1. There were nineteen significant correlations with the

TABLE 2. Habitat variables measured within 25 m of each point which were used in the analyses of breeding bird communities

|  | Median | Range |
|---|---|---|
| Maximum height (m) | 33.0 | 22–42 |
| Basal area (m$^2$ ha$^{-1}$) | 18.0 | 0–40 |
| Stem count (No. ha$^{-1}$) | 61.2 | 10–326 |
| Fraction of trees oak | 0.57 | 0–1.0 |
| % Canopy cover | 50 | 5–85 |
| % Shrub cover 4–10 m | 25 | 0–95 |
| % Shrub cover 2–4 m | 15 | 0–100 |
| % Shrub cover 0.5–2 m | 10 | 0–75 |
| % Ground cover—grass | 10 | 0–90 |
| % Ground cover—litter | 75 | 0–100 |

TABLE 3. Correlations between the overall breeding bird density, species diversity and individual species counts for fifteen groups of ten points ranked by the fraction of oak trees within 25 m of the census point. Results are given for only nineteen species with a total count greater than twenty-five. Scientific names are given in Table 1. *$P < 0.05$; **$P < 0.01$

|  |  |
|---|---|
| Total bird density | 0.04 |
| Species diversity | 0.00 |
| Jay | 0.43 |
| Great tit | 0.53* |
| Crested tit | 0.51 |
| Blue tit | −0.27 |
| Marsh tit | 0.36 |
| Nuthatch | 0.44 |
| Short-toed treecreeper | 0.56* |
| Wren | −0.16 |
| Song thrush | −0.13 |
| Blackbird | −0.76** |
| Robin | −0.09 |
| Blackcap | −0.69** |
| Garden warbler | −0.79** |
| Willow warbler | −0.82** |
| Chiffchaff | −0.61* |
| Wood warbler | −0.70** |
| Hawfinch | −0.14 |
| Chaffinch | 0.50 |
| Great spotted woodpecker | 0.29 |

total count. Eight species were significantly correlated with the score on axis 1: song thrush, robin, willow warbler and wood warbler positively so, and nuthatch, short-toed treecreeper, wren and chaffinch negatively so. On axis 2 there were eleven species with significant correlations: blackbird, blackcap, garden warbler, willow warbler, chiffchaff and wood warbler positively so, and great tit, crested tit, nuthatch, short-toed treecreeper and chaffinch negatively so.

TABLE 4. Loadings of the habitat variables on the first two Principal Component axes

| | Axis 1 | | Axis 2 |
|---|---|---|---|
| % Shrub 4–10 m | 0.92 | % Litter | 0.71 |
| % Shrub 2–4 m | 0.87 | % Shrub 0.5–2 m | 0.45 |
| % Shrub 0.5–2 m | 0.66 | Maximum height | 0.43 |
| % Litter | 0.26 | % Shrub 2–4 m | 0.33 |
| Fraction trees oak | 0.12 | % Shrub 4–10 m | 0.20 |
| Basal area | 0.01 | % Canopy cover | 0.12 |
| Stem count | −0.06 | Stem count | −0.54 |
| % Grass | −0.22 | Basal area | −0.55 |
| % Canopy cover | −0.56 | Fraction trees—oak | −0.66 |
| Maximum height | −0.57 | % Grass | −0.82 |

TABLE 5. Correlations of overall breeding bird density, species diversity and individual species counts for fifteen groups of ten points ranked by their scores on Principal Component axis 1 and axis 2. Scientific names are given in Table 1. *P < 0.05; **P < 0.01; ***P < 0.001

| | Axis 1 | Axis 2 |
|---|---|---|
| Total bird density | 0.38 | −0.36 |
| Species diversity | 0.70** | 0.29 |
| Jay | 0.04 | −0.04 |
| Great tit | −0.18 | −0.63* |
| Crested tit | 0.27 | −0.65** |
| Blue tit | −0.38 | 0.17 |
| Marsh tit | −0.45 | −0.36 |
| Nuthatch | −0.71** | −0.60* |
| Short-toed treecreeper | −0.59* | −0.62* |
| Wren | −0.71** | 0.38 |
| Song thrush | 0.77** | 0.02 |
| Blackbird | −0.29 | 0.71** |
| Robin | 0.89** | 0.44 |
| Blackcap | 0.34 | 0.96*** |
| Garden warbler | −0.07 | 0.80** |
| Willow warbler | 0.52* | 0.78** |
| Chiffchaff | 0.38 | 0.80** |
| Wood warbler | 0.52* | 0.77** |
| Hawfinch | 0.01 | 0.48 |
| Chaffinch | −0.51* | −0.75** |
| Great spotted woodpecker | −0.38 | 0.35 |

Axis 2 was the one most closely related to the simple fraction of oak trees as used in Fig. 1 and Table 3. The correlations with bird counts found in the two cases were also similar. It was clear from the loadings that axis 2 also contained considerable weighting for the shrub cover up to 4 m. It would therefore appear, that for this particular data set, the fraction of oak was inseparable from some of the structural measures. Even when the main open canopy/closed canopy variables were removed there still remained a residual relationship between the fraction of oak and the shrub cover—plots with high oak fraction having low shrub cover, particularly below 4 m.

For some of the species with significant correlations on axis 2, such as great tit, nuthatch and treecreeper, it is probable that they were associated with a high fraction of oak. Whereas for others, such as blackbird, blackcap, garden warbler and willow warbler, the key factor was probably the higher shrub coverage associated with the beech stands.

## DISCUSSION

This study illustrates well some of the difficulties in addressing the simple question of the effect of tree species mixtures. The differences in the breeding birds of conifers and broad-leaves are well described (Yapp 1952; Simms 1971; Adams & Edington 1973; Moss 1978b) but, surprisingly, the more subtle differences between particular broad-leaved or coniferous species are less well understood (Moss 1978b; French, Jenkins & Conroy 1986). This study attempted to differentiate between two dissimilar broad-leaved species, but was unable to separate the structural from the species effects. For a particular tree species the exact form and structure of a woodland will depend on the details of its history, management, soil type and climate. The results of a particular study are therefore likely to be site specific and replicates are required before conclusions can safely be generalized. Ideally, experimental manipulations should be carried out to test hypotheses but, particularly for birds, which can have large home ranges, such experiments would be prohibitively expensive. Thus most studies, although well designed, have been, and will continue to be, forced to make use of whatever stands are available. The choice of this study site in north-west France was dictated by the need to find large areas of broad-leaved woodland under relatively uniform management. Although some of the bird species found in continental woodlands are different (for instance crested tit replaces coal tit, *Parus ater*, and short-toed treecreeper replaces treecreeper) the overall trends and relationships with measures of the habitat within the study are likely to be similar to those in British woods.

With the advent of extensive conifer plantations in the uplands (Forestry Commission 1984) there is the real prospect of being able to study the effects of crop mixtures and clearly considerable relevance in doing so. There have, however, been surprisingly few studies of birds of upland plantations and it is by no means certain which tree species are preferred by which bird, much less the effects of crop and non-crop mixtures (Avery & Leslie 1990).

A subject which has received some attention is the impact of non-crop broad-leaves in conifer plantations. French, Jenkins and Conroy (1986) used Common Birds Census methods (Marchant 1983) to study nine such conifer/broad-leaf mixtures and concluded that those with admixed trees in clumps rather than evenly dispersed had higher numbers of bird species. Their conclusion differed from that of Bibby, Aston and Bellamy (1989) who used point counts to study the impact of broad-leaved patches in extensive conifer plantations of upland Wales. They found that a number of bird species of conservation interest were associated with the broad-leaves and that the highest numbers of these species would be found in a forest with an intimate mix of broad-leaves. This latter study, being based on a far larger and more widespread sample than the former, would be expected to give a more reliable conclusion.

In another recent study of birds of conifer plantations, Peck (1989), using point counts, found the highest densities and numbers of bird species in compartments with large numbers of tree species, again suggesting that dispersed admixtures of broad-leaves would support higher bird numbers than discrete blocks. Peck supplemented the density estimates by making direct observations on the foraging behaviour of the most common species in her study and found that differential usage of different tree species was one of the main sources of niche separation between the different bird species.

Both of these studies suggest that broad-leaves are likely to have maximum impact when they are dispersed throughout the forest rather than in large clumps. There has not, however, been any investigation on whether broad-leaved trees are best as part of ride management or dispersed within the interior of compartments. This is a question that requires further study but it must be borne in mind that ride management is likely to be determined by the needs of plants and animals other than birds (Warren & Fuller 1990).

There has been very little work on the practice, which occurs widely in the lowlands, of adding conifers to predominantly broad-leaved stands. In an unpublished study, using point counts in Chiltern beechwoods of southern England, the author found that the addition of conifers increased the overall bird density, by adding bird species which favour conifers such as coal tit and goldcrest without reducing the numbers of other species. This appears to be the converse of the effect found by Bibby, Aston and Bellamy (1989) in upland conifer plantations.

There are clearly still many questions to be answered but the facts that are known already suggest that there may be considerable scope to improve forests for birds by the judicious use of tree species mixtures.

## ACKNOWLEDGMENTS

The census data from the Forêt de Crécy were collected by B. Averis, G. Balanca and D. Burges. An earlier draft was greatly improved by referees comments.

## REFERENCES

Adams, M.W. & Edington, J.M. (1973). A comparison of songbird populations in mature coniferous and broadleaved woods. *Forestry*, **46**, 191–202.

Avery, M.I. & Leslie, R. (1990). *Birds and Forestry*. T. & A.D. Poyser, London.

Bibby, C.J., Phillips, B.N. & Seddon, A.J.E. (1985). Birds of restocked conifer plantations in Wales. *Journal of Applied Ecology*, **22**, 619–33.

Bibby, C.J., Aston, N. & Bellamy, P.E. (1989). Effects of broadleaved trees on birds of upland conifer plantations in north Wales. *Biological Conservation*, **49**, 17–29.

Cooley, W.W. & Lohnes, P.R. (1971). *Multivariate Data Analysis*. Wiley, New York.

Erdelen, M. (1984). Bird communities and vegetation structure I. Correlations and comparisons of simple and diversity indices. *Oecologia*, **61**, 277–84.

Forestry Commission (1984). *Census of Woodland and Trees, Great Britain*. Forestry Commission, Edinburgh.

French, D.D., Jenkins, D. & Conroy, J.W.H. (1986). Guidelines for managing woods in Aberdeenshire for song birds. In *Trees and Wildlife in the Scottish Uplands* (Ed. by D. Jenkins), pp. 129–43. Institute of Terrestrial Ecology, Abbots Ripton.

Gibb, J. (1954). Feeding ecology of tits, with notes on treecreeper and goldcrest. *Ibis*, **96**, 513–43.

Hamilton, G.J. (1975). *Forestry Commission Booklet No. 39—Forest Mensuration Handbook*. HMSO, London.

James, F.C. & Warner, N.O. (1982). Relationships between temperate forest bird communities and vegetation structure. *Ecology*, **63**, 159–71.

MacArthur, R.H. & MacArthur, J.W. (1961). On bird species diversity. *Ecology*, **42**, 594–8.

Marchant, J. (1983). *BTO Common Birds Census Instructions*. British Trust for Ornithology, Tring.

Millar, R.G. (1974). The jackknife—a review. *Biometrika*, **61**, 1–15.

Moss, D. (1978a). Diversity of woodland song-bird populations. *Journal of Animal Ecology*, **47**, 521–7.

Moss, D. (1978b). Song-bird populations in forestry plantations. *Quarterly Journal of Forestry*, **72**, 5–13.

Muller, Y. (1985). L'avifaune forestière nicheuse des Vosges du Nord, sa place dans le contexte médio-européen. Thesis, University of Dijon.

Muller, Y. (1986). Ecologie des oiseaux nicheurs de la Forêt de Haguenau (Alsace); comparaison des peuplements aviens de quatre formations boisées agées. *Ciconia*, **10**, 69–90.

Peck, K.M. (1989). Tree species preferences shown by foraging birds in forestry plantations in northern England. *Biological Conservation*, **48**, 41–57.

Shannon, C.E. & Weaver, W. (1949). *The Mathematical Theory of Communication*. University of Illinois Press, Urbana.

Simms, E. (1971). *Woodland Birds*. Collins, London.

Smith, K.W. (1988). Breeding bird communities of commercially managed broadleaved plantations. *RSPB Conservation Review*, **2**, 43–6.

Smith, K.W., Averis, B. & Martin, J. (1987). The breeding bird community of oak plantations in the Forest of Dean, Southern England. *Acta Oecologica*, **8**, 209–17.

Southwood, T.R.E. (1961). The number of species of insects associated with various trees. *Journal of Animal Ecology*, **30**, 1–8.

Southwood, T.R.E., Moran, V.C. & Kennedy, C.E.J. (1982). The richness, abundance and biomass of the arthropod communities on trees. *Journal of Animal Ecology*, **51**, 635–49.

Tomialojc, L., Wesolowski, T. & Walankiewicz, W. (1984). Breeding bird community of a primeval forest (Bialowieza National Park, Poland). *Acta Ornithologica*, **20**, 241–310.

Ulfstrand, S. (1975). Bird flocks in relation to vegetation diversification in a south Swedish coniferous plantation in winter. *Oikos*, **26**, 65–73.

Warren, M.S. & Fuller, R.J. (1990). *Woodland Rides and Glades: their Management for Wildlife*. Nature Conservancy Council, Peterborough.

Willson, M.F. (1974). Avian community organisation and habitat structure. *Ecology*, **55**, 1017–29.

Yapp, W.B. (1952). *Birds and Woods*. Oxford University Press, Oxford.

# Are tree species mixtures too good for grey squirrels?

R.E. KENWARD[*], T. PARISH[†] AND P.A. ROBERTSON[‡]

[*] Institute of Terrestrial Ecology, Furzebrook Road, Wareham, Dorset BH20 5AS, UK; [†] Institute of Terrestrial Ecology, Monks Wood, Abbots Ripton, Huntingdon, Cambridgeshire PE17 2LS, UK; and [‡] The Game Conservancy, Fordingbridge, Hampshire SP6 1EF, UK

## SUMMARY

The introduced North American grey squirrel causes problems for foresters by stripping bark from young trees, and for wildlife conservation by displacing the native British red squirrel. Trapping surveys at thirty damage-prone beech and sycamore sites showed that squirrel density was highest where tree diversity was greatest. Surveys of damage accumulated over 5–9 years at fifty-three sites showed that bark-stripping increased with the number of seed-bearing tree species near the vulnerable crop, confirming that vulnerable trees should not be planted close to mature seed-bearing trees, or with a nurse crop of those conifer species which may provide food for squirrels in winter. Young sycamores were more prone to damage than beech, especially in plantations and where pheasants were fed in winter, but damage was reduced where there was a high percentage of ground cover and where Warfarin was used to kill squirrels. Conceptual models are presented to show how factors interact to increase the risk of bark-stripping, and of red squirrel replacement by grey squirrels through the presence of oaks in tree mixtures. Further work is needed (i) to determine how small or isolated plantations should be to minimize risk of damage, (ii) to define why damage reduction is associated with extensive ground cover, (iii) to discover whether tree growth can be managed economically to reduce damage, and (iv) to determine whether the species content of woodland can be managed to ensure the survival of red squirrels in mainland Britain.

## INTRODUCTION

Except under extreme conditions, natural woodland in Europe contains mixtures of tree species (Rackham, this volume). In plantations on poor soils, mixtures are beneficial for tree growth, perhaps partly as a result of mycorrhizal associations. Mixtures are also desirable for nature conservation, at the least by promoting species diversity among ground flora and birds (Peck 1989).

Further support for maintaining tree species mixtures comes from the observation that animal populations are most stable in diverse ecosystems (Elton 1958), leading to a view that diversity may reduce pest outbreaks. However, there is

243

little evidence that insect pest outbreaks are less frequent or smaller in mixed stands (Watt, this volume). Indeed, where the diversity of mixtures provides a more stable food supply for pests, there may be more risk of damage associated with high population density.

The introduced North American grey squirrel (*Sciurus carolinensis*) is another mammal which should benefit from tree mixtures, because the squirrels depend mainly on tree seeds for winter food (Gurnell 1983, 1987) and the more tree species in an area the greater the chance that one will produce a good seed crop. Indeed, Sanderson *et al.* (1976) found that drey densities were highest in areas with the highest tree diversity, and Don (1985) has shown that drey density reflects squirrel density.

The grey squirrel causes damage by stripping bark, especially from plantations of beech (*Fagus sylvatica*) and sycamore (*Acer pseudoplatanus*) which are 15–40 years old (Shorten 1957; Rowe & Gill 1985). Squirrels eat the phloem tissue, which is especially voluminous during the main damage period, in June–July. The extent of fresh damage, which can be severe enough to kill most of the young trees at a site, depends mainly on the average phloem volume at the time (Kenward 1983; Kenward & Parish 1986). However, bark-stripping is also most likely to start when squirrel density is high, especially the density of spring young (Kenward & Parish 1986), probably because there are then many agonistic encounters to trigger gnawing behaviour (Taylor 1966, 1969). Once triggered, damage tends to be repeated the next year, in a pattern which suggests that squirrels have learned the habit (Kenward *et al.* 1988).

On this basis, it has been suggested that areas of 'potential high density' squirrel populations, with abundant mature seed-bearing trees, should be especially vulnerable to damage, especially in years with good seed crops (Gurnell 1987, 1989; Gurnell & Pepper 1988). This paper shows, using data from trapping and tree damage surveys, that grey squirrel density and bark-stripping intensity is greatest where there is high species diversity of seed-bearing trees. These results, together with recent findings on the comparative ecology of red and grey squirrels, suggest ways of designing new woodlands both to minimize the risk of damage and to conserve red squirrels (*Sciurus vulgaris*).

## METHODS

Squirrel densities were estimated by trapping sessions in March–April, and again in June–July, at thirty English midland woods which contained young beech and sycamore. Grids of twelve to fifty-six small Legg multiple-capture live-traps, which were regularly spaced 70–100 m apart, were used during 1979–83 for 1 year at twenty sites, 2 years at nine sites and 3 years at one site, to give forty-one separate annual density estimates. For density estimates where the grid covered part of a wood, the area trapped was considered to extend 100 m beyond the grid at edges within the woodland, because this was the average range radius of female squirrels radio-tracked in four woods. Spring densities were estimated as twice the female density, because male ranges varied in size with courtship activity: this often gave rise to an excess of males being trapped, despite a 1:1 sex ratio in five woods covered completely by trap grids (see Kenward & Parish (1986) for further details).

Fresh bark-stripping in each area was assessed during early July on a 6-point scale, with 0 where there was no sign of bark-stripping. Sometimes squirrels removed small ($<5\,cm^2$) flakes of bark: such areas scored 1. Where hand-sized patches ($<250\,cm^2$) were removed from a small minority of trees the score was 2: such damage was unimportant. Stripping of larger patches (sometimes affecting the whole trunk) from $<5\%$ of trees (scoring 3) or from 5 to 50% of trees (scoring 4) caused loss of the trees concerned, and stripping of large patches from a majority of trees (scoring 5) ruined the crop. The 6-point damage index correlated strongly with a more objective estimate, based on the proportion of attacked trees in transect counts during 1983 ($r = 0.880, n = 24, P < 0.001$). At fifty-three beech and sycamore woods, including twenty-two of the midland woods where squirrels were trapped and twenty-seven without trapping in Dorset, bark-stripping was assessed for 5–9 years. The worst annual damage score was used as an index of cumulative damage at these long-term sites.

At the 30 woods with squirrel trapping, one to three species of seed-bearing trees (typically oak *Quercus robur*, beech, sweet chestnut *Castanea sativa*, Scots pine *Pinus sylvestris*, Norway spruce *Picea abies* or larch *Larix decidua*) were common within 200 m of the trap grid. Radio-tracking indicated that squirrels from further away would not frequently have visited the grid. Similarly, one to three other seed-bearing tree species were common within 200 m of the vulnerable species at forty-seven of the fifty-three long-term survey sites, and none close to the vulnerable crop at six sites. Variables recorded at these sites also included tree age, tree spacing, the presence or absence of a conifer nurse crop, of pheasant feeding and of squirrel control, the percentage of canopy closure and of cover within 30 cm (squirrel 'height') of the ground (Robertson 1988), maximum, minimum and average phloem volumes, and the soil drainage quality (from 0 = sand to 9 = clay, see Kenward *et al.* 1988).

## RESULTS

Although squirrels were trapped in more than 1 year at ten of the thirty survey sites, the densities varied greatly between years, so results from all forty-one site-years were treated independently in the analysis. At fourteen sites, all with two or three seeding tree species, squirrels had been killed by shooting or trapping the previous year. Taking squirrel control into account in a rectangular (2 × 2) analysis of variance with unequal cells, there was a highly significant tendency for sites with three seeding species to have higher squirrel densities than sites with two species ($F_{2/36} = 7.11, P < 0.01$), and density at the site with only one seeding tree species was lower than at the others without squirrel control (Table 1).

TABLE 1. The mean number of grey squirrels per hectare (± S.E.) in spring in areas with 1–3 species of seed-bearing trees nearby, and with or without squirrel control the previous year

| | Number of seeding tree species within 200 m of trap grid | | | | | |
| | 1 | | 2 | | 3 | |
| | $n$ | Density | $n$ | Density | $n$ | Density |
|---|---|---|---|---|---|---|
| Squirrels shot or trapped during the previous year | | | 5 | 0.68 (0.23) | 9 | 1.45 (0.37) |
| No control of squirrels during the previous year | 1 | 0.74 | 16 | 1.14 (0.25) | 10 | 1.87 (0.37) |

There was no sign of bark-stripping at three of six long-term survey sites which had no seeding trees nearby, and serious damage at only one site. However, some bark was stripped in at least one year from all but two of the remaining forty-seven sites, and there was serious damage at thirty-five. The crop was ruined only at sites with at least two seeding tree species (Fig. 1).

Since the strength of this relationship differed between beech and sycamore trees, and sycamore crops attracted more damage than beech ($t = 2.955$, $n = 53$, $P < 0.01$), the two species were separated for further multivariate analysis. Using a regression-based analysis of variance and covariance, in which the best-fit relationship between damage and tree diversity was a logistic curve with asymptote at damage level 3, 68% of variation in the long-term damage at thirty-one beech plantations was explained by three factors. Damage was reduced by having few seeding tree species within 200 m of the vulnerable plantation, and also by good

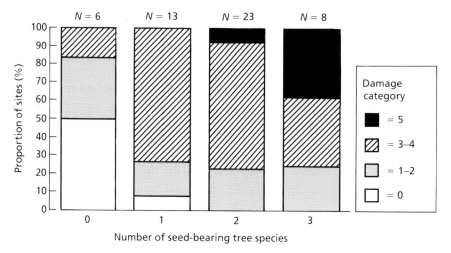

FIG. 1. Long-term damage scores (from 0 = none to 5 = most trees destroyed) for bark-stripping by grey squirrels at fifty-three sites which contained young beech or sycamore trees and had 0–3 species of seed-bearing trees nearby.

TABLE 2. Factors associated with the extent of bark-stripping damage which accumulated during 5–9 years in young beech plantations. *$P < 0.05$; **$P < 0.01$; ***$P < 0.001$

|  | Factor | Student's $t$ |
|---|---|---|
| All cases | | |
| $n = 31$ | Number of seeding trees within 200 m of site | +5.401*** |
| $R = 0.82$*** | % Herb cover within 0.3 m of ground | −2.238* |
|  | Whether or not Warfarin hoppers were used | −2.083* |
| Excluding sites where Warfarin hoppers used | | |
| $n = 25$ | Number of seeding trees within 200 m of site | +3.896*** |
| $R = 0.80$*** | % Herb cover within 0.3 m of ground | −2.903** |

TABLE 3. Factors associated with the extent of bark-stripping damage which accumulated during 5–9 years in young sycamore woods. †$P = 0.06$; *$P < 0.05$; **$P < 0.01$; ***$P < 0.001$

|  | Factor | Student's $t$ |
|---|---|---|
| All cases | | |
| $n = 17$ | Trees in plantation, not selfed | +3.813** |
| $R = 0.83$*** | Presence of pheasant feeding nearby | +2.298* |
|  | Number of seeding trees within 200 m of site | +2.052† |
| Including only plantation sites | | |
| $n = 8$ | Number of seeding trees within 200 m of site | +2.857* |
| $R = 0.91$** | % Herb cover within 0.3 m of ground | −2.606* |

herb cover and by the use of Warfarin to kill squirrels (Table 2). When the analysis was confined to twenty-five sites at which Warfarin was not used, 64% of the variation was explained by the combination of food-tree presence and herb cover.

There were sufficient data for a full multivariate analysis at only 17 sycamore sites. Among these, the most important factor in an analysis of variance and covariance was whether they were planted or self-seeded (Table 3). Only four of nine self-seeded sites suffered appreciable damage (two in category 3, two in category 4), whereas all eight sycamore plantations were damaged at least this heavily, with three in category 5 (i.e. commercial write-offs). Damage tended to be high at sites with pheasant feeding, with the contribution from food-tree species not quite significant at the 5% level. Although the first two variables differed from those at the beech sites, just three variables again explained 68% of variation in the damage score. Warfarin was not used at any of the sycamore sites, and, if the analysis was confined to sycamore plantations (a comparable sample to the twenty-five beech sites where Warfarin was not used, since all beech sites were plantations), once again the damage was most strongly related to the

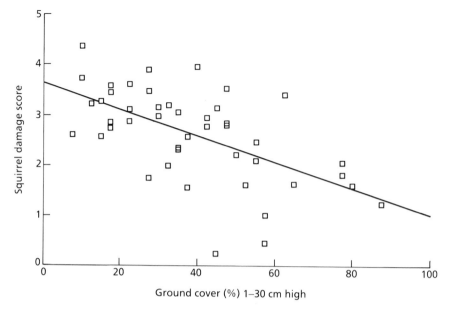

FIG. 2. Residual long-term damage scores for bark-stripping by grey squirrels (after correcting for stand type, use of Warfarin, pheasant feeding and seeding tree presence) in relation to % ground cover in young beech or sycamore woods.

number of food-tree species and the extent of herb cover. In the sample of only eight sites, these two variables explained 83% of variation in bark-stripping.

The slope of the relationship between damage and herb cover did not differ significantly between beech and sycamore sites ($P > 0.05$). Figure 2 shows the regression of herb-cover on damage-score residuals at all sites, after taking into account the effects of tree species, origin, proximity to seeding trees, pheasant feeding and Warfarin use. The addition of other factors, including tree age, tree spacing, canopy closure, soil quality, phloem volume and presence or absence of a conifer nurse crop did not significantly improve any of the regression analyses shown in Tables 1 and 2.

## DISCUSSION

### Tree mixtures and bark-stripping

At the fifty-three long-term sites, a high proportion lacked bark-stripping only when there were no seeding trees nearby (Fig. 1), the best-fit relationship between seed-bearer diversity and damage being a logistic curve. The density data from the thirty squirrel-survey sites (Table 1) indicate why. At the same sites, the strongest relationship between damage and squirrel density was also a logistic relationship, such that damage was always at level 2 or above when there were more than 0.35 young squirrels per hectare (Kenward & Parish 1986; Kenward *et al.* 1988). Areas with a similar density of non-juvenile squirrels can produce this many young squirrels in years with good seed crops, which are likely to occur at least once in nearby seeding trees during the vulnerable 20–30 years of a beech or sycamore crop. Areas with at least one species of seeding trees tend to have densities above 0.35 grown squirrels per hectare in spring (Table 1). Lower densities are likely only with no seed-bearing trees nearby.

Such isolation can result from a plantation being surrounded either by open country, or by trees which provide little food for squirrels in winter. The trees may be conifers (e.g. hemlock *Tsuga* or *Thuja* species), which are also suitable as a nurse crop, or small-seeded deciduous species (Gurnell 1989). However, data are still needed on squirrel densities and breeding in relatively food-less habitats. For example, how large can such plantations be, and how isolated for possible commuters, before squirrels will breed there? These are important questions for present re-afforestation projects, because grey squirrels have been little studied in areas with densities below 1 per hectare (Don 1983; Gurnell 1983). Unfortunately, the thirty squirrel-survey sites were all selected either because damage was already present, or because squirrels were known to be abundant enough to get reasonable density estimates from the trap grids. All thirty sites contained at least one seeding tree species, either pine or spruce nurse crops, which tended to seed at an early age (especially Scots pine), or mature trees in adjacent stands. Beech themselves eventually produce winter food for squirrels, but not usually to a great extent before the bark on their trunks is thick enough to deter squirrels.

Aldhous (1981) reported that young beech were often damaged when planted among conifers, and also suggested that the presence of sycamore might protect neighbouring beech. Since mixture with conifers at the long-term sites was not associated with extra damage after the conifers' role as seed-bearers was included, their importance may be mainly as a food source. There were no intimate beech/sycamore mixtures in the survey, and although sycamores were the more prone to damage, learning to bark-strip sycamore might as much trigger as deflect the stripping of beech.

### Damage and other site factors

At the fifty-three long-term sites, reduction in damage was associated not only with lack of nearby seed-bearing trees, but also with use of Warfarin, with absence of pheasant-feeding and with trees being self-set. These results confirm the findings of other surveys. For example, a questionnaire returned by members of the Royal Forestry Society (RFS) showed that only Warfarin, but not shooting or trapping, was effective for reducing damage (Kenward *et al.* 1988). Killing squirrels in this way presumably acts not only by reducing density, but also removes squirrels which have learned to strip bark. The RFS survey also recorded most damage where pheasants were fed through the winter. Squirrels regularly exploit the pheasant food and have higher densities in such areas than elsewhere.

Reduced damage to self-seeded trees probably reflected low phloem volume in them. In a survey during 1982, only 4% of self-set trees in British woods exceeded the $0.3\,\mathrm{ml\,cm}^{-2}$ threshold for serious damage (categories 3–5), compared with 23% in plantations. Similar low sap volumes measured during 1986 in North American woods, which are typically self-seeding, probably explain why there is so little bark-stripping there (Kenward 1989). American woods tend to be rich in seed-bearing species and can harbour dense grey squirrel populations (Flyger 1959; Mosby 1969). The lack of a direct relationship between accumulated damage and other factors, some of which were related to bark-stripping on an annual basis (e.g. phloem volume), probably reflected a tendency for conditions which are not favourable for bark-stripping every year nevertheless to be suitable at times during the life of each crop. For example, most plantations were thinned at some point during the study, thus affecting tree-spacing, canopy closure and phloem volume (Kenward 1989).

The relationship between bark-stripping and herb cover deserves further study. Perhaps the cover reduced agonistic encounters by squirrels at ground level, where bark-stripping often starts, because they could not see each other. Or perhaps squirrels were more reluctant to visit the ground where extensive cover increased risk of ambush by predatory mammals. Although it is also possible that cover was a correlate of an independent factor which affected the triggering or profitability of bark-stripping, any involvement of phloem volume, canopy closure or soil drainage can be excluded.

### Tree mixtures and red squirrels

The value of tree mixtures for grey squirrels is also relevant to the survival of the red squirrel (*Sciurus vulgaris*) in mainland Britain. Deciduous woodland which contains oaks gives grey squirrels an advantage, probably because a relative inability to detoxify tannins prevents red squirrels from exploiting acorn crops (Kenward & Holm 1989). In Scots pines, on the other hand, red squirrels not only achieve much higher densities than in deciduous woodland, but also higher densities and a lower mortality than grey squirrels in Scots pines (R.E. Kenward & S.S. Walls unpubl.). In this case it is particular components of tree mixtures which are important, with oaks favouring grey squirrels and pines perhaps giving red squirrels an advantage.

### Management implications

The risk of damage during the life of a crop, which is affected by several factors, can be represented as in Fig. 3. Risk should be low if new woods are planted and managed to keep them close to the top-right foreground of the three axes, i.e. with maximal isolation, minimal seeding tree presence and maximal ground cover (assuming the latter factor has a causal relationship with damage). However, the same three factors may also affect the suitability of woods for other considera-

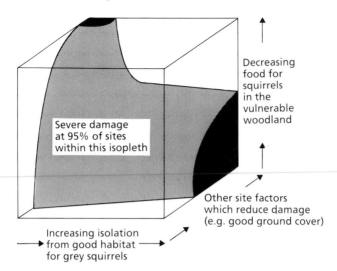

Decreasing
food for
squirrels
in the
vulnerable
woodland

Severe damage
at 95% of sites
within this isopleth

Other site factors
which reduce damage
(e.g. good ground cover)

Increasing isolation
from good habitat
for grey squirrels

FIG. 3. A model for the risk of economically significant bark-stripping by grey squirrels, in relation to habitat quality, isolation and other factors within areas of vulnerable young trees.

tions, such as nature conservation and game management. Refining the shape of the risk isopleths should allow more sophisticated management decisions to be made, especially where there are conflicts between different requirements.

At the least, it is important to note that the $y$-axis is one of food supply, not of tree diversity *per se*. Grey squirrels benefitted from mixtures of tree species because these usually either contained species which seeded at an early age (e.g. pine or spruce nurse crops) or were planted close to mature seed-bearers. It is probably unnecessary for vulnerable plantations to be monocultures, and thus poor for other wildlife, provided that mixtures are of species (or of late-seeding or sterile strains?) which provide little food for squirrels in winter. More than the observed 66% of damage in beech plantations might have been explained if the analysis had been based not only on the number of different tree species, but also on their density and the food value of their seeds. For planning purposes, the $z$-axis could include tree growth conditions as well as ground cover. In amenity woodland, at least, it may be worth planting beech as understorey saplings, or leaving dense stands unthinned, so that crowding reduces phloem volume until bark thickens with age enough to prevent serious damage. Further work is needed to show whether plantations could be managed economically by favouring height growth to girth growth, thus probably minimizing phloem volume, until the trees are old enough for bark thickness to deter squirrel damage.

A conceptual model like Fig. 4 can be used to illustrate conditions for the persistence of red squirrels. The $x$-axis, as in Fig. 3, is of increasing isolation from good habitat for grey squirrels, and the $y$-axis of food factors which favour the two squirrel species differentially. Mixed woodland containing oaks would be at the base of this axis, with conifers (especially mature Scots pine) at top, favouring

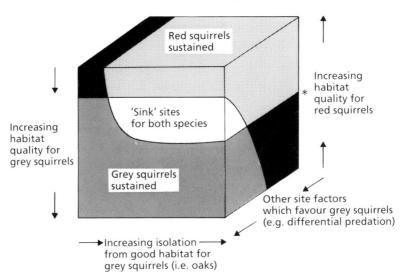

FIG. 4. A model of the conditions which may sustain red squirrels (pale grey volume), in sites away from competition with grey squirrels (black volume) or other factors which may militate against red squirrel survival whether there is competition from grey squirrels (dark grey volume) or not (white volume). * marks the hypothetical change in competitive advantage on the habitat axis.

red squirrels. The $z$-axis is for other site factors which may favour one squirrel species over the other. Differential predation might act in this way, since eleven of twenty-two radio-tagged goshawks (*Accipiter gentilis*) killed red squirrels in Swedish lowland areas, whereas only one of eleven hawks killed the larger grey squirrel in lowland Britain (Kenward 1981; Kenward, Marcström & Karlbom, 1981). In this case, the most favourable woodland for red squirrels would be that dense enough to hinder goshawk attacks. The crucial aspect of this figure is the presence of the boundary (*) on the $y$ axis, above which red squirrels are presumed to have a feeding advantage over grey squirrels. Below this line red squirrels are replaced by competition alone (black volume), or (in the smaller grey volume) by competition plus a possible third factor (e.g. predation) which also prevents their persistence in some sites above the boundary (white volume). As grey squirrels continue their advance through mainland Britain, there is an urgent need for research to define whether any composition of woodland, mixed or monoculture, can give red squirrels a competitive advantage.

## ACKNOWLEDGMENTS

We thank Deene, Elton, Lulworth, Mapledurham, Morden, Rempstone, Southill, Trigon and Woburn Estates, the Forestry Commission, the Nature Conservancy Council and the Northamptonshire Naturalists Trust for permission to work in their woods. The Royal Forestry Society and the Forestry Commission kindly provided some of the equipment. S. Adair, J. Holm, N. Payne, J. Oliver and R. Buxton helped with trapping. We are also grateful to J. Goss-Custard, M.G. Morris and the anonymous referees for comments and advice.

## REFERENCES

Aldhous, J.R. (1981). Beech in Wessex—a perspective on present health and silviculture. *Forestry*, **54**, 197–210.
Don, B.A.C. (1983). Home range characteristics and correlates in tree squirrels. *Mammal Review*, **13**, 123–32.
Don, B.A.C. (1985). The use of drey counts to estimate grey squirrel populations. *Journal of Zoology (London)*, **206**, 282–6.
Elton, C.S. (1958). *The Ecology of Invasions by Animals and Plants*. Wiley, New York.
Flyger, V.F. (1959). A comparison of methods for estimating squirrel populations. *Journal of Wildlife Management*, **23**, 220–3.
Gurnell, J. (1983). Squirrel numbers and the abundance of tree seed supplies. *Mammal Review*, **13**, 133–48.
Gurnell, J. (1987). *The Natural History of Squirrels*. Christopher Helm, London.
Gurnell, J. (1989). Demographic implications for the control of grey squirrels. *Mammals as Pests* (Ed. by R.J. Putman), pp. 131–43. Chapman & Hall, London.
Gurnell, J. & Pepper, H. (1988). Perspectives on the management of red and grey squirrels. *Wildlife Management in Forests* (Ed. by D.C. Jardine), pp. 92–109. Institute of Chartered Foresters, Edinburgh.
Kenward, R.E. (1981). Goshawk re-establishment in Britain—causes and implications. *The Falconer*, **7**, 304–10.
Kenward, R.E. (1983). The causes of damage by red and grey squirrels. *Mammal Review*, **13**, 159–66.
Kenward, R.E. (1989). Bark-stripping by grey squirrels in Britain and North America: why does the damage differ? *Mammals as Pests* (Ed. by R.J. Putman), pp. 144–54. Chapman & Hall, London.
Kenward, R.E. & Holm, J.L. (1989). What future for British red squirrels? *Biological Journal of the Linnean Society*, **38**, 83–9.
Kenward, R.E. & Parish, T. (1986). Bark-stripping by grey squirrels (*Sciurus carolinensis*). *Journal of Zoology (London)*, **210**, 473–81.
Kenward, R.E., Marcström, M. & Karlbom, M. (1981). Goshawk winter ecology in Swedish pheasant habitats. *Journal of Wildlife Management*, **45**, 397–408.
Kenward, R.E., Parish, T., Holm, J. & Harris, E.H.M. (1988). Grey squirrel bark-stripping. I. The roles of tree quality, squirrel learning and food abundance. *Quarterly Journal of Forestry*, **82**, 9–20.
Mosby, M.S. (1969). The influence of hunting on the population dynamics of a woodlot gray squirrel population. *Journal of Wildlife Management*, **33**, 59–73.
Peck, K.M. (1989). Tree species preferences shown by foraging birds in forest plantations in Northern England. *Biological Conservation*, **48**, 41–57.
Robertson, P.A. (1988). Pheasant management in small broadleaved woodlands. *Wildlife Management in Forests* (Ed. by D.C. Jardine), pp. 25–33. Institute of Chartered Foresters, Edinburgh.
Rowe, J.J. & Gill, R.M.A. (1985). The susceptibility of tree species to bark-stripping damage by grey-squirrels (*Sciurus carolinensis*) in England and Wales. *Quarterly Journal of Forestry*, **79**, 183–90.
Sanderson, H.R., Healey, W.M., Pack, J.C., Gill, J.D. & Ward, T.J. (1976). Gray squirrel habitat and nest tree preference. *Proceedings of the Annual Conference of the Southeastern Associated Game and Fish Committee*, **30**, 609–16.
Shorten, M. (1957). Damage caused by squirrels in Forestry Commission areas 1954–56. *Forestry*, **30**, 151–72.
Taylor, J.C. (1966). Home range and agonistic behaviour in the grey squirrel. *Symposia of the Zoological Society of London*, **18**, 229–35.
Taylor, J.C. (1969). *Social structure and behaviour in a grey squirrel population*. Ph.D. thesis, University of London.

# Effect of mixed-species tree planting on the distribution of soil invertebrates

J. BUTTERFIELD AND J. BENITEZ MALVIDO

*Department of Biological Sciences, University of Durham, Durham DH1 3HP, UK*

## SUMMARY

Mixed-species tree planting, whether broad-leaf/conifer or of more than one species of conifer, resulted in higher densities of soil-surface invertebrates than conifer monocultures. There were fewer species of ground beetle (Carabidae) in conifer stands than in oak woodland and the dominant species were different. Patches of broad-leaf woodland within conifer plantation, or mixtures of broad-leaves and conifers, resulted in higher numbers of carabid species and the presence of some of the species of deciduous woodland. Invertebrate species diversity and the conservation potential of conifer plantations was enhanced by the presence of broad-leaves.

## INTRODUCTION

Between 1924 and 1987 forest cover in Britain increased from $1180 \times 10^3$ to $2265 \times 10^3$ ha, or from 5.3 to 10% of the land area (Forestry Commission 1987). Most afforestation has been in the uplands with a considerable loss of moorland (a reduction of 41% in Wales, 22% in England and 11% in Scotland between 1946 and 1981). Additionally, there has been a 31% loss of rough pasture and 67% of this has been to forestry (Woods & Cadbury 1987). Change of this magnitude has far reaching effects not only on the landscape but also on the flora and fauna (Hill 1986; Ratcliffe & Thompson 1988). The loss of open habitat for ground-nesting birds and for wide-ranging scavengers and predators such as ravens and eagles has been a particular source of concern (Marquiss, Newton & Ratcliffe 1978; Marquiss, Ratcliffe & Roxburgh 1985; Thompson, Stroud & Pienkowski 1988). At the invertebrate level, new species associated with trees (Bevan 1987) have been introduced and species associated with open moorland have declined in afforested areas (Butterfield in press).

In 1985, the Forestry Act 1967 was amended by the Wildlife and Countryside (Amendment) Act to place a duty on the Forestry Commission to endeavour to achieve a reasonable balance between the interests of forestry and those of the environment; the idea being to operate management practices that maximize benefit to wildlife while maintaining the profitability of the timber crop (Ratcliffe

& Petty 1986). As the plant and animal communities of moorland and upland grassland are inevitably disrupted when forestry is introduced, the aims must be either to retain viable fragments of the original or to create favourable habitats for new and desirable species, from the conservation point of view.

One management option has been the planting of mixed tree species rather than single-species stands. Native deciduous trees harbour a greater diversity of insect species than do the imported conifer species used in afforestation (Southwood 1961; Kennedy & Southwood 1984) and it can be assumed that the retention of some of the self-seeded birch, *Betula pendula*, and rowan, *Sorbus aucuparia*, will increase the diversity of insect species in a plantation. Planting a mixture of non-indigenous conifer species could be expected to have a much smaller effect.

Planting of mixed-species stands has already been undertaken for purely commercial purposes and there have been a number of studies of its effect on bird distributions. Von Haartman (1971) and Newton and Moss (1977) found that mixed deciduous/conifer plantations were richer in both species and numbers of individuals than conifer plantations, while Peck (1989) found that plantations of mixed conifer species, as well as deciduous/conifer mixtures, attracted a higher density of insect-eating birds than monocultures. She associated the greatest densities of birds with the highest abundance of insect food, but this does not necessarily imply a high insect diversity. Sycamore, *Acer pseudoplatanus*, was favoured by all bird species purely because of the large biomass of aphids it produced.

In the present study we compared the numbers of surface-active invertebrates caught in mixed-species plantations with those caught in single-species conifer plantations and in an adjacent wood-pasture. The species distribution of the ground beetles, Carabidae, was analysed in relation to tree-species composition and other environmental variables to assess the importance of tree-species mixtures to one taxon within the mesofauna at the soil surface.

## SITE DESCRIPTION AND METHODS

### Study area

The study was carried out in a Forestry Commission plantation, Hamsterley Forest (NZ0530), and the adjacent Adder Wood, in County Durham, north-east England. The forest comprises c. 2000 ha lying between 150 and 400 m above sea level with peat moorland to the north and rough pasture to the south. Two streams, Euden Beck and Spurlswood Beck which join to form Bedburn Beck, flow through the area and the soils within the forest range from deep peat to brown earths, with alluvial soils along the stream sides. The majority of sites used in the present study lay on areas of shallow peat over sandstone. Adder Wood (NZ1231) lies 2 km to the east of the forest and is an ancient wood-pasture dominated by mature oak, *Quercus petraea*.

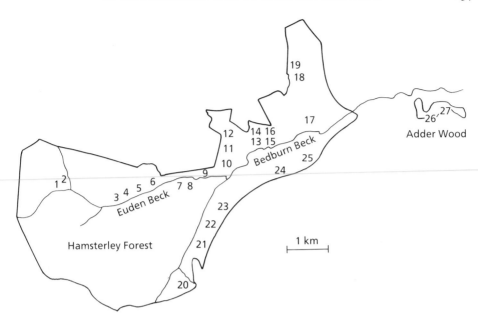

FIG. 1. Map of Hamsterley Forest and Adder Wood, in north-east England, showing distribution of sites.

## Site descriptions

Twenty-seven sites were selected to represent a range of soil types and tree species. Figure 1 shows their distribution within the woods and Table 1 shows the site characteristics. Measurements are the means of five replicates and samples for pH and organic content determination were taken within 60 mm of the litter layer. The complexity of the field layer was assessed on a scale of 0–3 as follows: 0 = needles only; 1 = sparse grass and moss, cover <50%; 2 = dense grass and other herbs, cover >90%; 3 = dense herb layer + shrubs, cover >90%.

The tree species within the forest ranged from birch and alder in relatively undisturbed natural woodland (Sites 4 and 6) through broad-leaf/conifer mixtures (sites 8, 9, 17, 18, 19 and 24) to pure conifer stands of one or two species. Four of the sites in single-species stands (10, 12, 20 and 21) were in Sitka spruce (*Picea sitchensis*), five (11, 13, 14, 22 and 23) in Scots pine (*Pinus sylvestris*) and two (1 and 7) in Norway spruce (*Picea abies*). Site 2 was in Japanese larch (*Larix kaempferi*) and Site 16 in western hemlock (*Tsuga heterophylla*). Sites 3, 5 and 25 were in mixed Sitka spruce/western hemlock plantations and Site 15 was in a Scots pine/Japanese larch plantation.

The peat depth and the organic content of the top 60 mm of the soil differed widely from site to site and the conifer plantations without broad-leaves did not differ significantly in these characteristics from the plantations with a broad-leaf component. However, the mean pH of the soil in the conifer-only plantations was

TABLE 1. Characteristics of twenty-seven sites within Hamsterley Forst and Adder Wood in north-east England. *o = no vegetation; 1 = herbs, cover < 50%; 2 = herbs, cover > 90%; 3 = herbs + shrubs, cover > 90%. †JL = *Larix kaempferi*, EL = *Larix decidua*, SP = *Pinus sylvestris*, SS = *Picea sitchensis*, NS = *Picea abies*, WH = *Tsuga heterophylla*, decid. = deciduous

| Sites | Age of stand (years) | % conifer | Field layer* | Peat depth (mm) | % organic content | pH | Litter depth (mm) | Tree species† |
|---|---|---|---|---|---|---|---|---|
| 1 | 35 | 100 | 0 | 20 | 10 | 3.8 | 10 | NS |
| 2 | 36 | 100 | 0 | 20 | 38 | 3.6 | 80 | JL |
| 3 | 39 | 100 | 0 | 20 | 33 | 3.7 | 13 | SS/WH |
| 4 | — | 0 | 2 | 0 | 1 | 4.1 | 0 | Birch |
| 5 | 45 | 100 | 0 | 20 | 12 | 3.8 | 26 | SS/WH |
| 6 | — | 0 | 1 | 0 | 11 | 4.0 | 0 | Birch |
| 7 | 45 | 100 | 0 | 0 | 7 | 3.7 | 0 | NS |
| 8 | 55 | 40 | 2 | 0 | 7 | 4.5 | 0 | SP/decid. |
| 9 | 14 | 80 | 3 | 10 | 12 | 4.5 | 10 | SS/birch |
| 10 | 17 | 100 | 0 | 0 | 4 | 3.9 | 30 | SS |
| 11 | 56 | 100 | 1 | 200 | 68 | 3.4 | 40 | SP |
| 12 | 56 | 100 | 0 | 500 | 94 | 3.4 | 50 | SS |
| 13 | 16 | 100 | 2 | 0 | 20 | 3.6 | 0 | SP |
| 14 | 16 | 100 | 2 | 0 | 11 | 3.6 | 0 | SP |
| 15 | 46 | 100 | 2 | 0 | 9 | 3.5 | 0 | SP/JL |
| 16 | 46 | 100 | 0 | 0 | 12 | 3.5 | 63 | WH |
| 17 | 49 | 50 | 3 | 0 | 12 | 4.5 | 0 | Oak/EL |
| 18 | 18 | 70 | 3 | 200 | 89 | 3.5 | 20 | SS/SP/birch |
| 19 | 16 | 70 | 3 | 50 | 22 | 3.6 | 10 | SS/birch |
| 20 | 22 | 100 | 0 | 0 | 11 | 3.7 | 40 | SS |
| 21 | 57 | 100 | 0 | 130 | 65 | 4.0 | 50 | SS |
| 22 | 57 | 100 | 3 | 80 | 79 | 3.5 | 20 | SP |
| 23 | 57 | 100 | 3 | 80 | 71 | 3.7 | 40 | SP |
| 24 | 16 | 60 | 3 | 0 | 43 | 4.5 | 0 | SS/decid. |
| 25 | 25 | 100 | 0 | 0 | 35 | 3.7 | 30 | SS/WH |
| 26 | — | 0 | 2 | 0 | 34 | 4.7 | 0 | Oak |
| 27 | — | 0 | 1 | 0 | 46 | 4.7 | 0 | Oak |

$3.7 \pm 0.06$, significantly lower than the $4.3 \pm 0.04$ mean of the broad-leaf and broad-leaf/conifer mixtures ($t = 4.2$, d.f. 11, $P < 0.01$). A further difference between plantation types lay in the field layer which was present in the plantations with a broad-leaf component and in the Scots pine stands but absent in all the other conifer plantations.

## Sampling

The surface-active mesofauna was sampled for 1 year by means of five pitfalls (plastic coffee cups, rim diameter 70 mm, partly filled with 2% formalin and detergent), placed at 2 m intervals in a straight line. The traps were in position from 1 April until the end of October and were emptied at fortnightly intervals.

## Analysis

An initial comparison between the numbers of individuals belonging to the major groups of surface-active arthropods caught in plantations of different tree-species composition was made. The ground beetles (Carabidae), which contributed about 25% of the surface-active invertebrates caught in pitfalls in the conifer plantations, were selected in order to compare the effects of tree species and edaphic factors on the species distribution. Identification was according to Lindroth (1974) and the forest sites were grouped according to the similarities of their carabid faunas using TWINSPAN (Hill 1979). The Canonical Correspondence Analysis option within the CANOCO program (Ter Braak 1988) was used to relate changes in the species composition to the environmental variables listed in Table 1 (with the tree species reduced to the format: o = single species, 1 = mixed). Pseudospecies (Hill 1979) were inserted at abundance levels of 6, 36 and 216 (giving a geometric progression assigning approximately a third of the species to pseudospecies). Carabids being active animals, occurrences of one individual of a species on a site have been ignored.

## RESULTS

### The relative abundance of invertebrates in mixed-species and single-species plantations

There was a wide variation between the numbers of surface-active invertebrates caught in pitfall traps at each site. An initial comparison has been made using the numbers of individuals caught in each of the invertebrate taxa shown in Table 3. The sites have been grouped into the following forest habitat types: broad-leaved woodland, mixed broad-leaf and conifers, mixed conifers, Scots pine, Sitka spruce and a group consisting of single-species stands of other conifers. The geometric means of the catches in each taxon for each group of sites have been calculated and the means for each of the eight taxa have been used as pairs in paired *t*-tests for a series of comparisons between one forest habitat and another. Table 2 shows the significance levels of paired *t*-tests comparing the numbers of surface-active invertebrates caught on each habitat type. The numbers caught in the broad-leaf woodland did not differ significantly from those caught in any of the other forest types, nor were there significant differences between the catches in the forest monocultures. However, the catches in both the broad-leaf/conifer mixtures and in the mixed conifer plantations were significantly higher in two comparisons out of three with the conifer monoculture groups. Accordingly, the numbers caught in each taxon in mixed- and single-species stands have been compared (Table 3). For all taxa the numbers caught in pitfalls in the mixed-species stands were greater than those in the single-species stands (significantly so in six of the eight cases). Of the numerically dominant groups, about twice as many harvestmen and rove-beetles and one and a half times the numbers of spiders and ground-beetles were

TABLE 2. Comparisons between the mean numbers of invertebrates caught in pitfalls in single and mixed species plantations. − indicates significantly smaller catches, $P < 0.05$ in the plantations in the left-hand column compared with those in the horizontal line, d.f. 7 in all cases

| Plantation type | No. of sites | | | | | |
|---|---|---|---|---|---|---|
| Broad-leaf | 4 | | | | | |
| Broad-leaf/conifer | 6 | N.S. | | | | |
| Mixed conifer | 4 | N.S. | N.S. | | | |
| Scots pine | 5 | N.S. | − | − | | |
| Sitka spruce | 4 | N.S. | N.S. | − | N.S. | |
| Other conifer | 4 | N.S. | − | N.S. | N.S. | N.S. |
| | | Broad-leaf | Broad-leaf/ conifer | Conifer mix | Scots pine | Sitka spruce |

TABLE 3. Geometric mean numbers (with 95% confidence limits) of invertebrates caught in mixed- and single-species plantations

| Invertebrate group | Mixed ($n = 10$) | Single ($n = 13$) | $t$ | $P$ |
|---|---|---|---|---|
| Isopoda | 12 (5–28) | 3 (2–5) | 3.25 | <0.01 |
| Diplopoda | 31 (18–55) | 16 (10–26) | 2.00 | N.S. |
| Staphylinidae | 204 (141–293) | 104 (69–158) | 2.68 | <0.02 |
| Carabidae | 197 (109–355) | 136 (95–195) | 1.18 | N.S. |
| Lepidoptera larvae | 10 (4–24) | 2 (1–4) | 2.87 | <0.01 |
| Homoptera | 14 (4–50) | 2 (1–6) | 2.54 | <0.02 |
| Araneae | 327 (196–546) | 193 (129–290) | 1.77 | N.S. |
| Opiliones | 104 (63–171) | 45 (28–74) | 2.59 | <0.02 |

caught in the mixed-species stands. In many cases pitfall catches give a poor indication of population densities as the magnitude of the catch also depends on activity (Southwood 1978) and vegetation round pitfalls may impede access (Briggs 1961; Luff 1975). In the present case the ground vegetation was dense in the deciduous/conifer stands but the catches were higher than in spruce or western hemlock, which lacked a field layer, and real differences in invertebrate densities between mixed-species stands and single-species stands could be inferred. Moreover, Table 2 indicates that mixed conifer stands, as well as mixed broad-leaf/conifer stands, had greater densities of surface-active invertebrates than single-species stands of Scots pine or Sitka spruce (the total mean catches of the invertebrate groups listed in Table 3 were in the ratios of 2.8:1.2:1 in mixed conifers, Scots pine and Sitka spruce, respectively).

## *The carabid communities in the forest plantations and their affinities with those of deciduous woodland*

TWINSPAN (Hill 1979) was used to classify the forest sites according to their carabid faunas. The first division produced a positive group of ten sites with a relatively high species richness and a negative group of fifteen sites with low diversity (mean numbers of species 16.6 ± S.E. 1.6 and 10.9 ± S.E. 1.0, respectively, $t = 3.02$, d.f.23, $P < 0.02$). The positive indicator species were *Pterostichus madidus*, *Amara plebeja*, *Loricera pilicornis* and *Carabus violaceus* and the negative indicators were *Notiophilus biguttatus* and *Calathus micropterus*.

Table 4 shows the comparison between the percentage distribution of the abundant species (forming 5% or more of the carabid catch on a group of sites) in the four groups of plantation sites formed at the second TWINSPAN level and the deciduous woodland sites. Four of the abundant species of deciduous woodland, *Agonum assimile*, *Calathus melanocephalus*, *Pterostichus melanarius* and *Nebria brevicollis* were rare or absent in all four plantations groups but *Abax parallepipedus*, *Pterostichus madidus* and *Carabus violaceus* were present in both the deciduous woodland and in Group 1 and Group 2 plantations. The catch in the plantations, except for Group 1, was dominated by *Carabus problematicus* or *Calathus micropterus* and *Trechus obtusus*. The species spectrum in Group 1, as

TABLE 4. Comparison between the percentage catch of the more abundant (5% or more) species of carabid in deciduous woodland and in the groups of forest sites formed at the second TWINSPAN division. − indicates numbers of less than 5% on less than half of the sites in a group; + numbers of less than 5% on at least half the sites

|  | Deciduous woodland ($n = 2$) | Forest plantations | | | |
|---|---|---|---|---|---|
|  |  | Group 1 ($n = 6$) | Group 2 ($n = 4$) | Group 3 ($n = 8$) | Group 4 ($n = 7$) |
| *Agonum assimile* | 5 | − | 0 | − | − |
| *Calathus melanocephalus* | 7 | 0 | − | 0 | 0 |
| *Pterostichus melanarius* | 23 | 0 | − | 0 | 0 |
| *Nebria brevicollis* | 17 | − | − | − | − |
| *Abax parallepipedus* | 5 | 9 | + | − | + |
| *Pterostichus madidus* | 3 | 12 | 11 | + | − |
| *Carabus violaceus* | 5 | 5 | 17 | + | − |
| *Leistus rufescens* | 0 | 6 | 4 | + | 8 |
| *Carabus problematicus* | 7 | 9 | 39 | 46 | 9 |
| *Calathus micropterus* | 2 | 18 | 10 | 19 | 34 |
| *Trechus obtusus* | 0 | 11 | + | + | 23 |
| Mean number of species | 20.5 | 16.2 | 17.3 | 11.9 | 9.9 |

TABLE 5. The groups of sites formed at the second level of a TWINSPAN analysis of the distribution of the common carabids in the forest plantations. Abbreviated names of tree species as in Table 1

| | Group 1 | | Group 2 | | Group 3 | | Group 4 |
| Site | Tree species | Site | Tree species | Site | Tree species | Site | Tree species |
|---|---|---|---|---|---|---|---|
| 23 | SP | 22 | SP | 21 | SS | 2 | JL |
| 24 | SS/birch | 11 | SP | 10 | SS | 1 | NS |
| 8 | SP/alder | 18 | SS/birch | 12 | SS | 3 | SS/WH |
| 6 | Birch | 19 | SS/birch | 13 | SP | 25 | SS/WH |
| 17 | EL/oak | | | 15 | SP/JL | 5 | SS/WH |
| 4 | Birch | | | 16 | WH | 7 | NS |
| | | | | 14 | SP | 20 | SS |
| | | | | | | 9 | SS/birch |

well as having affinities with the wood-pasture of the present study, was similar to other oak woodland in the north of England (Walker 1985) and the shift in species distribution between Group 1 and Group 4 suggested a sequence from broad-leaf to conifer woodland. This was substantiated in Table 5 which shows the distribution of sites in the four TWINSPAN groups together with their tree species. Group 1 consisted of the two birch woodland sites and the three mixed broad-leaf/conifer sites with the highest proportion of broad-leaves, together with a mature Scots pine stand. Group 2 consisted of two mixed broad-leaf/conifer sites and two mature Scots pine plantations while the other conifer plantations fell into Groups 3 and 4 with no apparent distinction between tree species or between mixtures and monocultures. Site 9, which formed an exception, being a mixed conifer/broad-leaf stand falling into Group 4, had an 80% tree cover of Sitka spruce, the highest proportion of conifer among the broad-leaf/conifer plantations.

In general, the ground beetle fauna of conifer plantations was less species rich than that of deciduous woodland and was dominated by different species. Broad-leaf woodland within conifer plantations and broad-leaf/conifer mixtures retained some of the species of deciduous woodland and were, on average, richer in species than the plantations.

### Carabid distribution in relation to environmental variables

The relationship between the carabid communities and other environmental variables, as well as the proportion of conifers present was analysed by canonical correspondence analysis, CANOCO (Ter Braak 1987). The first four canonical axes accounted for 77% of the variance and Fig. 2 shows site scores on axis 2 plotted against axis 1. The direction and strength of the most important environmental gradients are represented by arrows superimposed on the plot. Axis 1 (eigenvalue 0.31, $P < 0.01$) was most closely related to pH and organic content of

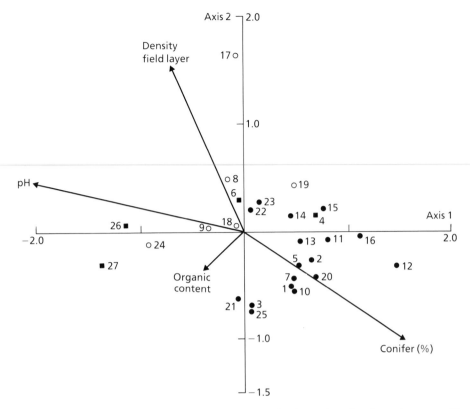

Fig. 2. Axis 2 CANOCO site scores, based on carabid distributions, plotted against axis 1 scores. The arrows represent the directions and strengths of the most important environmental gradients.

the soil, while axis 2 (eigenvalue 0.20) was most closely related to the density of the ground vegetation. The proportion of conifers present showed the highest correlation with axis 3 (eigenvalue 0.14) and was also closely correlated with axis 2. Axis 1, therefore, reflected a gradient from mineral to peat soils, while axis 2 was related to a gradient from the dense herb layer of the deciduous and open canopy sites to the unvegetated, litter layer of the closed canopy conifer stands.

## DISCUSSION

In the present study, the presence of more than one tree species, whether conifers or a mixture of conifer and broad-leaved species, increased the density of invertebrates, as measured by pitfall trap catches. Pitfall catches are dominated by active predatory species (Southwood 1978) and it is possible that these species were responding to an increased food supply, either in the soil or falling from the trees above. Although the number of insect species associated with non-indigenous conifers such as Sitka spruce or western hemlock is low compared to indigenous broad-leaved species (Southwood 1961), addition of just a few extra species associated with the second tree may buffer fluctuations in food supply to the ground predators and allow greater densities to be maintained.

TWINSPAN analysis (Hill 1979) of the carabid species distributions indicated that, at the species level, these predators were not influenced by whether one or more conifers were planted together, but by the presence of a broad-leaved component. However, in a multivariate analysis (Ter Braak 1988) the first canonical axis was closely related to pH and organic content of the soil. In addition, the abundant species of the more impoverished plantations in the present study were the same as those of the adjacent moorland (Butterfield, in press). Thus, the main gradient appeared to be from the relatively fertile soil of deciduous woodland to peat and to have little to do with the trees. Despite this, the type of vegetation growing on non-calcareous upland soils is known to influence pH and fertility and, as broad-leaved species are associated with a relatively high pH (Miles 1985, 1988) their presence, as well as the original soil status, probably contributed to the gradient. The second canonical axis was most closely related to the complexity of the field layer and the proportion of conifers present, and thus indicated the importance of the broad-leaf component more directly. Newton and Moss (1977) could have been writing about carabid communities when they wrote 'In general, the variety and density of birds is greater in woods (a) of birch and other broad-leaved trees than of pine (b) on rich than poor soils . . . (d) which have much, rather than little undergrowth'.

Species richness in birds and ground beetles, taxa of widely differing habitat and size range, shows similar increases in response to the presence of broad-leaved tree species in conifer plantations. A single management option, to leave some of the patches of broad-leaf seedlings that appear at the planting stage of the second rotation, would greatly increase the conservation potential of conifer plantations for at least these groups. Mixed planting of conifer species favours bird species diversity (Peck 1989) and appears to allow increased densities of surface-active invertebrates but further investigation is needed to determine whether species diversity of the latter is increased.

## ACKNOWLEDGMENTS

We are grateful to the Forestry Commission for permission to work at Hamsterley and, especially, to G. Simpson and B. Walker for their help and enthusiasm and to Mr A Thompson for permission to work in Adder Wood. We would also like to thank B. Huntley and N. Aebischer for advice and constructive criticism.

## REFERENCES

**Bevan, D. (1987).** *Forest Insects: A Guide to Insects Feeding Trees in Britain.* Forestry Commission Handbook, No. 1. HMSO, London.

**Briggs, J.B. (1961).** A comparison of pitfall trapping and soil sampling in assessing populations of two species of ground beetle (Col.: Carabidae). *Report East Malling Research Station 1960,* 108–12.

**Butterfield, J.** (in press). The effect of conifer plantations on the invertebrate communities of peat moorland. In *Peatland Ecosystems and Man: an Impact Assessment* (Ed. by O.M. Bragg, P.D. Hine, R.A. Robertson & H.A.P. Ingram). International Peat Society. Cambridge University Press, Cambridge.

**Forestry Commission (1987).** *Forestry Facts and Figures.* Forestry Commission, Edinburgh.

**Haartman, von L. (1971).** Population dynamics. In *Avian Biology* (Ed. by D.S. Farner & J.R. King), Vol. 1, pp. 391–459. Academic Press, London.

**Hill, M.O. (1979).** *TWINSPAN—a Fortran Program for Arranging Multivariate Data in an Ordered Two-way table by Classification of the Individuals and Attributes.* Ecology and Systematics, Cornell University, New York.

**Hill, M.O. (1986).** Ground flora and succession in commercial forest. In *Trees and Wildlife in the Scottish Uplands* (Ed. by D. Jenkins), pp. 71–8. ITE, Cambridge.

**Kennedy, C.E.J. & Southwood, T.R.E. (1984).** The number of species of insects associated with British trees; a re-analysis. *Journal of Animal Ecology*, **53**, 455–78.

**Lindroth, C.H. (1974).** *Carabidae. Handbooks for the Identification of British Insects*, Vol. IV (2). Royal Entomological Society of London, London.

**Luff, M.L. (1975).** Some features influencing the efficiency of pitfall traps. *Oecologia*, **19**, 345–57.

**Marquiss, M., Newton, I. & Ratcliffe, D. A. (1978).** The decline of the raven, *Corvus corax*, in relation to afforestation in southern Scotland and northern England. *Journal of Applied Ecology*, **15**, 129–44.

**Marquiss, M., Ratcliffe, D.A. & Roxburgh, L.R. (1985).** The numbers, breeding success and diet of golden eagles in southern Scotland in relation to changes in land use. *Biological Conservation*, **34**, 121–40.

**Miles, J. (1985).** The pedogenic effects of different species and vegetation types and the implications of succession. *Journal of Soil Science.* **36**, 571–84.

**Miles, J. (1988).** Vegetation and soil change in the uplands. In *Ecological Change in the Uplands* (Ed. by M.B. Usher & D.B.A. Thompson), pp. 57–70. Special Publication No. 7 of The British Ecological Society. Blackwell Scientific Publications, Oxford.

**Newton, I. & Moss, D. (1977).** Breeding birds of Scottish pinewoods. In *Native Pinewoods of Scotland* (Ed. by R.G.H. Bunce & J.N.R. Jeffers), pp. 26–31. ITE, Cambridge.

**Peck, K.M. (1989).** Tree species preferences shown by foraging birds in forest plantations in northern England. *Biological Conservation*, **48**, 41–57.

**Ratcliffe, D.A. & Petty, S.J. (1986).** The management of commercial forestry for wildlife. In *Trees and Wildlife in the Scottish Uplands* (Ed. by D. Jenkins), pp. 177–87. ITE, Cambridge.

**Ratcliffe, D.A. & Thompson, D.B.A. (1988).** The British uplands: their ecological character and international significance. In *Ecological Change in the Uplands* (Ed. by M.B. Usher and D.B.A. Thompson), pp. 9–36. Special Publication No. 7 of The British Ecological Society. Blackwell Scientific Publications, Oxford.

**Southwood, T.R.E. (1961).** The number of species of insect associated with various trees. *Journal of Animal Ecology*, **30**, 1–8.

**Southwood, T.R.E. (1978).** *Ecological Methods: with Particular Reference to the Study of Insect Populations.* Chapman & Hall, London.

**Ter Braak, C.J.F. (1988).** *CANOCO—a Fortran Program for Canonical Community Ordination by (Partial) (Detrended) (Canonical) Correspondence Analysis, Principle Components Analysis and Redundancy Analysis (Version 2.1).* Agricultural Mathematics Group, Ministry of Agriculture and Fisheries, Wageningen.

**Thompson, D.B.A., Stroud, D.A. & Pienkowski, M.W. (1988).** Afforestation and upland birds: consequences for population ecology. In *Ecological Change in the Uplands* (Ed. by M.B. Usher & D.B.A. Thompson), pp. 237–59. Special Publication No. 7 of the British Ecological Society. Blackwell Scientific Publications, Oxford.

**Walker, M. (1985).** A pitfall trap study on Carabidae and Staphylinidae (Col.) in County Durham. *Entomologists Monthly Magazine*, **121**, 9–18.

**Woods, A. & Cadbury, C.J. (1987).** Too many sheep in the hills. *RSPB Conservation Review*, **1**, 65–7.

# Insect pest population dynamics: effects of tree species diversity

A.D. WATT

*Institute of Terrestrial Ecology, Bush Estate, Penicuik, Midlothian EH26 0QB, UK*

## SUMMARY

The evidence to support the view that insect pest outbreaks occur more frequently in forest monocultures than in mixed-species stands is inconclusive, partly because of the lack of evidence and partly because factors other than stand-species composition strongly affect insect pest status. Several factors may make monocultures more or less prone to insect attack; these include lack of natural enemies, greater concentration of host plants, the absence of alternative hosts and the development of closer coincidence between insect and plant phenologies. More research is required before it becomes clear whether tree species composition can be effectively manipulated to reduce the incidence of pest outbreaks in forests.

## INTRODUCTION

There is a widespread belief that forest and agricultural monocultures experience outbreaks of insect pests much more frequently than mixed-species forests and other natural ecosystems, a belief which is closely tied to the view that the population fluctuations of insects (and other species) are more stable in diverse ecosystems (e.g. Elton 1958). Indeed the occurrence of insect pest outbreaks in forest monocultures is often cited as support for the stability–diversity hypothesis (e.g. Krebs 1985). The topic of diversity and stability goes much further than insect pest outbreaks, and is probably not the best basis for discussing the population dynamics of individual species (pests or non-pests) (see Young, this volume). This paper will therefore be restricted to addressing the following largely applied, questions.

1  Do outbreaks of insect pests occur more frequently in forest monocultures than in mixed-species stands?

2  What factor(s) make monocultures more or less prone to insect pest outbreaks?

3  Can the tree species composition of forests be manipulated to reduce the frequency of pest outbreaks?

These questions should ideally be answered with reference to extensive surveys and properly formulated experiments which compare the population dynamics of insect pests in pure and mixed-species stands of trees. Such surveys and experiments do not exist, so these questions have largely to be answered through studies on comparisons of natural, semi-natural and plantation forests of various types, and through relevant work in agro-ecosystems.

## THE ROLE OF MONOCULTURE IN CAUSING INSECT PEST OUTBREAKS IN FORESTRY

There are many insect pests causing damage throughout the world to plantation forestry (e.g. Gibson & Jones 1977; Speight & Wainhouse 1989). However, it is one thing to state that forest monocultures are prone to attack from insect pests, it is quite another to conclude that pest outbreaks are more prevalent in pure rather than mixed-species stands. The evidence, even for the same pest species, such as the eastern spruce budworm (*Choristoneura fumiferana*) is inconclusive. Spruce budworm outbreaks in eastern Canada have increased in severity during the twentieth century (Blais 1983). It is widely believed that budworm outbreaks are more likely to occur in extensive stands of mature balsam fir (*Abies balsamea*) than in areas of forests with a mixture of tree species (e.g. Blais 1968, 1983). To a large degree, man's actions appear to have been responsible for the upsurge in budworm damage through forest-felling practices, fire control measures and pest control itself (Blais 1983). The felling of white pine (*Pinus strobus*) and clear-felling of fir/spruce and fir/spruce/hardwood stands have led to an increase in the proportion of fir. Likewise fire protection measures have meant that pioneer tree species have declined in the area they cover. These species (such as trembling aspen (*Populus tremuloides*), paper-bark birch (*Betula papyrifera*), Jack pine (*Pinus banksiana*) and black spruce (*Picea mariana*)) become established after fire, and tend to form the typical mixed boreal forest with balsam fir and white spruce which only gradually succeed the pioneer species. Pest control measures which prevent balsam fir mortality also prevent the establishment of other tree species. The eastern spruce budworm therefore appears to be a classical pest of forest monoculture, growing in significance with the decline in mixed forest stands. However, a rather different view of the ecology of this budworm emerged from an analysis of the 1967–74 outbreak in Quebec (Hardy, Lafond & Hamel 1983). The Quebec outbreak originated from seven epicentres which, instead of being dominated by balsam fir and other host-species, were areas of mixed forest where non-host-tree species outnumbered host-species, host-trees forming an average of only 12% of the forest cover. These epicentres were also characterized by abundant pioneer-tree species suggesting that ecological disturbance led to an invasion by pioneer-species and, in turn, the establishment of transition-species like balsam fir in areas where fir is normally scarce. In a similar study on the Douglas fir tussock moth (*Orgyia pseudotsugata*), the basal area of white fir (*Abies concolor*), a preferred host of this insect, was found to be lower in outbreak areas than non-outbreak areas (Williams *et al.* 1979). These findings are difficult to explain. Perhaps insect pest species in these epicentres found themselves in enemy-free space (Jeffries & Lawton 1984) for long enough for an outbreak to be initiated, or perhaps the foliage of their host plants was more suitable nutritionally (see below) in these areas. Whatever the reason, those examples illustrate the difficulty in making generalizations about the influence of forest stand composition on insect pest outbreaks.

## FACTORS INFLUENCING THE OCCURRENCE OF OUTBREAKS IN MONOCULTURES AND MIXED STANDS

There are various factors which can influence the abundance of forest insect pests through stand species composition. The main reasons for concluding that forest monocultures experience greater levels of abundance than mixed forest stands are:

1  natural enemies: mixed forest stands may contain a greater abundance or diversity of natural enemies, or both;

2  resource concentration: the mortality of dispersing stages of phytophagous forest insects may be at a minimum in a monoculture;

3  insect–plant relationships: insect pest abundance may be affected by the presence or absence of alternate hosts, or by coincidence between insect and plant phenology; and

4  incidental effects: the incidence of pest damage may be affected by factors associated with, but not directly related to, monocultural forestry.

### Natural enemies

The reasons for concluding that natural enemies are of greater importance in diverse ecosystems are outlined by Root (1973) and Russell (1989):

1  there is a greater diversity of phytophagous insects and therefore potential prey species in diverse ecosystems;

2  this permits relatively stable populations of generalist natural enemies to take advantage of a greater variety of phytophagous insects which vary in their spatial and temporal availability within the habitat;

3  specialized natural enemies are buffered because the presence of prey refuges in complex ecosystems prevents them from being faced without a food resource;

4  diverse ecosystems are more likely to provide critical requirements for some natural enemies such as nectar and pollen; and

5  these factors together act to inhibit the development of insect pest outbreaks.

There is very little evidence from studies of forest insects to support or refute these ideas, but Reeves, Dunn and Jennings (1983) showed that carabid beetle diversity in New Hampshire forests was greatest in stands with the greatest tree species diversity. The role of natural enemies in diverse rather than monocultural ecosystems has received considerably more attention in agricultural than forest ecosystems, but even in agro-ecosystems where the ability to set up complex experiments is much greater than in forests, the number of studies is surprisingly small. Russell (1989) reviewed eighteen studies that tested the 'enemies hypothesis'. He found that in studies where mortality from predation or parasitism was recorded, there was a higher mortality in mixed systems in nine studies, a higher mortality rate in monocultures in two, and no difference in two.

The principles underlying the 'enemies hypothesis' and the agricultural studies which test it, may have some general relevance to forests of different species compositions, but Way (1977), in reviewing the role of natural enemies in mono-cultures and diverse ecosystems in general, made the important point that pest control by natural enemies is brought about by a limited number of key species. Thus, factors which may bring about an increase in the diversity and abundance of natural enemies are less important, from an applied viewpoint, than factors which increase the action of a small number of key natural enemies. In some cases, nonetheless, tree species diversity may promote the abundance of important natural enemies of forest pests. Wood ants are significant predators of some temperate forest pests (Adlung 1966). They have been reported to be most abundant in mixed woods, probably because as aphid honeydew feeders they derive more consistent feeding conditions from the different periods of honeydew production of the various aphid species in mixed woodland (Laine & Niemela 1989).

### Resource concentration

An alternative reason for predicting that monocultures contain higher numbers of pest species is that they may be expected to more easily locate their host-plants in a habitat where their food resource is concentrated (Root 1973; Kareiva 1983; Russell 1989). This 'resource-concentration hypothesis' considers three aspects of the food resource (Kareiva 1983):
1   density, the distance between individuals of the same host-plant species;
2   patch size, the size of the host-plant stand (perhaps including non-host species); and
3   stand diversity, the species composition of the stand and the frequency of each species.

Studies on agricultural and similar natural ecosystems have shown that increasing plant density almost invariably leads to a decrease in the number of herbivores per plant: all seven studies reviewed by Kareiva (1983) found that herbivore numbers per plant decreased as plant density increased. Studies on the effect of host-plant patch size have shown that herbivore abundance tends to increase with host-plant patch size: nine out of nineteen studies have reported this, two found the opposite effect and eight studies gave no evidence of an effect of host-plant patch size (Kareiva 1983). Unfortunately, these results are difficult to relate to the forest level: it is difficult to separate the effects of host-plant spacing from host-plant quality (Kareiva 1983) and agricultural studies tend to be of a short duration, perhaps adequate for annual agricultural pests but not for insects which typically take several years to reach pest status. Similarly, the conclusions from studies on the effects of host-plant species diversity on agri-cultural ecosystems should be taken with caution. Nevertheless, they are worth consideration because of the principles they have exposed, some of which are undoubtedly relevant to forest ecosystems. Searching herbivores may be unable to find their host plants in a mixture of host- and non-host-plants because their hosts are physically hidden (Rausher 1981); host plants may be difficult to find against a background of non-host plants (Cromartie 1981); and herbivores may be confused by the chemical odours of non-host plants (Tahvanainen & Root 1972). Even when a host has been located there is evidence that non-host plants interfere with adult oviposition (Coaker 1980). Larval insects suffer high levels of mortality when dispersing and this is likely to be worst among passive dispersers and those which are unable to detect their hosts from a distance, when larvae must locate a second host plant, and when host plants are dispersed among non-host plants (Kareiva 1983).

Although research on agro-ecosystems has led some to conclude that problems associated with host location form the major reason why herbivores are less abundant in host–plant mixtures than in monocultures (Risch, Andow & Altieri 1983), there is no background of studies in forestry to permit a similar conclusion. Nevertheless, there are a few experimental studies, and there is enough circumstantial evidence, to indicate that the resource concentration hypothesis is important in forest ecosystems.

Host location by adult insects is particularly relevant to bark-feeding and wood-boring insects. These insects tend to be restricted to logs and unhealthy trees: the defences of healthy host plants are usually sufficient against attack, but these defences may be overcome when trees are attacked by large numbers of insects (Berryman 1982). Thus, there are many examples of this type of insect becoming serious pests of healthy trees when grown in monoculture plantations (Gibson & Jones 1977).

Dispersal by first or second instar larvae is an important part of host colonization by defoliating forest pests, particularly among Lepidoptera with flightless females such as the gypsy moth, Douglas fir tussock moth and the winter moth (*Operophtera brumata*). Kemp and Simmons (1979) found that survival by second instar larvae of the eastern spruce budworm during spring dispersal increased as the degree of crown closure increased and was lower in a mixture of host and non-host trees than in stands largely composed of host species.

Finally, it should be stated that the 'enemies hypothesis' and the 'resource concentration hypothesis' are not mutually exclusive: the species composition of forest stands may affect insect abundance both through the action of natural enemies and by affecting host location.

## Insect–host-plant relationships

In addition to the above factors, the species composition of a stand can have an influence on insect abundance through the provision of alternate host plants. For example, the Douglas fir woolly aphid (*Adelges cooleyi*) requires both a spruce and a fir host to complete its full life cycle with both sexual and asexual stages.

Mixtures of these two host plants are particularly vulnerable to outbreaks (Speight & Wainhouse 1989).

Another aspect of insect–host-plant relationships is the coincidence between insect and host-plant phenology, particularly between egghatch and budburst. Poor coincidence can dramatically affect insect survival (Watt 1987) and is thought to be the main reason why species like the winter moth fluctuate so markedly in abundance (Feeny 1976). Some indication of the impact of stand species composition on such species comes from research on the green oak tortrix (*Tortrix viridana*). The survival of the newly hatched larvae of this species is closely dependent on the growth stage of its host plants (du Merle & Mazet 1983). The date of maximum suitability of oak buds is 1–3 weeks later for *Quercus ilex* than *Quercus pubescens*, but du Merle (1983) found that eggs laid on *Q. ilex* hatch 8–16 days later than those laid on *Q. pubescens*. The development of host-specific traits does, however, depend upon stand structure: it occurs in pure but not mixed stands (Mitter *et al.* 1979). Thus, although a stand mixture of two or more species which are equally suitable for *T. viridana* or similar species would appear to have no effect on insect abundance, it is likely to do so by hindering the development of host-specific traits.

## Incidental effects

Although there are sound reasons why pest status is affected by the species composition of forests, it is also clear that there are occasions when it is not monoculture *per se* that has led to an increase in pest problems but factors associated with monoculture forestry. Some of these factors are far removed from 'monoculture', others can be regarded as factors closely associated with monoculture.

The main factor which is confused with monoculture forestry is the use of exotic tree species. For example, outbreaks of the pine beauty moth (*Panolis flammea*) on lodgepole pine (*Pinus contorta*) in Scotland did not occur until plantations of this tree were established in the late 1950s and early 1960s (Watt & Leather 1988). However, this pest did not emerge because of factors which directly result from planting this tree in a monoculture: nearby plantations of Scots pine (*Pinus sylvestris*) remain free from outbreaks. Outbreaks also occur in single-species plantations when they are planted in areas susceptible, for some reason, to insect outbreaks. For example, outbreaks of the pine beauty moth are limited to lodgepole pine monocultures on deep unflushed peat (Evans *et al.* 1991). In cases similar to this, the link between soil type and insect abundance appears to be the effect that site conditions can have on host-plant quality (e.g. White 1974), but in this example it appears that an impoverished natural enemy fauna in plantations on deep peat plays a more straightforward role in causing pine beauty moth outbreaks (Watt, Leather & Evans 1991).

Forestry practices associated with monoculture plantations can also lead to insect pest problems. For example, the scolytid beetle (*Xyleborus mascarensis*) is a pest of African mahogany (*Khaya ivorensis*) and okoumé (*Aucoumea klaineana*) because weed clearance and accumulation of woody debris provide breeding sites for this insect (Gibson & Jones 1977). Thus, pest problems arise because of plantation practices rather than because of the use of forest monoculture.

## PEST MANAGEMENT AND SPECIES MIXTURES

In considering species mixtures as a method of pest management, both hope and caution are appropriate. The caution is needed because, as discussed above, the outbreaks of many pests in forest monocultures occur because of factors related more to plantation forestry than monoculture forestry, and because manipulation of susceptibility to forest pests is limited by factors such as the impact of soil conditions on host-plant quality and the abundance of soil-dwelling predators. This point is well illustrated by the varied benefit of forest understorey in Polish forests (Szujecki 1987). The addition of an undergrowth or understorey can have an impact on the abundance of predators (Carabidae and Staphylinidae) (see also Butterfield & Malvido, this volume) and parasites (Ichneumonidae) but this is secondary to the effect of soil conditions. An understorey has a greater impact in forests growing in rich rather than poor soils (Szujecki 1987). Thus, for example, a pine/birch stand on a poor site will not benefit greatly from an oak or a beech understorey.

## CONCLUSIONS

There is very little direct information on the relative abundance of insect pests in monocultures and mixed-species stands. But circumstantial evidence and a few direct studies plus the experience of research in agro-ecosystems demonstrate that tree species-rich forests can be relatively pest-free. However, this is not an absolute rule: the opposite may be true for some pests. In implementing species mixtures as a method of pest management it must first be appreciated that pest problems develop in forest monocultures for many reasons other than those which directly follow from the planting of large stands of the same tree species, i.e. a direct result of monoculture. The planting of tree mixtures in such circumstances is not very likely to alleviate the problem. The second factor to be borne in mind is that some of the problems often directly associated with forest monocultures, such as a lack of natural enemies, cannot always be solved by increasing tree species diversity, because of the effect that site conditions may have on host-plant quality and natural enemy abundance.

Clearer conclusions await further research on this topic. Although research on the effects of plant species diversity on insects in agro-ecosystems is much further advanced than that in forest ecosystems, that research is not without its faults (Kareiva 1983; Russell 1989). As well as learning from the research in agro-ecosystems, the research on forest ecosystems should be directed toward factors which may be more appropriate to the more gradual build up of insect numbers typical of forest pests. Thus, attention must be paid to the dynamics of insect–natural enemy interactions over the long term: research on the resource concentration hypothesis should be directed more towards pest-colonization dispersal than initial colonization; and emphasis should be placed on the consequences of aspects of insect–plant relationships, such as phenological coincidence.

## REFERENCES

**Adlung, K.G. (1966).** A critical evaluation of the European research on use of red wood ants (*Formica rufa* group) for the protection of forests against harmful insects. *Zeitschrift für angewandte Entomologie*, **57**, 167–89.

**Berryman, A.A. (1982).** Biological control, thresholds, and pest outbreaks. *Environmental Entomology*, **11**, 544–9.

**Blais, J.R. (1968).** Regional variation in susceptibility of eastern North American forests to budworm attack based on history of outbreaks. *Forest Chronicle*, **50**, 19–21.

**Blais, J.R. (1983).** Trends in the frequency, extent, and severity of spruce budworm outbreaks in eastern Canada. *Canadian Journal of Forest Research*, **13**, 539–47.

**Coaker, T.H. (1980).** Insect pest management in *Brassica* crops by intercropping. *Integrated Control Brassica Crops, International Organization for Biological Control West Palearctic Regional Section Bulletin*, **3**, 117–25.

**Cromartie, W.J. (1981).** The environmental control of insects using crop diversity. *Handbook of Pest Management* (Ed. by D. Pimentel), pp. 223–51. Chemical Rubber Company in Agriculture, Boca Raton, Florida.

**Elton, C.S. (1958).** *The Ecology of Invasions by Animals and Plants*. Methuen, London.

**Evans, H.F., Stoakley, J.T., Watt, A.D. & Leather, S.R. (1991).** Development of an integrated approach to control of pine beauty moth in Scotland. *Forest Ecology and Management*, **39**, 19–28.

**Feeny, P. (1976).** Plant apparency and chemical defense. *Biochemical Interactions Between Plants and Insects, Recent Advances Phytochemistry* (Ed. by J.W. Wallace & R.L. Mansell), Vol. 10, pp. 1–40. Plenum Press, New York.

**Gibson, I.A.S. & Jones. T. (1977).** Monoculture as the origin of the major forest pests and diseases. *Origins of Pest, Parasite, Disease and Weed Problems* (Ed. by J.M. Cherrett & G.R. Sagar), pp. 139–61. Blackwell Scientific Publications, Oxford.

**Hardy, Y.J., Lafond, A. & Hamel, L. (1983).** The epidemiology of the current spruce budworm outbreak in Quebec. *Forest Science*, **29**, 715–25.

**Jeffries, M.J. & Lawton, J.H. (1984).** Enemy free space and the structure of ecological communities. *Biological Journal of the Linnean Society*, **23**, 269–86.

**Kareiva, P. (1983).** Influence of vegetation texture on herbivore populations: resource concentration and herbivore movement. *Variable Plants and Herbivores in Natural and Managed Systems* (Ed. by R.F. Denno & M.S. McClure), pp. 259–89. Academic Press, New York.

**Kemp, W.P. & Simmons, G.A. (1979).** Influence of stand factors on survival of early instar spruce budworm. *Environmental Entomology*, **8**, 993–6.

**Krebs, C.J. (1985).** *Ecology: The Experimental Analysis of Distribution and Abundance*, 3rd edn. Harper & Row, New York.

**Laine, K.J. & Niemela, P. (1989).** Nests and nest sites of red wood ants (Hymenoptera, Formicidae) in Subarctic Finland. *Annales Entomologici Fennici*, **55**, 81–7.

**Merle, P. du (1983).** Phenologies comparées du chêne pubescent, du chêne vert et de *Tortrix viridana* L. (Lep., Torticidae). Mise en évidence chez l'insecte de deux populations sympatriques adaptées chacune à l'un des chênes. *Acta Œcologica Œcologia Applicata*, **4**, 55–74.

**Merle, P. du & Mazet, R. (1983).** Stades phénologiques et infestation par *Tortrix viridana* L. (Lep., Torticidae) des bourgeons du chêne pubescent et du chêne vert. *Acta Œcologia Œcologia Applicata*, **4**, 47–53.

**Mitter, C., Futuyma D.J., Schneider J.C. & Hare J.D. (1979).** Genetic variation and host plant relations in a parthenogenic moth. *Evolution*, **33**, 777–90.

Rausher, M. (1981). The effect of native vegetation on the susceptibility of *Aristolochia reticulata* (Aristolochiaceae) to herbivore attack. *Ecology*, **61**, 905–17.

Reeves, R.M., Dunn, G.A. & Jennings, D.T. (1983). Carabid beetles (Coleoptera: Carabidae) associated with the spruce budworm, *Choristoneura fumiferana* (Lepidoptera: Torticidae). *Canadian Entomologist*, **115**, 453–72.

Risch, S., Andow, D. & Altieri, M. (1983). Agroecosystem diversity and pest control: data, tentative conclusions and new research directions. *Environmental Entomology*, **12**, 625–9.

Root, R.B. (1973). Organisation of a plant–arthropod association in simple and diverse habitats: the fauna of collards (*Brassica oleracae*). *Ecological Monographs*, **43**, 95–124.

Russell, E.P. (1989). Enemies hypothesis: a review of the effect of vegetational diversity on predatory insects and parasitoids. *Environmental Entomology*, **18**, 590–9.

Speight, M.R. & Wainhouse, D. (1989). *Ecology and Management of Forest Insects*. Clarendon Press, Oxford.

Szujecki, A. (1987). *Ecology of Forest Insects*. Polish Scientific Publishers, Warszawa.

Tahvanainen, J. & Root, R.B. (1972). The influence of vegetational diversity on the population ecology of a specialized herbivore, *Phyllotreta cruciferae* (Coleoptera: Chrysomelidae). *Oecologia*, **10**, 1205–14.

Watt, A.D. (1987). The effect of shoot growth stage of *Pinus contorta* and *Pinus sylvestris* on the growth and survival of *Panolis flammea* larvae. *Oecologia*, **72**, 429–33.

Watt, A.D. & Leather, S.R. (1988). The impact and ecology of the pine beauty moth in upland pine forests. *Ecological Change in the Uplands* (Ed. by M.B. Usher & D.B.A. Thompson), pp. 261–72. Blackwell Scientific Publications, Oxford.

Watt, A.D., Leather, S.R. & Evans, H.F. (1991). Outbreaks of the pine beauty moth on pine in Scotland: the influence of host plant species and site factors. *Forest Ecology and Management*, **39**, 211–21.

Way, M.J. (1977). Pest and disease status in mixed stands vs. monocultures; the relevance of ecosystem stability. *Origins of Pest, Parasite, Disease and Weed Problems* (Ed. by J.M. Cherrett & G.R. Sagar), pp. 127–38. Blackwell Scientific Publications, Oxford.

White, T.C.R. (1974). A hypothesis to explain outbreaks of looper caterpillars, with special reference to populations of *Selidosema suavis* in a plantation of *Pinus radiata* in New Zealand. *Oecologia*, **16**, 279–301.

Williams, C.B., Jr., Wenz. J.M., Dahlsten, D.L. & Norick, N.X. (1979). Relation of forest site and stand characteristics to Douglas-fir tussock moth (Lep. Lymantriidae) outbreaks in California. *Bulletin de la Société Entomologique Suisse*, **52**, 297–307.

# Conserving insect communities in mixed woodlands

## M.R. YOUNG

*Department of Zoology, Aberdeen University, Tillydrone Avenue, Aberdeen AB9 2TN, UK*

## SUMMARY

Conservation of insect communities in woodland has received little attention, although there are several studies relating to individual species, almost all butterflies. Furthermore mixed woodlands, either as stands containing more than one tree species or a mixture of stands of separate species, have received very little consideration. Conservation strategies have too often been either uncritical continuation of previous management or concentration on one species in the hope that others will also benefit. Nevertheless collation of the few studies that exist provides some basis for successful woodland management for insect conservation, and emphasizes the special need to provide for non-tree features and a varied structure. So far it seems that the woodland edges, rides, glades and other discontinuities are the most important features needing attention. Consequently if a commercial crop, even a monoculture, can be made 'mixed', in the sense of adding locally native tree species to it, by providing the extra trees and other features at its edges and in rides and glades, then this may well prove the type of 'mixed', commercial woodland that is most practicable for insect conservation. Even so, some insects can only be conserved effectively in extensive semi-natural woodlands with mature and senescent native trees.

## INTRODUCTION

Much work has focused on specific insect pests in commercial woodland, for example on the pine beauty moth, *Panolis flammea*, in lodgepole pine (*Pinus contorta*) woods in Scotland (Leather & Barbour 1987; Watt, this volume), or on the ecology of individual woodland insects, such as the wood white butterfly *Leptidea sinapis* (Warren 1985) or the hover fly, *Callicera rufa* (Rotheray & MacGowan 1990). However, there have been very few general studies on insect communities or their conservation in woodlands, presumably because of the bewildering diversity of species involved, many of which are imperfectly known taxonomically, and the difficulties of sampling adequately across the range of 'ecological types'. Some interesting work has included all the species caught by a particular sampling method, such as the use of insecticide fogging to sample the canopy fauna of oaks (*Quercus robur, Q. petraea*) in England (Barnard, Brooks & Stork 1986), or the species which share a special habit, such as leafminers in coppice woodland (Sterling & Hambler 1988). However, it has proved more or less impossible to devise an integrated sampling programme to include even the majority of species within a woodland and it is most unlikely that it will ever be possible to achieve that. As Disney (1986) points out, basic inventories of species, even from very limited areas, would consume an unrealistic sampling and identification effort and so a common limitation has been to concentrate on one taxonomic group, so that a reasonably complete account can be made of that. Owen (1989), for example, has recently completed a lengthy survey of the Coleoptera of a Scots pine (*Pinus sylvestris*) wood in the Spey valley in Scotland, but even so he does not claim that it is complete.

Against this background it can be appreciated why there have been no comprehensive studies which have set out to compare the overall insect communities of mixed- and single-species woodlands, and why there are almost no satisfactory published data sets from particular woodland types, which can be abstracted and contrasted.

Despite this it is possible to find some studies, with a taxonomic limitation, which do make a comparison between different woodland types, and examples are Magurran (1985), Young (1986) and Waring (1988, 1989a), all of which concentrate on Lepidoptera. Although none of these studies specifically address the topic of mixed-species woods, they do have direct relevance to them and so are discussed further below.

In the context of the lack of directly relevant studies, this review first considers general approaches to the conservation of insect communities in woods, so as to set the scene, and then relates this to mixed woodlands. My own taxonomic preference means that there is an emphasis on Lepidoptera.

## CONSERVING WOODLAND INSECTS

It is impossible to imagine developing a strategy for the collective conservation of a woodland insect community based on detailed knowledge of habitat requirements of even a proportion of the species present. Owen (1989) recorded 820 species of Coleoptera in a Scottish pine forest, which is generally regarded as a rather impoverished woodland type, and obviously the total insect fauna must be several times that number. Of all these, some are even now of questionable taxonomic status, many are exceedingly difficult to sample and for almost none are the habitat requirements known. It would be possible to begin to list larval foodstuffs (as done for Lepidoptera in Bernwood Forest by Waring, 1989b), but even this simple piece of information is imperfectly known for many species.

Since these data have not been collected, two approaches have generally been used in the conservation of insect communities in woodlands. The first is to continue existing management or to change to a management regime that is believed to have favoured insects in other woods. The second is to take 'key' species, about which something is known, and to manage for those, hoping that most or all other species will also benefit.

## Continuing or adapting existing management

It is obviously true that today's insect community has been established under yesterday's management regime and so continuing this regime should provide a safe route to the conservation of the community. Despite drawbacks described below this has been a popular choice of strategy, largely because it is easy and reduces the prospect of catastrophic losses.

To succeed there must be clear, detailed and accurate prescriptions of past management and the first problem is that these are rarely available. During work on insect communities in native pine woods, M. Young, G. Armstrong and A. Edgar (unpubl.) found that even on estates such as Glen Tanar, Aberdeenshire, where records are regarded as exceptionally complete, vital data were missing on such relevant activities as juniper (*Juniperus communis*) removal, the fate of dead timber or the exact provenance of newly planted trees. Although Thomas (1988) was able to record the general changes in management at Bernwood Forest, she was not able to list the detailed prescriptions used over the years, such as would be needed to duplicate management accurately.

A second problem is that woods develop and alter and are also subject to changing outside influences, so that it is not neccessarily appropriate or adequate to persist with precisely similar management techniques. It may be possible to continue coppice rotation in large woodland blocks, for example, (as described for Bradfield Woods, Essex (Rackham 1980)), but in smaller areas, which are developing towards a high forest crop, management has had to change, at least to some extent, to meet various constraints and goals. For conservation to succeed, some assessment must be made of the key management features for insects, so that these can be prioritized and retained where possible.

In practice it has been more common for woodland conservation managers to alter previous management regimes at least slightly, in the hope that the changes will benefit insects, usually reverting to what was thought to be a previous situation, and this has sometimes led to depressing results. Two classic examples are firstly Monkswood National Nature Reserve, Huntingdon, which was originally exceptionally favoured by butterflies with over forty species in the early years of this century. When bought for conservation purposes, management was changed from coppice with standards to a more non-interventionist system, with the result that rides shaded over and blackthorn (*Prunus spinosa*) and other shrubby species became scarce. In consequence many butterflies, including the nationally rare black hairstreak (*Satyrium pruni*) disappeared from the wood. In recent years the re-establishment of ride management and some coppicing, plus some reintroductions, have redressed some of the early losses (Steele & Welch 1973; Thomas 1984; Morris & Thomas 1989).

At Blean Woods, Kent the heath fritillary (*Mellicta athalia*) vanished from areas which were treated with especially relaxed management regimes designed to favour it, but fortunately it survived in adjacent commercial woodland until regular short-cycle coppice was reintroduced in the reserve area as a result of a detailed study of its requirements (Warren 1987), so that it has returned and is now thriving.

General changes in woodland management have often been implicated in changes in insect communities, for example, there has been a substantial localization of butterflies to the ride system in the Bernwood Forest complex, as the commercial plantations have matured (Peachey 1980) and the fritillary butterflies may well have been generally lost from southern and eastern British woods because of the overall decline of coppicing (Heath, Pollard & Thomas 1984).

## Managing for key species

Certain insect species, usually butterflies or dragon flies, catch the eye and attract disproportionate conservation attention. Often woodland management is devoted primarily to their survival, with the implied justification that if management favours them then it will also be favouring myriad other 'unseen' species (at its most simplistic this is the 'if one looks after the plants, the invertebrates will look after themselves' approach, debunked by Sterling and Hambler (1988) amongst others). In fact it is almost never known whether the chosen species have any 'key' status in an ecological sense, and very rarely known whether many other species do also benefit. Nevertheless, because butterflies and other large conspicuous species can be used as a popular attractant for conservation effort, and because they are certainly acting as a 'monitor' of habitat quality to some extent, it seems prudent to encourage management for them, whilst simultaneously trying to widen the base for conservation measures for the future.

It has often been found that certain rather simple factors govern the survival of individual species, so simplifying their conservation. The white admiral (*Ladoga camilla*) needs semi-shaded honeysuckle (*Lonicera periclymenum*) and a ready nectar source, but is otherwise found in a range of woodland types (Pollard 1979); the purple hairstreak (*Quercusia quercus*) can survive on almost isolated mature oaks on the edge of a sea of conifers (Thomas 1975); the Kentish glory moth (*Endromis versicolora*) requires young, sheltered but uncrowded birch (*Betula pendula*) bushes on which to lay its eggs (Young & Barbour 1989); the robber fly (*Laphria flava*) needs prominent, horizontal dead trees on which it perches in hunting behaviour (I. McLean, unpubl.); *Callicera rufa*, a rare hover fly, seems to be found in all pine wood relicts where deep leaf-filled rot holes remain (Rotheray & MacGowan 1990); and so on.

In some cases, of course, one species's key factor is another species's disadvantage. Coppice for heath fritillaries will lead to a general reduction in leaf-mining microlepidoptera for example (Sterling & Hambler 1988).

A strategy for avoiding the limitations imposed by the choice of one species as a 'key' species, is to widen the choice and try to accommodate a deliberately contrasting range; for example a butterfly of open rides, a dead wood beetle, a fungus-feeding fly, a tree-canopy herbivore, etc. The practical problem is finding sufficient species whose ecological requirements are known. A more usual strategy is to choose nationally rare species for attention, with the problem that so little is known of the distribution and status of the more cryptic orders that some truly rare species may be overlooked.

A more forward-looking alternative might be to try and choose species which can be described as 'key ecological indicators'. Parasites (with their extra direct dependence on their insect hosts) might be such a choice, or one of several species known to respond very sensitively to a predicted change in woodland structure.

Despite the difficulties and limitations discussed above, it is essential to try and conserve insects. There is no hope in a plea for inaction because insufficient is known and mistakes may be made if something is attempted; and when a nationally rare species is known to be present its presumed conservation requirements may be the only ones available for consideration.

### Action points for conservation

Examination of the data from recent studies, accumulation of general wisdom from past failures and successes, and paying due attention to insect biology, has led to the identification of a series of principles and common actions for conserving woodland insect communities. Questions remain, but a general pattern is emerging which is fortunately broadly similar to conservation plans for other groups, such as developed in Peterken (1981) and noted in Speight and Wainhouse (1989).

This section sets out some general principles and points, leaving specific consideration of their role in mixed woodlands until later.

Knowledge of general insect ecology leads to the following realizations.

1  Insects may not survive even very short unfavourable periods. With at least one generation a year and no 'resistant' stage (such as buried seed-banks for plants), it is essential that there is *no* break in the continuity of favourable conditions.

2  Insects may have extremely specific requirements so that (for example) the beetle *Salpingus castaneus*, which requires dead wood in which to breed, uses only 1–2 cm diameter pine twigs (Hunter 1977). It is not good enough to ensure just that some (unspecified) dead wood is present. Similarly a north-facing tree may not be as suitable, however luxuriant its foliage, as one that faces south. Almost every species of insect could offer its own example here.

3  Insect populations may be viable in very small areas, but this is sometimes overstressed. Thomas (1984) makes a guestimate of the areas needed by woodland butterflies and some of the larger species may need up to 50 ha. However, the whole area does not necessarily have to be made up of the key survival

TABLE 1. British woodland butterflies, their status and conservation requirements. Data principally from Emmet and Heath (1989). *Data from Thomas (1984). Note that some of these species also occur outside woodland

| Species | Larval food-plant | British status | Reason for change of status | Area needed for colony* | Conservation needs |
|---|---|---|---|---|---|
| *Carterocephalus palaemon* Chequered skipper | Scotland: *Molinia* England: *Brachypodium* | Stable, well established Extinct —1976 | 'Changes' in woodland management | ? | Not yet established |
| *Leptidea sinapis* Wood white | Usually *Lathyrus* (or *Lotus*) | Historical decline; more recent slight spread | Shading over of rides; recently favoured by commercial forest rides | 1–2 ha | Sheltered but sunny rides with abundant *Lathyrus* |
| *Thecla betulae* Brown hairstreak | Blackthorn —low on sunny bushes | Significant decline; less so in S W | Loss of unkempt hedges and woodland edge | 10–50 ha (colonies very dispersed) | Retention of low growing blackthorn |
| *Quercusia quercus* Purple hairstreak | Oak | Stable but some local extinctions | Some loss of oak woodland | <1 ha (single or few trees) | Retention of some large oaks |
| *Satyrium w-album* White-letter hairstreak | Elm—in woodland edge and hedgerow | Significant but scattered losses | Loss of elm due to disease | <1 ha (single or few trees) | Retention of some elm |
| *S. pruni* Black hairstreak | Blackthorn —on long established stands | Now stable— previous losses. Always local | Some loss of blackthorn thicket. Very low dispersal ability | <1 ha | Retention of blackthorn scrub in long woodland rotation |
| *Celastrina argiolus* Holly blue | Holly and ivy—woodland edge species | Stable (high population fluctuation) | — | Open population | Retention of mature holly and ivy |
| *Hamearis lucina* Duke of Burgundy fritillary | Primrose and cowslip | Severe decline | Loss of food-plant in shaded woodlands | <1 ha | Provide coppice and open woodland glades |
| *Ladoga camilla* White admiral | Honeysuckle —in semi-shade | Stable; some spread this century | Favoured by increased shade in woodland | 10–50 ha | Retain semishaded honeysuckle and nectar sources |

| *Apatura iris* Purple emperor | Sallow | ?Stable; some historical losses | Some loss of woods | >50 ha | Retain extensive woodland mosaic with sallow in rides |
|---|---|---|---|---|---|
| *Boloria selene* Small pearl-bordered fritillary | Violets in open, damp areas | Severe decline in E. Stable in N | Decline in coppice, shading of woods | 5–10 ha | Reintroduce coppice or open up woodland rides and glades |
| *B. euphrosyne* Pearl-bordered fritillary | Violets in very open, short vegetation | Severe decline in E. Stable in N | Decline in coppice, shading of woods | 5–10 ha | Reintroduce coppice or open up woodland rides and glades |
| *Argynnis adippe* High-brown fritillary | Violets in open, sunny areas | Very severe decline— almost all woodland populations extinct | Decline of coppice, shading of woods | ? | Reintroduce coppice or open woodland rides and glades |
| *A. paphia* Silver-washed fritillary | Violets in clumps within light woodland | Decline in E. Stable or slight-loss in W | Some woodland loss, increased woodland shade | ? | Open up woodlands |
| *Mellicta athalia* Heath fritillary | *Melampyrum* in woods | Very severe decline but recent stabilization | Loss of coppice | <1 ha | Reintroduce coppice |
| *Pararge aegeria* Speckled wood | Broad-bladed grasses | Some increase this century | Increased woodland— tolerates shade | ? | Retain wood |

features and in general it is true that insects can use smaller reserves than can larger animals (see Table 1).

**4** Insects have low ability to survive broad environmental changes. Clear-felling part of a wood may not affect the presence of insect habitats in the remainder but it may alter the microclimate beyond the tolerance of some species. Management must be carried out with consideration for such effects.

**5** Insects have characteristically wide population fluctuations with several consequences (Varley, Gradwell & Hassell 1973; Pollard, Hall & Bibby 1986). First of all monitoring their numbers uncritically may lead to inappropriate optimism or gloom about their progress in any one year. Secondly, if extra adverse factors coincide with natural low points this may lead to unexpected extinction. Con-

TABLE 2. General list of larval food resources of British woodland Lepidoptera. Data collated from Allan (1949), Skinner (1984), Emmet & Heath (1989) and Emmet (1988). N.B. Identification of which species to include as 'woodland', and the placement of polyphagous species, are difficult and somewhat arbitrary, non-tree species are especially likely to be excluded in error

| Food resource | No. of associated species of Lepidoptera | |
|---|---|---|
| *Trees* | | |
| Oak | 96 | |
| Birch | 76 | |
| Sallow | 51 | |
| Elm | 23 | |
| Alder | 21 | |
| Other deciduous trees | 73 | |
| Polyphagous on deciduous trees | 86 | |
| Pine | 26 | |
| Spruce | 10 | |
| Other coniferous trees | 11 | |
| Polyphagous on coniferous trees | 7 | 480 |
| Shrub and understorey species | 80 | 80 |
| *Other plants* | | |
| Grasses | 43 | |
| Herbs | 114 | |
| Lichens | 12 | 169 |
| *Other resources* | | |
| (dead wood, withered leaves, fungi, etc.) | 41 | 41 |
| Overall total | | 770 |

TABLE 3. Larval food sources of Lepidoptera breeding in Tillyfourie oak wood site of special scientific interest, Aberdeenshire (after Young 1986)

| | Categories of food resource | | | |
|---|---|---|---|---|
| | Trees | Shrubs | Herbs, grasses, etc. under canopy | Herbs, grasses, etc. in clearings |
| No. of spp. of Lepidoptera associated with each resource | 100 | 44 | 42 | 146 |

versely environmental improvement may lead to rapid increase and spread, as has been seen in many successful cases where correct management has recently been applied (e.g. the adonis blue, *Lysandra bellargus*, Emmet & Heath (1989)).

6 Despite the flying ability of most adult insects, they may actually be highly sedentary, with very low recolonization ability, even over very small distances. Thomas (1980) exemplifies this for the black hairstreak butterfly (*S. pruni*). Within a wood this means that newly created habitat patches must be within easy reach of existing ones.

**7**  Each life cycle stage of an insect may require a different set of habitat features and the essential feature may not be (for example) the obvious nectar source for the adult, but the unseen protected pupation site or the apparency of the larval food plant.

**8**  For many species it has now been found that the structure of the habitat may be more important than the precise vegetation composition. Warren (1985) found that it was the ride structure, providing shelter but allowing some sun to get to the vetch food plants of the wood white butterfly (*Leptidea sinapis*), that was important, rather than the identity of the trees providing the shelter.

**9**  Many woodland insects depend on the associated shrubs, herbs, fungi and other features, rather than the actual trees. Although oak provides a pabulum for very many species, at least an equivalent number will feed on the other components of an oak wood. Consequently woodland management for conservation must be as much to do with all these other resources as with the trees (see Tables 2 and 3). (Lists such as those in Tables 2 and 3 carry with them many problems. It is difficult to ascribe correctly the food-plants of polyphagous species, some of which feed on both trees and herbs; it is also difficult to know exactly which species to include as 'woodland'. Species of clearings are often also found in more open habitats. Consequently the data in Tables 2 and 3 can only be regarded as broadly indicative.)

## MIXED WOODLANDS AS INSECT HABITATS

Before detailed consideration is given to conserving insect communities in mixed woodlands, it is necessary to consider the general effect of different woodland types and different management regimes.

It is a truism that intensively managed forests of non-native trees harbour fewer insects than largely unmanaged forests of native trees. Young (1986) illustrates this for two woods in Aberdeenshire (Table 4) and it is due to two broad factors.

Firstly, Kennedy & Southwood (1984) (and many other authors) have noted that in general native tree species have more insect species associated with them than non-native trees. This is complicated by many factors. For example, native yew (*Taxus baccata*) has very few associated insects because of its effective complement of toxic chemical defenses; non-native larch (*Larix decidua*) (introduced around 400 years ago) has already accumulated far more species than non-native sweet chestnut (*Castanea sativa*) (introduced in Roman times); recently established southern beech (*Nothofagus* sp. ) has quickly been colonized by some polyphagous insects (Welch 1986), whereas native ash (*Fraxinus excelsior*) has surprisingly few. However it remains true that in general the common native forest trees, such as oak and birch, act as host to more insect species than the common non-native commercial species such as larch (*Larix* spp.) or Sitka spruce (*Picea sitchensis*) (Winter 1983; Emmet 1988). Furthermore, and of relevance to conservation, the insects of non-native commercial species are predominantly either pests or common

TABLE 4. Comparisons between insect communities in deciduous and coniferous woodlands

| Study authors | Methods used | Comparison | | |
|---|---|---|---|---|
| | | **Tillyfourie** | | |
| Young (1986) | Light trap; direct search for adults, larval signs, Aberdeenshire | Oak wood 340 spp. of Lepidoptera | Coniferous woods 100 spp. of Lepidoptera | |
| | | **Abernethy** | | |
| M.R. Young, G. Armstrong & A. Edgar (unpubl.) | Pitfalls; water traps; window traps some use of 'fogging', suction traps, D-Vac, light traps, Inverness-shire | Semi-natural pine forest 344 spp. of all orders (N.B. many 'tourist' spp. present) | Pine plantation 404 spp. of all orders | |
| | | **Ynys-Hir** | | |
| Kearns & Majerus (1987) | Light trap (single occasion), Dyfed, Wales | Mixed deciduous wood 36 spp. of Lepidoptera | Coniferous woods 20 spp. of Lepidoptera | |
| | | **Bernwood Forest** | | |
| Waring (1988, 1989b) | Light traps; some larval searching, Oxfordshire | Neglected coppice 4141 individual Lepidoptera 183 spp. of Lepidoptera (N.B. many 'tourist' spp. present) | Coppice 1809 individual Lepidoptera 177 spp. of Lepidoptera | Coniferous plant 2694 individual Lepidoptera 181 spp. of Lepidoptera |
| | | **Banagher** | | |
| Magurran (1985) | Light trap and some direct recording, N. Ireland | Oak wood Plant diversity $\lambda$ log normal $= 129$ Lepidoptera diversity $\lambda$ log normal $= 163$ | Coniferous plantation Plant diversity $\lambda$ log normal $= 68$ Lepidoptera diversity $\lambda$ log normal $= 97$ | |

polyphagous species, whereas the native tree faunas may well contain rarer species, considered more worthy of conservation (Shirt 1987).

Secondly, intensively managed forests will be less varied, with a consequent reduction in the microhabitats available for insects (Heliovaara & Vaisanen 1984; Watt this volume). Hunter (1977) divides up a pine tree into many individual microhabitats and lists the beetles associated with each, and many of these would not be present in a rather uniform tree which had developed in a commercial forest environment.

Despite this general relationship between insect species richness and tree and management type, there are embarrassingly few studies which provide adequate data to confirm it. It has apparently been too simple a question to attract attention. Two exceptions are the recent studies in Northern Ireland (Magurran

1985, 1988) and workers at Bernwood Forest, as exemplified by Waring (1988, 1989a) (see Table 4). Magurran compared an old oak woodland with adjacent areas of Sitka spruce, using light traps to catch Lepidoptera, recording plant data and measuring environmental variables. She used various diversity indices to illustrate the comparison and found that the spruce plantation had a significantly lower diversity and lower species richness of both plants and Lepidoptera than the oak wood. However, although she chose her trapping sites carefully, so that they were equally exposed, she was admittedly comparing an old, heterogeneous oak wood, of various tree ages and densities, and with an admixture of other native, deciduous trees such as birch, with a spruce plantation of trees that were only 45 years old. She correlated the Lepidoptera and plant diversity by pointing out that, since most Lepidoptera are broadly oligophagous, then more plant species provide more potential food-plants and so more Lepidoptera. The factor which she believed to be of overriding importance was that shade was more dense in the plantation, so that many plants were eliminated.

Anderson (1979) has pointed out that plantations of conifers can be managed so that they have a lower crown density throughout the crop rotation, allowing many of the plants to survive, and if this system of forestry were to be widely adopted it might prove very beneficial to insects. However, at present most coniferous plantations do rapidly shade out most of the ground flora.

Waring (1988, 1989a) also used light traps and direct searching for larvae to compare the lepidopterous fauna of coniferous, coppiced and neglected-coppice areas of Bernwood Forest. Despite the complication that light traps catch many non-resident casual species, which just happened to have strayed into any area, he found that the greatest diversity was in the neglected coppice, with the least in the coniferous plantation, and he also attributed the differences principally to the difference in the presence of plants, which itself was caused by the density and subsequent shade of conifers and the influence of coppicing. Only 3% of the species he recorded at Bernwood fed directly on conifers, the remainder caught in the plantations were either strays or were feeding on the various plants and deciduous trees that were still present amongst and around the crop. In fact 77% of the birch-feeding species found in the overall study were at some time found feeding on birch trees in the plantation areas. Unfortunately Waring does not say specifically where these birches occurred or what position was most favourable for the insects feeding on them, although his diagrams imply that they bordered the rides.

A further small-scale observation was reported by Kearns and Majerus (1987), who ran a light trap in adjacent blocks of coniferous and mixed deciduous woodland, recording twenty and thirty-six species of macrolepidoptera in these respective locations. They did not claim much from this one observation but noted that it would be regarded as typical of the experience of lepidopterists.

In contrast, M.R. Young, G. Armstrong and A. Edgar (unpubl.) found little difference in species richness of all insects between ancient semi-natural pine wood stands and neighbouring plantations. However, in this case the plantations were also of pine of local provenance and they were immediately surrounded by the older woods. This study was also affected by a by-catch of casual 'tourist' species in each area and it was noticed that, at times of strong winds, the traps really only worked in the densely sheltered plantation, so artificially boosting the catch there.

An alternative approach has been to compare old lists with more recent ones and to try to correlate any differences with changes in woodland management. This was done by Peachey (1980), who noted the loss of butterfly species from Bernwood Forest, Oxfordshire this century, as well as localization of the survivors, consequent on the changes in the forest from predominantly old deciduous and coppiced trees to even-aged coniferous plantation. Hadley (1982) and Parsons and Hadley (1987) contrasted old and recent lists of macrolepidoptera from Abbots Wood, Sussex, where the mature oak was mostly felled in the 1939–45 period and replaced with conifers. They discuss the difficulties caused by trying to assess whether the methodology used at the two times was sufficiently comparable, but conclude that at least 25–30% of species have disappeared from the wood between 1930 and 1980. More importantly, all of what are now regarded as 'Red Data Book' species (Shirt 1987) have gone. In this case the actual change in tree species has been of great importance, whereas the butterflies in Bernwood Forest are mainly dependent on herbs and grasses and have been affected at least as much by the increase in shading in the rides and the failure to open up new areas on a regular basis.

Waring (1990) has made a similar analysis of the current and past status of the rarer species of moth found in Bernwood Forest, noting a decline or extinction for many species in the last 30 years. The small black arches (*Meganola strigula*) has been recorded only once since the early 1960s, for example, whereas it remains common in a nearby wood where mature oak woodland survives. He also attributes the loss of moth species both to the replacement of mature native trees with younger plantations of exotic species and to the loss of suitable rides and glades in the early years of coniferization, before conservation management agreements were enacted.

Unfortunately equivalent studies for other insect groups do not seem to exist, apart from some rather casual comments on the loss of some beetle species (especially dead wood species) from coniferized woods.

Recently, as noted by Thomas (1988), management of substantially coniferized woods has allowed the retention of more fringing deciduous trees and it may well be that the insect communities of the 'best' of these forests, with varied age and species composition, and with continued sympathetic management, approach those of some woods of more natural character. This bears directly on the question of whether forests of exotic conifers, with an additional mixture of native trees and their associated microhabitats, can be a satisfactory habitat for the conservation of insects. There is obviously still a clear need for better studies of the insect communities of a range of woodland types, bearing in mind the problems and provisos of the earlier section.

A further requirement is for studies which illustrate the proportionate role of woodland edge, woodland canopy and woodland interior in the overall insect community. In an open-structured, varied woodland, such as the semi-natural pine wood stands studied by M. Young, G. Armstrong and A. Edgar (unpubl.), these distinctions may have less separate identity, but in a closed-canopy wood they are important. In tropical forests there is a general agreement that canopy dwellers are of great importance (Sutton 1983), whereas in temperate woodlands, at least until the insecticide fogging of Barnard, Brooks and Stork (1986), it has been sometimes assumed that many species in a woodland insect community are associated with rides and glades, i.e. woodland edge. The canopy has been assumed to be too exposed, and the woodland interior too shaded, to allow for high insect species richness. The reality of these assumptions for temperate woodlands needs to be tested because of the implications for conservation in mixed woodlands. If it is true that woodland edge is of primary significance, then it may be possible to manage rides and glades sympathetically, even in the context of a mainly commercial forest, and so achieve satisfactory conservation results. Hall and Greatorex-Davies (1989) have recently been providing advice to the Nature Conservancy Council on the best ways to manage rides in commercial forests, directly touching on this point.

Since conservation action is urgently needed now, the Forestry Commission (FC) has produced its own advice (Carter & Anderson 1987), based on published work on butterflies and on internal FC studies, setting out the basic principles for ride management. This concentrates on achieving sheltered but sunny ride sides and a varied height and composition in the semi-natural fringe of vegetation and some FC woods are now managed according to the prescriptions, with obvious advantages. This is a development greatly to be welcomed.

The Nature Conservancy Council has also recently produced advice on the management of woodland rides and of coppice (Fuller & Warren 1990; Warren & Fuller 1990), emphasizing the way in which practical management can be fitted into the commercial use of the woods. They also discuss the relative advantages and disadvantages of different management regimes to various types of insects, with suggestions for ways in which mixed management can benefit many groups.

## APPLYING CONSERVATION TO MIXED WOODLANDS

It is possible to apply the arguments and considerations above to the general topic of whether mixed woodlands are more suitable for insect communities than single-species woodlands, and how they should be arranged and managed, so as to maximize any advantage.

1   Since many herbivorous insects are at least broadly oligophagous, it must follow that each new tree species will permit the establishment of new insect species. Even a single isolated rowan tree (*Sorbus aucuparia*), for example, will

eventually act as host for one or more of its specialist Nepticulid leaf-mining Lepidoptera. For most insects more than a single tree will be needed, but in almost all cases, the exact number or area of trees required is not known. Furthermore whether the trees are grouped together, their age and condition, their aspect and shelter, and many other factors are also of importance. In any case, conservation aims are not as indiscriminate as to value every new species, regardless of its status at that site—there is little worth in adding birch trees to an oakwood (on a non-birch site), just because they would add some insects to the total faunal list.

In practice it is likely that there will be a great benefit in adding locally native tree species to a woodland from which they are currently absent, especially when these are arranged in such a way as to maximize their effectiveness. There will be less value from single, isolated trees, scattered through the wood, than from grouping them in appropriate places and allowing them to develop their natural growth forms and associated ground flora. The clumps should also be sufficiently frequent to allow interchange of insects between them, and this would be facilitated by a satisfactory ride system. It seems possible that this system will carry more benefits than interplanting within the commercial crop, but it has to be stressed that no studies have attempted to assess the value of an interplanted mixture for conservation. As always the retention of existing semi-natural features will be of more value than their replacement with new planting.

2  Similar arguments apply to the provision of other habitat features, with the general guideline that variety is preferable, provided that each microhabitat is of natural local occurrence, is of sufficient area or size, and is accessible to colonizing insects. Such features could include: plants other than trees, low and bushy areas, shaded and unshaded glades, bare ground, wet areas, dead wood of various sizes, ages and positions, fungi, etc.

3  An important principle embodied in points 1 and 2 is that a varied woodland 'architecture' is of primary importance, especially at woodland edges. Southwood, Brown and Reader (1979) stress the value of varied architecture to the fauna of an area.

4  Most ecotonal features are transitory and in nature would be continually renewed by events like fresh tree-fall, soil erosion, grazing or natural fires. These factors may not be of common occurrence in commercial forests, so recurrent ride and glade management will be needed to provide the renewal. This renewal must be progressive so that the features are always present, whilst being adequately renewed in adjacent areas.

5  The insect communities and their required habitat features must survive commercial management practices, including thinning but especially final felling, and this can often be achieved if the original plans allow for it. However a thin fringe of deciduous trees will probably not retain their value for insects if the main crop is entirely removed from behind them. The changes in microclimate will probably be too great unless sufficient shelter is left.

6  Although it has so far been assumed that insects will spread effectively along

rides, acting as 'corridors', to newly available areas, this is not yet adequately investigated.

**7** A resource very commonly absent in commercial woodland is overaged trees and dead wood and so, in the context of the management plan to achieve habitat renewal, some provision must also be made to allow some trees to achieve old age and to die naturally. Mixed-species woodland must also become mixed-age woodland.

**8** Interactive effects must be considered. The purple emperor butterfly (*Apatura iris*) not only requires sallow (*Salix* spp.) for its caterpillers but also uses tall trees as assembly points (Heslop, Hyde & Stockley 1964); the holly blue butterfly (*Celastrina argiolus*) has alternate generations, usually on holly (*Ilex aquifolium*) and ivy (*Hedera helix*) (but occasionally on other plants (Plant 1990)) and so lives only where the right combination are present (Emmet & Heath 1989); most aphids have different summer and winter hosts; many species need nectar plants as well as larval food-plants. These examples illustrate clearly that 'mixed' woodlands are often essential (even if this is not strictly mixed tree-species woodlands).

**9** Whether the higher species richness and diversity of insects in 'mixed' woodlands will be of benefit in reducing the likelihood of insect pest outbreaks is discussed by Watt (this volume).

**10** A negative consideration is whether potential tree crop pests will be encouraged by the extra diversity of trees present (see Watt, this volume), or by the presence of such features as nectar sources. Equally likely is that nectar sources will also benefit insect predators and parasites, both by the provision of food and by acting as assembly points where mating is facilitated (C. Carter, unpubl.).

**11** With the exception that some insects, such as the beetle *Magdalis phlegmatica*, which were originally confined to native pine woods, have now been able to spread more widely into the nearby Scots pine plantations (Hunter 1977), there are few examples of a direct conservation benefit to insects being derived from the establishment of commercial plantations of non-native trees. A few species (such as the moth *Lithophane leautieri*) have spread into Britain since the establishment of their food-plant (Heath & Emmet 1983) (in this case *Cupressus* or *Chamaecyparis*), but few rare British species have increased because of the plantations, although the juniper berry-feeding bug, *Cyphostethus tristriatus*, now probably survives in southern England on *Chamaecyparis lawsoniana*, following the decline in native juniper (Carter & Young 1974). However, in a more general sense forest plantations have extended Britain's very meagre woodland cover and some insect species have extended with them, by living in the rides and glades.

In summary it seems likely that the most fruitful method of marrying the requirements of commercial forestry and of insect conservation may be to establish the presence of desirable features at the edges of the commercial stock, that is along rides, in widened ride areas as glades, along natural features such as stream gullies and steep areas and on the actual edge (rather than using interplanting or a mixture of species in blocks). Provided that these features are managed appropriately, are of sufficient size and proximity, and with a favourable aspect, then they may go a considerable way to improving commercial woodlands for insects, even if they can never be a full substitute for some, such as the dark crimson underwing moth (*Catocala sponsa*), which seems to survive only in very extensive forests of mature, native trees (Heath & Emmet 1983).

There are some habitat features that cannot usually be accommodated in this way in the early years of a forest crop rotation, however, such as an extensive shrub layer under the light shade of an open mature woodland, or the presence of a coppice rotation. Rides and glades are permanent open areas, which tend to 'grass over', whereas coppice opens a succession of new areas, providing opportunities for different plants and insects. Very short rotation conifer crops, such as for Christmas trees (*Picea abies*), may in some circumstances have a similar effect to coppicing, but this also needs further study.

The keys to success will be (i) the production of a realistic set of aims, related as closely as possible to a range of insect species whose habitat requirements are understood; (ii) the initial production of an establishment prescription and a management plan to provide continuity of the desired features through all the commercial stages of the forest, including eventual felling; (iii) some form of monitoring scheme which will indicate whether the prescriptions are meeting the aims; and (iv) the flexibility to respond to changing conditions and achievements and increased knowledge.

## REQUIREMENTS FOR FUTURE WORK

It will be clear that much further work is required before there is a proper understanding of the way in which insect communities respond to the establishment and management of the various types of mixed woodlands. In particular there have been very few experimental studies, so that cause and effect has almost never been established and yet this might surely be possible in a commercial setting.

Specific requirements are for the following:

1 More general studies which enumerate insect faunas of varied woodland types (if these can overcome the difficulties noted earlier).

2 More research on individual species of insect. (However these should be chosen not just for their rarity or general attraction but for the light they can shed on the other questions set out below.)

3 Assessment of the comparative abundance and diversity of insects in different areas within a wood (for example the canopy, the interior, or the woodland edge).

4 Assessment of the 'size' and 'shape' of habitat areas needed to harbour viable populations of a range of different species. In woodlands this may as well be the

length of a linear feature, the area of a block, or the number of a resource such as a dead tree.

**5** Assessment of the rate of spread of various insects within a wood, along rides, between glades or from adjacent woodlands.

**6** Research specifically designed to illustrate whether rides do act as 'corridors' between habitat patches, and for what types of insect species this applies. Assessment of the best design for such corridors.

**7** Research on whether there are many other 'unseen' species which benefit from the conservation of a 'key' species, as often claimed.

**8** The development of effective but simple monitoring schemes for woodland insects—overcoming the inherent problems of the great variability of population size and inconspicuousness of most species.

**9** Experimental manipulation of rides to establish clearer relationships between different management regimes and the success of insect populations. (This should substantiate and refine the design features set out in the management guidelines which exist (e.g. Carter & Anderson 1987; Warren & Fuller 1990).)

**10** Assessment of whether interplanting of additional tree species carries any benefit for insects, as compared with block or ride planting.

**11** Experimental assessment of whether management regimes, such as the use of low initial planting density or of line-thinning, carry benefits for insects.

**12** Further assessment of the pest–predator interactions between the tree crop and the deciduous admixture of trees and non-tree resources such as nectar plants.

**13** Consideration of more successful ways of developing long-term management plans, so as to avoid the undue disruption caused by commercial operations during the rotation of the crop.

## CONCLUSIONS

Currently it seems that the most fruitful way of 'mixing' woodlands, for the benefit of insect populations, is to introduce native tree and shrub species (with their associated flora) along the edges and in rides, glades and natural features. Here they may be designed so as to produce an approximation to the natural variety of native woodlands, whilst taking up only a small proportion of the available area, so that the wood remains commercially viable. Because of the limitations of size and position these areas must be managed frequently and carefully, to provide continuity and interconnection of habitat and so as to survive the intervention of commercial operations. Even so other strategies will be needed, if features such as coppice, or high canopy are to be included, although mixing by interplanting is unlikely to provide the desired results. It seems unlikely that even the best-managed commercial forests can be a full substitute for semi-natural native woodlands, although so few of these now exist that in general insects may well benefit from the increased forest area due to commercial planting.

## ACKNOWLEDGMENTS

I am most grateful to the many entomologists and foresters with whom I have had useful discussions, including D. Barbour, A. Watt, C. Carter and especially P. Waring.

## REFERENCES

**Allan, P.B.M. (1949).** *Larval Foodplants.* Watkins & Doncaster, London.

**Anderson, M. (1979).** The development of plant habitats under exotic tree crops. *Ecology and Design in Amenity Land Management* (Ed. by S.E. Wright & G.P. Buckley). Wye College and Recreation Ecology Research Group.

**Barnard, P.C., Brooks, S.J. & Stork, N.E. (1986).** The seasonality and distribution of Neuroptera, Rhapidioptera and Mecoptera on oaks in Richmond Park, Surrey, as revealed by insecticide knock-down sampling. *Journal of Natural History*, **20**, 1321–31.

**Carter, C.I. & Anderson, M.A. (1987).** *Enhancement of Lowland Forest Ridesides and Roadsides to Benefit Wild Plants and Butterflies.* Forestry Commission. Research Information Note 126.

**Carter, C.I. & Young, C.W.T. (1974).** *Chamaecyparis lawsoniana* (Murray) Parlatore, another host for *Cyphostethus tristriatus* (F.) (Hem. Acanthosommatidae). *Entomologists Monthly Magazine*, **109**, 180.

**Disney, R.H.L. (1986).** Inventory surveys of insect faunas: discussion of a particular attempt. *Antenna*, **10**, 112–16.

**Emmet, A.M. (1988).** *Field Guide to the Smaller British Lepidoptera*, 2nd edn. British Entomological and Natural History Society, London.

**Emmet, A.M. & Heath, J. (Eds) (1989).** *The Moths and Butterflies of Great Britain and Ireland*, Vol. 7. Harley Books, Colchester.

**Fuller, R.J. & Warren, M.S. (1990).** *Coppiced Woodlands: Their Management for Wildlife.* Nature Conservancy Council, Peterborough.

**Hadley, M. (1982).** The decline of the indigenous macro-lepidoptera of Abbot's Wood, East Sussex. *Entomologists Record*, **94**, 92–6.

**Hall, M.L. & Greatorex-Davies, J.N. (1989).** *Management Guidelines for Invertebrates, especially Butterflies, in Plantation Woodland.* Nature Conservancy Council Contract Report, Peterborough.

**Heath, J. & Emmet, A.M. (Eds) (1983).** *The Moths and Butterflies of Great Britain and Ireland.* Harley Books, Colchester.

**Heath, J., Pollard, E. & Thomas, J.A. (1984).** *Atlas of Butterflies in Britain and Ireland.* Viking, Middlesex.

**Heliovaara, K. & Vaisanen, R. (1984).** Effects of modern forestry on north western European forest invertebrates: a synthesis. *Acta Forestalia Faunica*, **189**, 4–29.

**Heslop, I.R.P., Hyde, G.E. & Stockley, R.E. (1964).** *Notes and Views of the Purple Emperor.* Southern Publishing Co., Brighton.

**Hunter, F.A. (1977).** Ecology of Pine Wood Beetles. *Native Pinewoods of Scotland*, (Ed. by R.G.H. Bunce & J.N.R. Jeffers), pp. 42–55. Natural Environment Research Council.

**Kearns, P.W.E. & Majerus, M.E.N. (1987).** Differential habitat selection in the Lepidoptera: a note on deciduous versus coniferous woodland habitats. *Entomologists Record*, **99**, 103–6.

**Kennedy, C.E.J. & Southwood, T.R.E. (1984).** The numbers of insects associated with British trees: a reanalysis. *Journal of Animal Ecology*, **53**, 455–78.

**Leather, S.R. & Barbour, D.A. (1987).** Associations between soil type, lodgepole pine (*Pinus contorta* Douglas) provenance, and the abundance of pine beauty moth, *Panolis flammea* (D. and S.). *Journal of Applied Ecology*, **24**, 945–52.

Magurran, A.E. (1985). The diversity of Macrolepidoptera in two contrasting woodland habitats at Banagher, Northern Ireland. *Proceedings of the Royal Irish Academy*, **85b**, 121–32.

Magurran, A.E. (1988). *Ecological Diversity and its Measurement*. Croom Helm, London.

Morris, M.G. & Thomas, J.A. (1989). Re-establishment of insect populations. *The Moths and Butterflies of Great Britain and Ireland* (Ed. by A.M. Emmet & J. Heath), Vol. 7, pp. 22–36. Harley Books, Colchester.

Owen, J.A. (1989). Beetles from pitfall-trapping in a Caledonian pinewood at Loch Garten, Inverness-shire. *British Journal of Entomology and Natural History*, **2**, 107–13.

Parsons, M. & Hadley, M. (1987). Abbot's Wood—a history of a woodland and its butterflies. *Entomologists' Record*, **99**, 49–58.

Peachey, C.A. (1980). *The Conservation of Butterflies in Bernwood Forest*. Nature Conservancy Council, Newbury.

Peterken, G.F. (1981). *Woodland Conservation and Management*. Chapman & Hall, London.

Plant, C.W. (1990). *Pyracantha* as a possible foodplant of the holly blue butterfly, *Celastrina argiolus* (Linn.) (Lep.: Lycaenidae) in the London area. *Entomologists Record*, **102**, 41–3.

Pollard, E. (1979). Population ecology and change in range of the white admiral butterfly *Ladoga camilla* L. in England. *Ecological Entomology*, **4**, 61–74.

Pollard, E., Hall, M.L. & Bibby, T.J. (1986). *Monitoring the Abundance of Butterflies 1976–1985*. Nature Conservancy Council, Peterborough.

Rackham, O. (1980). *Ancient Woodland*. Arnold, London.

Rotheray, G.E. & MacGowan, I. (1990). Re-evaluation of the status of *Callicera rufa* Schummel in the British Isles. *The Entomologist*, **109**, 35–42.

Shirt, D.B. (1987). *British Red Data Books: 2. Insects*. Nature Conservancy Council, Peterborough.

Skinner, B. (1984). *Colour Identification Guide to Moths of the British Isles*. Viking Press, Middlesex.

Southwood, T.R.E., Brown, V.K. & Reader, P.M. (1979). The relationships of plant and insect diversities in succession. *Biological Journal of the Linnean Society*, **12**, 327–48.

Speight, M.R. & Wainhouse, D. (1989). *Ecology and Management of Forest Insects*. Oxford Science Publications, Oxford.

Steele, R.C. & Welch, R.C. (1973). *Monkswood—A Nature Reserve Record*. Nature Conservancy, Monkswood, Huntingdon.

Sterling, P.H. & Hambler, C. (1988). Coppicing for conservation: do hazel communities benefit? *Woodland Conservation and Research in the Clay Vale of Oxfordshire and Buckinghamshire* (Ed. by K.J. Kirby & F.J. Wright), pp. 69–80. Nature Conservancy Council, Peterborough.

Sutton, S.L. (1983). The spatial distribution of flying insects in tropical rain forests. *Tropical Rain Forest: Ecology and Management* (Ed. by S.L. Sutton, T.C. Whitmore & A.C. Chadwick), pp. 77–91. Blackwell Scientific Publications, Oxford.

Thomas, J.A. (1975). Some observations on the early stages of the purple hairstreak butterfly, *Quercusia quercus* (Linnaeus) (Lep. Lycaenidae). *Entomologists Gazette*, **26**, 224–6.

Thomas, J.A. (1980). The extinction of the large blue and the conservation of the black hairstreak butterflies (a contrast of failure and success). *Annual Report of the Institute for Terrestrial Ecology*, **1977**, 19–23.

Thomas, J.A. (1984). The conservation of butterflies in temperate countries: past efforts and lessons for the future. *The Biology of Butterflies* (Ed. by R. Vane-Wright & P. Ackery), pp. 333–53. Symposium of the Royal Entomological Society, London. Academic Press, London.

Thomas, R.C. (1988). Historical ecology of Bernwood Forest Nature Reserve 1900–1981. *Woodland Conservation and Research in the Clay Vale of Oxfordshire and Buckinghamshire* (Ed. by K.J. Kirby & F.J. Wright), pp. 20–9. Nature Conservancy Council, Peterborough.

Varley, G.C., Gradwell, G. & Hassell, M. (1973). *Insect Population Ecology*. Blackwell Scientific Publications, Oxford.

Waring, P. (1988). Responses of moth populations to coppicing and the planting of conifers. *Woodland Conservation and Research in the Clay Vale of Oxfordshire and Buckinghamshire* (Ed. by K.J. Kirby & F.J. Wright), pp. 82–113. Nature Conservancy Council, Peterborough.

Waring, P. (1989a). Comparison of light trap catches in deciduous and coniferous woodland habitats. *Entomologists Record*, **101**, 1–10.

Waring, P. (1989b). *Moth Conservation Project—News Bulletin 2*. Nature Conservancy Council,

Peterborough.

**Waring P. (1990).** The status in Bernwood Forest of moth species which are recognised as nationally uncommon. *Entomologists Record*, **102**, 233–8.

**Warren, M.S. (1985).** The influence of shade on butterfly numbers in woodland rides with special reference to the wood white, *Leptidea sinapis*. *Biological Conservation*, **33**, 147–64.

**Warren, M.S. (1987).** The ecology and conservation of the heath fritillary butterfly *Mellicta athalia* III. Population dynamics and the effect of habitat management. *Journal of Applied Ecology*, **24**, 499–514.

**Warren M.S. & Fuller, R.J. (1990).** *Woodland Rides: Their Management for Wildlife.* Nature Conservancy Council, Peterborough.

**Welch, R.C. (1986).** What do we know about insects in Scottish Woods? *Trees and Wildlife in the Scottish Uplands* (Ed. by D. Jenkins), pp. 95–100. ITE Report 17, NERC.

**Winter, T.G. (1983).** *A catalogue of phytophagous insects and mites on trees in Great Britain.* (Forestry Commission Booklet 53) Forestry Commission, Edinburgh.

**Young, M.R. (1986).** The effects of commercial forestry on woodland Lepidoptera. *Trees and Wildlife in the Scottish Uplands* (Ed. by D. Jenkins), pp. 88–94. ITE Report 17, NERC.

**Young, M.R. & Barbour, D.A. (1989).** *Ecology of the Kentish Glory Moth* (Endromis versicolora). Nature Conservancy Council contract report, Aberdeen.

# Discussion

EDITED BY M.G.R. CANNELL

## INTRODUCTION

The papers published in these proceedings were discussed at the end of symposium in six discussion groups, each consisting of ten to thirty people. The task given to each group was to highlight the main issues, cover any major omissions in the papers, consider any points of contention, and identify priorities for future research.

The following reports were drafted by the group chairmen to reflect the balance of views at the symposium. They form a valuable addition to the presented papers and provide a statement of current thinking on the ecology of tree species mixtures.

## COMPETITION, YIELD PREDICTION AND SILVICULTURE

### H.E. Burkhart

### Competition

Competition is an elusive concept. However, we do know that trees in stands share resources of light, water and nutrients necessary for their survival and growth. Because resources are limited, there is interference between trees. A better understanding is needed of the different types of interference between neighbouring trees of the same and different species.

### Yield modelling and experimentation

Yield prediction for mixed-species stands should emphasize tree-level approaches; stand-level models cannot provide information on the contribution of different species to stand structure that is needed. Empirical relationships from past studies in monocultures and mixtures can be incorporated into growth and yield prediction models to develop 'first approximations' for various mixtures of interest. However, it will be necessary to establish additional experiments and field trials with mixtures in order to validate existing yield prediction models and to extend and refine yield prediction capabilities.

*Silviculture*

Silvicultural prescriptions for mixtures can be developed only in the context of specific objectives. However, for a wide range of land management objectives, it is clear that silvicultural systems are needed (i) to achieve desired regeneration in mixed-species stands, and (ii) to convert mixed stands of undesirable composition to stands of the desired species composition. As a corollary to the need for silvicultural systems to change species composition, one could add the need to develop silvicultural systems to maintain desired mixtures.

*Recommendations*

The body of knowledge developed from the study of mixed-species stands should be summarized and made readily available. This summary requires a two-pronged effort. First, a compilation of yield and silvicultural information should be undertaken. Such a compilation would require a clearly articulated purpose and format for authors from many countries. Second, a computer-based decision support system—perhaps in the form of an Expert System—should be developed for management of mixed-species stands. Any Expert System would need to be made specific to local conditions, but a well developed prototype should lessen the development time required for others that follow.

## CANOPY INTERACTIONS IN MIXTURES

P.G. Jarvis

*Definitions*

The initial definition of canopy mixtures for the discussion included the vertical strata, horizontal patches, conifers and broad-leaves.

It was presumed that mixtures would affect canopy architecture and physiology, including canopy leaf area index and roughness; crown shape, dimensions and leaf area density distribution; and leaf size, shape, optical and physiological properties.

It was expected that there might be consequences for over- and under-storey microclimate; canopy processes, including radiation absorption, carbon assimilation, evaporation and transpiration; and tree growth and stand production.

*Evidence for effects of mixed-species stands on canopy processes*

1 Silviculture and stand management. There is much *empirical evidence* from long-term studies in uneven-aged, selection forest and replacement mixtures, especially in central Europe, that there are differences in tree growth and stem-wood production between mixtures and monocultures.

These results *suggest* that canopy interactions may be involved especially with respect to leaf phenology and tree stability.

However, there has been *little detailed analysis* to show *how* or *if* canopy structure and processes are involved. Other variables, such as the heterogeneity and patchiness of soils, or the definition of proper controls for comparison, combine to confound the observations.

2  Stand ecology and canopy processes. There has been *much informed speculation* that mixtures have different canopy structures, i.e. different degrees of non-randomness in the leaf area density distribution, that lead to differences in light transmission, absorption and utilization efficiency, and possibly other processes.

However, there have been *no detailed analyses* of processes in canopies of mixtures compared with monocultures.

### Conclusions

Despite assertions made, based on the two types of evidence identified above, there is no hard evidence that a mixed canopy is responsible for the different performance of trees in a mixture, rather than, say, root, soil or mycorrhizal effects, nor for synergism based on canopy structure and canopy processes.

### Recommendations

The following research needs were identified:
1  the canopy structures of mixtures need to be explicitly defined;
2  the consequences of different canopy structures need to be analysed;
3  interaction effects that do lead to mixed canopies being more (or less) effective than the sum of their parts need to be identified;
4  models need to be used—parameterized to include interaction terms—to investigate the performance of different mixtures;
5  model predictions need to be tested against field observations and experiments.

## NUTRIENT-RELATED SYNERGISMS

### O.W. Heal

### Introduction

It is clear that nutrient-mediated interactions between tree species can result in positive (and possibly negative) effects on the survival and growth of the species involved. These interactions are important for forest management, particularly in tropical and subtropical areas, where nutrient supplies often limit tree growth. The productivity of large areas of forest might be significantly enhanced by manipulating the nutrient dynamics of tree mixtures.

### Factors and mechanisms involved in synergism

A wide range of mechanisms are involved in nutrient-related synergism. The relative importance of individual mechanisms varies among sites, depending on climate, soil, species and management conditions. No single generalizations or prescriptions can be made and quantitative information is currently inadequate to allow useful predictions to be made. However, there is a general body of know-ledge based on theory and management experience from a wide range of situations throughout the world which provides a basis for research and development. The major mechanisms of interaction include the following.

**1**  Ground vegetation. The amount, composition and phenology of the ground vegetation is partly determined by stand composition. Although relatively small amounts of nutrients are circulated through the ground vegetation, they can influence nutrient supply to trees by (i) providing a short term sink or source of readily available nutrients, (ii) modifying the soil surface microclimate, and (iii) competing with the trees, especially in the young stages.

**2**  Decomposition. Tree mixtures differ from single-species stands in the quantity, quality and timing of litter input. Nutrient release from litter and humus can be enhanced (or retarded) as a result of (i) amelioration of the microclimate, (ii) microbially mediated nutrient transfer, carbon utilization and breakdown of inhibitory organics, and (iii) faunal diversification in response to substrate heterogeneity.

**3**  Nitrogen fixation. $N_2$-fixation has potentially the greatest effect on the nutrient economy as a result of (i) enhanced nitrogen availability, (ii) competition for other nutrients, (iii) acidification through leaching of 'excess' nitrate and produc-tion of strong organic acids, (iv) enhanced mineralization of phosphorus by exoenzymes, and (v) chelation of other elements.

**4**  Roots. The rooting characteristics of different tree species and their associated microflora, potentially interact by affecting (i) the availability of carbon for microbial transformation of nutrients, (ii) differential uptake of nutrient elements, (iii) the spatial separation and partitioning of soil space, (iv) the temporal sep-aration of nutrient uptake, and (v) the direct transfer of nutrients through mycor-rhizal connections between trees.

**5**  Leaching. The leaching of inorganic and soluble organic fractions can be affected by tree species mixtures as a result of (i) a modified soil moisture regime, (ii) an extended period of nutrient uptake, and (iii) the control of ionic balance. Changes in the intensity of leaching can result in irreversible acidification and accelerated pedogenesis.

## Conclusions

**1**  The most important synergisms between tree species involve the process of decomposition (of both surface litter and humus) and root interactions (including mycorrhizas).

**2**  Symbiotic nitrogen fixation is also a key process, and has important effects on the availability of other elements and on the ionic balance in soils, which have long-term effects on leaching and pedogenesis, including profile differentiation.

**3**  Nutrient interactions are primarily between individual trees and are usually effective over distances of less than 20 m, related particularly to root distribution. A general feature of synergy is to reduce spatial heterogeneity in nutrient supply.

4  The capacity of mixtures to extend the seasonal period of nutrient utilization, by demanding nutrients over different periods, is an important hypothesis which is particularly lacking in data.

5  Nutrient-related synergism is strongly influenced by modification of the microclimate in mixtures. Alternation of the soil moisture regime by the trees can be a key determinant of the degree of nutrient synergy, particularly in wet and dry climatic extremes.

6  The selection of tree species in managed mixtures is largely based on empirical experience. In tropical areas knowledge of the growth strategies and phenologies of species is needed to assist in the selection of potentially synergistic species, e.g. by allowing accelerated succession and complementary resource utilization.

## Recommendations

Observations and data need to be synthesized into mechanistic models which can be used to explore the relative importance of different mechanisms under varying environmental conditions, and which can assist in experimental design.

Small-scale experiments are needed in the laboratory and field, which explore the localized nature of nutrient synergism and which can quantify the effects of particular factors or processes. These experiments can be opportunistic, exploiting both natural and managed mixtures and assessing, for example, microclimatic effects, nutrient transfer by tracers, and root separation.

Larger-scale field experiments are an expensive and valuable resource. Replication is essential. Whilst full use should be made of existing experiments, new experiments are needed.

# MIXTURES AS HABITATS FOR PLANTS

## M.O. Hill

### Introduction

This discussion considers, primarily, mixtures where at least the major component is planted. Such mixtures predominate in Britain, where natural regeneration is rarely employed as a means of maintaining tree cover.

### Conclusions

Plant communities under mixtures are not necessarily equivalent to an average of those under pure stands, and various mechanisms can account for this. Typically, there are gradients in temperature, humidity, light, soil moisture and litter quality.

Some plants may require an intermediate position on the gradient for optimum development or even survival. There are also secondary interactions dependent on primary interactions in other components of the ecosystem. For example, if one component of the mixture grows better than it would in a pure stand, then plants growing under it may respond to its improved nutrition.

Plant species often depend on refuges for their long-term survival. Mixtures may provide refuges, especially when one canopy species casts a shade so dense that it eliminates the ground vegetation. Even a small admixture of a broad-leaved species in a dense conifer plantation can allow shade-bearing woodland species to survive. Most shade-bearers cannot find a refuge in the seed-bank, because their seeds are short-lived and often sparsely produced. The actual proportions of the tree species making up a mixture may be relatively unimportant for short-term survival, but if plants are confined to sparse refuges for part of the tree crop cycle then they must rapidly spread out from them when conditions improve.

*Recommendations*

Experiments, both with mixtures in the field and with plants grown in mixed litters in pots, would add usefully to our knowledge, but we are still at the stage where observation and survey will be most rewarding. We still do not know whether beneficial effects attributed to mixtures are really due to mixtures or whether they are the result of the silvicultural practices used to establish mixed stands.

At a practical level, we should continue our search for understorey-friendly combinations of trees, perhaps by planting a shrub layer such as hazel. The practice of planting a wide variety of broad-leaved tree species contributes nothing to the conservation value of a wood. In order to sharpen our practical recommendations, we must improve our theoretical understanding. Theory requires not only data on species' short-term responses to mixtures but also more information on their capacity to spread during the favourable period of a tree crop's growth. From such data we can infer the effect of differing tree planting patterns and patch sizes, although factoring out the effect of differences in canopy structure will be difficult.

# MIXTURES AS HABITATS FOR VERTEBRATES

P.R. Ratcliffe and D.J. Jardine

*Introduction*

Food availability is probably the most significant factor determining the presence, absence, productivity and performance of animal populations. Most vertebrates are not dependent on particular tree species for food and therefore tree species mixtures generally have much less significance for them than for the many invertebrates which are specialist feeders.

*Conclusions/recommendations*

The discussion group was unconvinced of the benefits of studying tree species mixtures as habitats for vertebrates, except in a few cases where the trees represent important food sources. This is because tree species mixtures represent only one component of a large assemblage of vertebrate requirements.

There is a wide variety of vertebrate groups (mammals, birds, reptiles, amphibians and fish) all of which have varying degrees of mobility. The capacity of an individual species to benefit (or suffer) from a mixed tree species stand will vary with the mobility of the organism (i.e. its home range and exploratory behaviour) and the geographical scale of the mixture involved. There remains a need to quantify the interface effects of two or more tree species on various vertebrate species or groups to discover the relative importance of intimate, row/row, subcompartment or compartment level mixtures. Such research should be combined with studies of other important habitat requirements such as structure of the field and shrub layers, and proximity to water and shelter.

While there is a significant body of information on the effects of diversity of tree species on the diversity and density of vertebrates, little is known about the effect of various scales of mixtures on their productivity, e.g. what is the minimum area of broad-leaved woodland required to allow successful breeding of redstart in upland spruce plantations?

In a number of studies it has been difficult to separate the relative importance of tree species and canopy structure. In view of limited research resources, this work should be undertaken before any work is done on the effect of mixtures *per se* on vertebrates.

Many of the answers being sought by woodland and forest managers can be provided by autecological studies on important species (e.g. woodlarks and red squirrels), or by community studies of important groups of species (e.g. bats and amphibians). Additionally, studies of food supplies that are common to a variety of animals can be cost effective. These approaches enable diversity to be produced by developing management systems which are sensitive to key species found in particular areas. Overall, they could be used to achieve management objectives taking into account all of the costs (e.g. marketing), and benefits of growing trees in mixtures.

## MIXTURES AND INSECT POPULATIONS

### H.F. Evans

*What is a mixture?*

Mixtures can be identified at a number of levels, encompassing both intimacy and scale. These factors are clearly pertinent to interactions between and within tree species but are not so clearly defined as far as interactions with insects are concerned. A subsidiary question is: where does a mixture matter to a given insect? In other words, is the interface between one tree species and another the critical component or is it better to consider numbers of individuals of a tree species per unit area?

### *Do mixtures have more insects per unit area than monocultures made up of the same tree species?*

Evidence to support the possible view that mixtures harbour a richer species diversity of insects than monocultures are provided by Watt (this volume). He considered herbivores such as the spruce budworm and, using published data, showed that outbreaks of the insect were more frequent in certain monocultures. However, for the same insect, but using a different set of data, he was able to show that the opposite conclusion could be reached; outbreaks were more frequent in mixtures where the favoured species were grown in small patches among non-susceptible trees. We concluded that there were insufficient data to give an overall view regarding insects associated directly with trees.

Mixtures, as discussed by Young (this volume) are valuable in enhancing the numbers of insect species of conservation interest associated with woodland habitats. This reflects the increased habitat diversity of a mixed woodland but, being generally associated with understorey or rideside vegetation, does not necessarily reflect the mixtures of the trees themselves.

### *How do we enumerate the insects in a mixture?*

Decisions have to be made on whether the purpose of enumeration is ultimately to track a key species or to help in establishing the greatest possible biodiversity on a given site. The effort required to study a key species is manageable, bearing in mind the resources likely to be available. By contrast, enormous efforts would be required to study maximum biodiversity. Thus, in developing sampling regimes, one has to consider how best to achieve a high return for a given effort. To do this, it is essential to define clearly the purpose of the work.

### *Type of forest management?*

Clearly, it will be simpler to manage for the particular habitat requirements of one or more key species. However, there may be costs in terms of biodiversity, because the habitat requirements may change to the detriment of other non-key species. Management for maximum biodiversity is probably not viable with our current state of knowledge of mixtures and insect populations.

### *Are we trying to establish general principles?*

This question arises directly from the preceding ones. To develop a more defined framework from which to determine the roles of mixtures, data on a number of aspects will have to be gathered. At a minimum the following would need to be addressed:

forest structure (tree species and other vegetation);
site characteristics (soil, water relations and nutrients);
monoculture versus mixtures;
temporal aspects (changing structure with time); and
movement of insects (ability to colonize components of the forest).

## Conclusion

Our knowledge base is poor concerning mixtures, mainly because the concepts have not been specifically addressed in a quantitative manner. Bearing in mind inevitable constraints in resources, the best approach would be a relatively broad-based study within existing forest structures, concentrating initially on insects at different trophic levels within the forest.

# Index

Page numbers in *italics* refer to figures and tables.